T0142306

# Advances in Information Security

Volume 83

**Series Editor**
Sushil Jajodia, George Mason University, Fairfax, VA, USA

 MyCopy

Dear MyCopy Customer,

This Springer Nature book is a high quality monochrome, digitally printed version of the eBook, which is accessible to you on SpringerLink. As a result of your library investing in licensing at least one Springer Nature eBook subject collection, this copy is provided to you at a highly discounted price from the usual softcover edition.

MyCopy books are only offered to patrons of a library with access to at least one Springer Nature eBook subject collection and are strictly for individual use only. Personal resale or library shelving of the MyCopy book is in breach of this agreement.

You may cite this book by referencing the bibliographic data and/or the DOI (Digital Object Identifier) found in the front-matter. This book is an exact but monochrome copy of the print version of the eBook on SpringerLink.
www.springernature.com/mycopy

More information about this series at http://www.springer.com/series/5576

Wei Chang • Jie Wu
Editors

# Fog/Edge Computing For Security, Privacy, and Applications

Springer

*Editors*
Wei Chang
Department of Computer Science
Saint Joseph's University
Philadelphia, PA, USA

Jie Wu
Computer and Information Sciences
Temple University
Philadelphia, PA, USA

ISSN 1568-2633          ISSN 2512-2193   (electronic)
Advances in Information Security

https://doi.org/10.1007/978-3-030-57328-7

© Springer Nature Switzerland AG 2021
This work is subject to copyright. All rights are reserved by the Publisher, whether the whole or part of the material is concerned, specifically the rights of translation, reprinting, reuse of illustrations, recitation, broadcasting, reproduction on microfilms or in any other physical way, and transmission or information storage and retrieval, electronic adaptation, computer software, or by similar or dissimilar methodology now known or hereafter developed.
The use of general descriptive names, registered names, trademarks, service marks, etc. in this publication does not imply, even in the absence of a specific statement, that such names are exempt from the relevant protective laws and regulations and therefore free for general use.
The publisher, the authors, and the editors are safe to assume that the advice and information in this book are believed to be true and accurate at the date of publication. Neither the publisher nor the authors or the editors give a warranty, expressed or implied, with respect to the material contained herein or for any errors or omissions that may have been made. The publisher remains neutral with regard to jurisdictional claims in published maps and institutional affiliations.

This Springer imprint is published by the registered company Springer Nature Switzerland AG
The registered company address is: Gewerbestrasse 11, 6330 Cham, Switzerland

# Preface

With the advance of computer technology and high-speed networks, we have witnessed the rise of a new technology in fog/edge computing in recent years. The past decade has witnessed a significant advance in cloud technology, Internet of Things (IoT), and 4G/5G wireless communications that expand the traditional computer technology at both macro and micro levels. In the era of big data, IoT includes many sensors, actuators, and mobile devices at the network edges. With the help of 4G and future 5G high-speed communication, data collected at IoT will be sent to the cloud for storage and processing. However, communication latency, due to the sheer volume of data collected at IoT to be transmitted to the cloud, poses a major challenge in cloud technology. Various efforts have been made to allow IoT to perform limited computation and storage; fog/edge computing goes one step further to bring some cloud service close to the network edge.

*Fog computing* is an architecture that makes use of edge devices to carry out a good amount of computation and storage. Their communication is done locally and/or routed over the Internet and/or 4G/5G. Fog computing supports IoT and consists of a control plane and a data plane. Like cloud computing, fog computing also provides computation, storage, and applications to end-users. However, fog computing is closer to end-users at the network edge (also known as *edge computing*, although these two terms are sometimes used interchangeably) and has wider geographical distribution. Note that rather than a substitute, fog/edge computing often services as the complement to cloud computing, and in many cases, works together with existing cloud technology, like cloudlet. There are many technical challenges in fog/edge network design, such as computation offloading which deals with delay minimization, energy minimization, a combination of both, and caching which decides both placing caching units and their contents. This book focuses on security and privacy in fog/edge computing.

Security and privacy in fog/edge computing pose some unique challenges as various services are distributed at the network edge. Security and privacy issues can be divided into two parts: system-level and service-level. System-level security and privacy deal with issues in the computing system itself such as modern network design using virtualization and special threats and attacks and their counter methods in intrusion and malware detections. Service-level security and privacy handle issues under a service, which can be broadly divided into authentication and trust, access control, data confidentiality and integrity, privacy preservation, and non-repudiation. Security and privacy issues can also be partitioned in another orthogonal way based on system functions, including service provisioning, data processing, data transmission, and data storage. Note that service decentralization in fog/edge computing offers a double-edged sword, compared to service centralization in cloud computing: fog/edge computing offers more flexibility in the system and network design; however, it also poses some additional challenges in ensuring security and privacy, especially in supporting mobility, device heterogeneity, location-awareness, and lightweight solutions.

To handle the security and privacy issues in fog/edge computing, many secure and privacy-preserved systems, architectures, and algorithms have been designed. Based on the target, we can classify the existing security and privacy solutions to fog/edge computing into four categories: user-centric, device-centric, application-centric, and end-to-end-centric. The user-centric methods focus on the roles of users that participate in the fog/edge computing, and the corresponding security and privacy mechanisms are determined based on the roles. The device-centric methods provide security and privacy solutions for each end device based on its resources, location, and the roles it plays in the applications. The application-centric solutions take full advantage of the power, flexibility, and performance of the existing fog/edge computing systems and consider how to apply policies to different applications to meet their unique security and privacy requirements. The last group, end-to-end-centric, emphasizes the secure and privacy-preserved communications among all participants, such as the remote cloud, edge devices, and end devices. Because of the heterogeneity of the participants and variety of security and privacy goals, more and more fog/edge computing-related schemes have been proposed in the past decade.

The goal of this book is to collect the state-of-the-art development on security and privacy of fog/edge computing, together with their system architectural support and applications. This book will be of special value to academics, researchers, government officials, practitioners, and business organizations (e.g., executives, system designers, and marketing professionals) who conduct teaching, research, decision-making, and designing fog/edge technology. The content of the book will be particularly useful for students studying computer science, computer technology, and information systems, but also applies to students in business, education, and economics, who would benefit from the information, models, and case studies therein.

This book is suitable for serving as a reference book for a graduate course in fog/edge computing, computer security, privacy, and applications, as well as for developers in the fog/edge technology industry. Our focus is to expose readers to the technical challenges in building fog/edge with security and privacy in design, together with various applications that are related to security and privacy, and to offer some ideas on how we might overcome them. This book is organized into five parts with a total of 15 chapters. Each area corresponds to an important snapshot, starting from the introduction of fog/edge computing and ending with the applications of fog/edge computing with a focus on security and privacy. This book complements several books that have emerged recently in the area, but none addresses all major issues and possible solutions.

- Part I: Overview of Fog/Edge Computing (Chaps. 1 and 2)
- Part II: Security in Fog/Edge Computing (Chaps. 3, 4, and 5)
- Part III: Privacy in Fog/Edge Computing (Chaps. 6 and 7)
- Part IV: Architectural Design in Fog/Edge Computing (Chaps. 8 and 11)
- Part V: Applications of Fog/Edge Computing (Chaps. 12, 13, 14, and 15)

Part I gives an overview of fog/edge computing, its definition, and relevant concepts. Chapter 1 presents an overview of fog computing, focusing on its relationship with cloud technology in terms of low latency and overviewing the future with the use of 5G communication. Especially, this chapter foresees a variety of potential security concerns and risks of fog computing on 5G. Chapter 2 focuses on edge computing as an open and distributed architecture that features decentralized processing power, enabling mobile computing technologies, as well as the IoT devices or local edge servers. Several applications of edge computing are also discussed.

Part II focuses on security in fog/edge computing. Chapter 3 describes secure storage and search services in cloud computing, with a focus on attribute-based encryption and searchable encryption. Discussion is also given on how to port this approach to fog/edge where distributed storage is used and protocols should be lightweight. Chapter 4 focuses on IoT-fog computing, and it discusses a special type of intrusion detection method based on collaboration, as the traditional and centralized cloud-based intrusion detection cannot be applied in fog/edge computing. Chapter 5 studies the feasibility of deploying Byzantine agreement protocols to improve the security of fog/edge computing in untrusted environments, emphasizing the consistency, availability, and partition-tolerance tradeoffs.

Part III deals with privacy in fog/edge computing. Chapter 6 investigates the unique privacy challenges in fog/edge computing as more edge servers and communication between edge servers and end devices bring more challenges to users' privacy. Since edge computing causes a tremendous exchange of a user's data, identity, and location to edge server compared to cloud computing, privacy concerns are more severe compared to that of the cloud. Chapter 7 emphasizes privacy on edge-based video analysis, a popular machine learning (ML) application on fog/edge. The authors discuss privacy issues that occurred during the model training process and propose federated learning, a special type of distributed ML, driven by a privacy-preserving model training framework.

Part IV takes on the architectural design of fog/edge computing. Chapter 8 gives a comprehensive overview of vulnerabilities in fog/edge computing within multiple architectural levels, including virtualization and integration into the 5G networks. Some feasible countermeasures are also recommended. Chapter 9 deals with security and intelligent management of fog/edge computing, emphasizing trust management, security isolation, unified data storage, and a smart resource partition.

Chapter 10 studies an efficient implementation of network-function-virtualization (NFV)-enabled multicasting in mobile edge computing. Network functions offer special architecture support, which includes several security functions, such as intrusion protection/detection system (IPS/IDS), firewalls, web filtering, and flow filtering. Chapter 11 overviews the blockchain technology as an architectural building block for trustworthy distributed applications in fog/edge computing. To support the development of dependable fog services, this chapter also discusses how to use the blockchain to realize security services such as authentication, secured communication, availability, privacy, and trust management.

Part V surveys applications of fog/edge computing. Chapter 12 starts with fog/edge computing in Industrial IoT (IIoT). The focus is on how the creation of the IIoT technology with fog/edge computing revolutionizes the industrial. Several industrial applications are presented, such as smart grids, agriculture, healthcare, and supply chain management. Chapter 13 deals with security problems in edge computing in applications of augmented reality (AR). Several potential impacts that edge computing can make on AR system security, including user authentication, data collection, transformation, and output verification, are given. The authors also present three open problems for future work. Chapter 14 studies the application of data streaming in fog/edge computing for the optimal deployment of data stream processing (DSP) applications, including security and privacy issues. The last chapter of Part V presents results in the fog/edge-based blockchain application for finite-lifetime blocks and discusses a special system call LiTiChain, which allows the deletion of expired transactions and blocks from the blockchain for Edge-IoT applications.

We would like to express our thanks to all the contributing authors. This book would not be possible without their generous contributions and dedications. Our special gratitude is given to the Springer managing editor, Susan Lagerstrom-Fife, who gave us both initial encouragement and continuous support during the book editing process. Finally, we want to thank our families in the USA and in China for their support and patience during this project. This book is dedicated to our parents for their unwavering support and understanding, especially during the difficult period of COVID-19 when the only means of interaction is through remote communication. Readers are encouraged to provide feedback to the contacts below. We hope readers will find this book useful in their study as well as in the workplace!

Philadelphia, PA, USA                                                        Wei Chang

Philadelphia, PA, USA                                                           Jie Wu

# Contents

## Part IV    Architectural Design in Fog/Edge Computing

## Part V    Applications of Fog/Edge Computing

# Part I
# Overview of Fog/Edge Computing

# Confluence of 4G LTE, 5G, Fog, and Cloud Computing and Understanding Security Issues

**Khaldoon Alshouiliy and Dharma P. Agrawal**

## 1  Introduction

Nowadays, if people want to buy a new cellphone, they might find that there are too many acronyms to wrap their heads around. With today's repertoire containing the names CDMA, GSM, LTE, WiMax, this is just the scratching the surface of what technology has molded into today. When one connects to the 'cloud,' which is a computing model in which data is stored on remote servers accessed from the internet, multimedia and private information can be saved safely. This is an important server that has changed people's lives significantly over the years and will continue to grow in innovative ways in the next few. Therefore, we can conclude that the Internet has become more than a desktop accessory—it has become a virtual brain that anyone can contribute to, take, or and spectate. The world relies on the Internet for virtually everything. GPS is used to trace or route distances and durations. Cloud access allows from making money from just the palm of your hand using a cellphone, to immediate entertainment through games and streaming platforms. The capabilities of communication have also evolved through applications such as WhatsApp, Skype, Viber, and Facetime—making the lives of people easier and more comfortable to see someone on the other side of the world on a screen right in front of them. No one can disclaim the impact of the Internet, telecommunications, cloud computing and how they have altered all aspects of life. As a result, mutuality of these technologies has, as a result, established a huge dataset which leads to control of money and power. Process, storage and protect costs a lot of efforts, specifically in terms of cloud computing, and they are prominent factors in the workplace internationally. Companies like Microsoft

K. Alshouiliy (✉) · D. P. Agrawal
University of Cincinnati, Cincinnati, OH, USA
e-mail: alshoukr@mail.uc.edu; agrawadp@ucmail.uc.edu

© Springer Nature Switzerland AG 2021
W. Chang, J. Wu (eds.), *Fog/Edge Computing For Security, Privacy, and Applications*, Advances in Information Security 83,
https://doi.org/10.1007/978-3-030-57328-7_1

3

and Amazon provide virtually the entire world with their cloud and services and have left a very reliant mindset upon the workforce, and thus the power and money aforementioned are granted as their very 'lifeforce.' That is why, it is important to consider these concepts and associated security issues and examine how they can change the future.

## 2  4G and LTE

To start with, the first question that comes is what is 4G? To explain simply, 4G is a newer generation of the 3G. The International Telecommunication Union (ITU-R) set standards for 4G connectivity, approved in March 2008, requires that all services described as 4G to adhere to a set of speed and connection standards. For mobile use, including smartphones and tablets, connection speeds need to have a peak of at least 100 megabits per second, and for more stationary uses such as mobile hot spots, at least 1 gigabit per second. When these standards were announced, these speeds were unheard of in the practical world, as they were intended as a target for technology developers—a point in the future that marked a significant jump over the current technology. Over time, the systems that power these networks have caught up as new broadcasting methods have found their way into the products, as previously established 3G networks have been improved to the point to which they can be classified as 4G [1]. The next question is "how does 4G work?" The best answer to this question is graph provided in Fig. 1.

All customers are connected to the terrestrial unit and then to the base station where everyone can access to the Internet and other services. If the cell phone doesn't support 4G, then there is no way to get any service. A following question would be, "how does 4G compare with 3G?" To answer this question, we explain

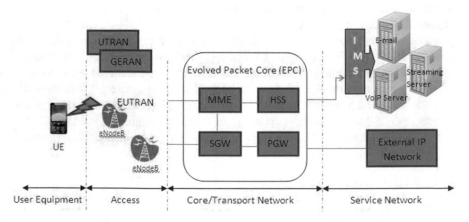

**Fig. 1** 4G network [2]

that in 3G, a network required to meet a set of technical standards for speed and reliability must offer peak data transfer rates of at least 200 kilobits per second. The first network that was able to reach this requirement was phased out in the U.S. in 2003, and with smartphones starting to achieve more mainstream adoption, there was a resulting rise in demand for better mobile broadband service. This drive for higher data speed moved the pace forward in only a few short years, and now 3G networks can be accessed from 200 kbps to thousands of times higher everywhere. A network will provide maximum data speeds of at least 100 megabits per second for high mobility communication (users in vehicles, trains, etc.) as well as at least 1 Gigabit per second for low mobility communication (walkers and stationary users) to be branded as 4G. Nevertheless, not all 4G networks are made together—they come with a range of different colors, and others are introduced quicker and larger than some. LTE, WiMAX, and HSPA+ are the most common deployments, but LTE is without question the most commonly deployed by major US carriers. Since we already have 4G, then what does LTE mean?

4G was created by researchers to become what we now know as Long Term Evolution (LTE). People use 4G as LTE but it's not the same, truthfully. The distinction between 4G and 4G LTE is that on a smartphone, the latter one operates more effectively, offering the highest efficiency and better network coverage. In addition, the 4G LTE is the fastest in terms of speed. The basic argument would be that 4G LTE networks will transmit information at speeds between 5 and 12 megabits per second—enough to broadcast live content seamlessly, with faster reaction time for playing video games online. By comparison, practical transmission rates for 4G networks can range from 3 to 8 megabits per second, relying on the latency—the cellular operator and the particular technologies used by its data network [3]. To follow this, what is next in the world of Telecommunications?

## 3  5G Network

"What is 5G? Do we need 5G? What is the difference between what we have now and 5G?" There are many questions that pop up in people's heads about 5G. To begin, lets understand the 5G concepts. 5G is the fifth generation of the telecommunication cellular network, but the question is, why do we need 5G? The answer is to fulfill our future needs since LTE soon will not be able to satisfy our requirements. 5G cellular network promises to offer sub-millisecond latency and 1 gigabit per second transmission speed to its users. Nonetheless, the new cloud-based storage and data distribution paradigm does not allow these QoS assurances to be easily implemented due to the amount of wired network hops between the 5G base stations and the cloud (different cables and wireless connections provide different data transmission), which contributes largely to a substantial improvement in latency. The forwarding directed to the cloud of all the data generated by the devices will impact the bandwidth and contribute to congestion. Therefore, it is important to house processing near the computers, close to the data source, so that data can be accessed, stored and sorted out by the time, it enters the cloud for the high-speed transfer of 5G. It brings up the computation, storage and networking services to the edge of the network and creates many new areas of study to extend Fog computation over the architecture of the cellular network. This segment addresses the implications of spreading cloud infrastructure to the edge by addressing use cases that can be realized by fog computing through 5G networks [4]. However, 5G contents five different things in it, thus earning its name. It has Millimeter waves, Massive MIMO, Small cells, beamforming, and Full duplex transmission. Each of these concepts works totally different from the earlier networks as follow:

- *Millimeter Waves*

   In 5G the electromagnetic signals distances from 1 mm to 1 cm are usually referred to mm-waves. From 5G of cellular connectivity networks, optimistic output is required to satisfy a range of applications (e.g. remote control, tracking, adaptive transport systems, and tactile interaction), with customer experienced data levels of up to 1 Gbps (500 Mbps) in downlink (uplink) and latency as low as 0.5 ms. Such goals cannot be seen easily by leveraging the bandwidth available for 4G systems, rather they require the introduction of additional frequencies. When the International Telecommunications Union (ITU) launched the international mobile telecommunications standard 2020 (IMT-2020) as early as 2012, it also called for a new distribution of bandwidth to cellular networks around the world. In 2015, the World Radio Communication Conference (WRC) defined different portions of spectrum for mobile communications from 24 and 86 GHz. It is estimated that tens of GHz will be made available in the mm-wave band, in compliance with different spectrum allocations per region [5].

- *Massive MIMO*

   Massive Multi-input multi-output (Ma-MIMO) integrated networking infras-tructure facilitates cellular connectivity improvements and integrates fixed broad-band technologies such as Wi-Fi and LTE. Another wireless communication technology is the small cell which plays a vital role in providing 5G communica-tion. In addition, the Ma-MIMO increases the spectral efficiency for the cellular networks by using antenna array at the macro base stations (MBSs) provided. Such MBSs comprise a significant number of active elements to perform the coherent transceiver operation. The antenna array ensures better signal paths and improve efficiency by increasing data transmission rate and communication reliability. This approach requires precise emphasis to be put on providing electricity for intended consumers, and the effect offers optimum energy output capacity [6].

- *Small Cells*

   Small cells are made up of complex technologies. This is a node with less regulated, wireless radio connectivity that can monitor for both licensed and unlicensed spectrum. This has a length from 10 m to less than a kilometer. It appears small, but they can individually increase their surface area relative to length. It will provide improved cellular coverage, power, and applications for the households, organizations, and rural areas. They can include various types of technology in it. Small-cell base stations are a community array, which plays a very significant role in increasing wireless network capacity. It can have consistency and has the potential to improve service quality at a highly competitive price. It's very useful for the consumers to incorporate a small-cell network. The number of small-cell towers will decline, too. It also provides a fine, cleaner signal with less power [7, 8].

- *Beamforming*

Beamforming is equivalent to providing wireless network traffic signals, to avoid data streams colliding and interfering with each other.

Beamforming technology allows base stations to target data streams on individual applications, allowing for improved performance and incoming data streams while reducing interruption. It can achieve so by monitoring individual signals as they bounce around, and using "signal-processing algorithms" to triangulate and map the optimal path back through each user unit through the air—effectively creating a "wave" of data that flows directly from the base station to the end of the user [9].

- *Full Duplex*

Due of the interference it creates, many base stations today cannot transmit and receive data synchronously at the same frequency. Instead, the sending and receiving of data will take turns at these base stations. Alternatively, certain base stations are capable of transmitting data over one frequency and receiving it over another, but this is not common.

Full Duplex technology will allow base stations to send and receive data at the same frequency—effectively doubling wireless network capability at their physical layer. This is important, as 5G networks with emerging applications will require increased data capacities. This technique has only recently become possible thanks to silicon transistors that mitigate the interaction of the signal with itself and the kinks are yet to be found but researchers are high expectations for the potential of full duplex technique [9]. However, one of the biggest questions is how 5G helps in terms of the cloud, fog networks, and other technologies?

## 4   Fog Computing on 5G Networks

Fog computing and 5G networks are two technologies that have distinct backgrounds but will eventually merge, as the promises provided by the 5G network is to make it possible to get data out to the edges.

## 4.1 Fog Computing: A Requirement of 5G Networks

The 5G mobile networks are scheduled to enter the market by 2020, but not a reality at the moment. Communications in 5G networks will indeed be associated with high-frequency signals that can have more bandwidth to deliver smoother, higher-quality video and multimedia content in the millimeter-wave frequency band. 5G network aims to have millisecond and sub-millisecond latency while at the same time delivering a maximum rate of over 1 gigabit per second [10]. The latency is so low that the likelihood of radio communication between the bottlenecks is excluded. Next-generation mobile networks are designed to accommodate communications that are not limited to human beings (where one can probably mask the latency) because they are often developed to facilitate secure and efficient machine-to-machine communication, a use-case that demands low latency in order to be successful. It has to support fog computing for 5G to be effective, otherwise low latency radio interfaces would be of no benefit. A standard 5G network would have smartphone users connecting to a base station, which in turn will be linked through wired connections to the main network. To eventually access the cloud servers, requests for a web-based service will go via the base station and the core network. In such a deployment, even though the low latency radio interface enables sub-millisecond communication between the mobile device and base station, sending the request from the base station to the cloud will lead to increased delay in the order of magnitude.

Fog computing will play a major role in addressing the demands of future 5G networks. The platform is marketed as a successful way of providing low latency offered by the 5G New Radio standard. With plans to phase out 5G globally by 2020, 5G and fog computing integration are considered to be an eventual result of getting computational activities closer to the edge of an industrial network [11].

## 5 Fog Computing in 5G Networks: An Application Perspective

Having the 5G networks more than just a networking system is imperative. If supplied by the network close to the machines, processing and storage facilities would allow applications to take advantage of low latency radio to have very quick end-to-end response time. This would significantly benefit both the consumers (by offering prompt responses) and the operator (by the backbone network loads). Cloud-to-edge networking shapes the concept of fog computing, and it would not be incorrect to say that 5G networks can't meet their commitments without fog computing. While most see it, fog computing is not a convenience but a required prerequisite for 5G networks to thrive. Small cells (pico and femto cells), also known as micro cells, are a core feature of the 5G network that enables fog computing. Small cells will relieve the pressure on base stations (macro-cells) at the roof top by allowing end points to attach to them. A unit can attach to a macro-cell or a micro-cell. This makes the architecture of 5G networks a hierarchical one—with at the apex the central network (cloud), followed by macro-cell base stations and base stations for micro-cells, and eventually, end users. Therefore, from the fog computing viewpoint, all macro and microcell base stations form the fog nodes, i.e., networking nodes that also provide computation and storage. Packets sent uplink by the systems should be analyzed at the base stations of micro-cells or macro-cells until they enter the main network. Another major development in technology is that 5G provides secure device-to-device connectivity. User data is transmitted directly to the receiver system from the sender computer, with the base station only handling control information from this switch. It allows communicating between devices to take place without burdening the base station, thereby embellishing fog systems with the scalability of managing multiple interacting devices. With projects containing multiple linked points and constant connectivity between these points, with an example, smart houses, this would be of categorical benefit. The remainder of this section addresses 5G network infrastructure, and how fog computing can be realized. Besides this, the fog applications architecture is often defined as segregation of application logic into components that can leverage the services offered by fog computing [4].

## 5.1   Physical Network Architecture

A fog network over 5G's physical network infrastructure will expand the state-of-the-art Heterogeneous Cloud Radio Access Networks architecture [12]. Both application processing activities are conducted in the conventional HCRAN architecture on the cloud within the core network, which requires billions of end-users to transmit their data to the core network. Such a large volume of connectivity may be crucial to the functionality of the fronthaul which could overburden the core network, which would have a negative effect on the end-user QoS.

An elegant solution to this issue is to pull down processing and storage capacities from the cloud near the edge, so that it is possible to eliminate the need to transfer all the produced data to the cloud by end-users, thus alleviating the fronthaul and the central network of the enormous traffic surge. Figure 2 indicates areas where this processing and storage offload can be carried out. The architecture of the fog network consists of three logical layers that are seen in Fig. 2. The machines in each layer are capable of hosting computation and storage so that complex computing offload policies could be developed.

- *Device Layer*: All terminal machines connected to the fog network are subsumed by the device layer. The products include IoT products such as cameras, gateways and so on, and handheld devices such as smartphones, laptops, etc. These devices can share data directly with the network or connect peer-to-peer with each other. Such systems are the lowest layer of the fog networks, being the source of

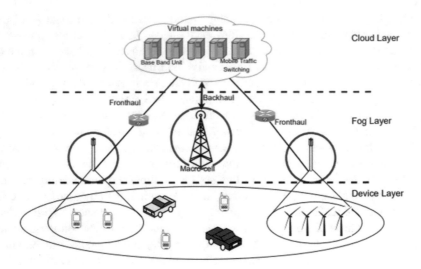

**Fig. 2** The network architecture of a fog network over 5G

all data accessing the network and the primary actuators executing the tasks. The application layer hosts computing either by embedded coding (for low-end equipment such as sensors) or as a program running on the computer's operating system.

• *Fog Layer*: The fog layer is comprised of intermediate network equipment situated in the device layer between the end devices and the cloud layer. The first phase of offloading in this layer is the Remote Radio Heads (RRHs) and small cells are connected to the main network by fiber fronthaul. Handling of incoming data would greatly reduce the pressure on fronthaul here. Macro-cells also form an offload processing point which sends processed data over backhaul links to the core network. Ethernet links and intermediate items, such as routers and switches, in the direction from the radio heads to the core form possible locations where processing and storage activities can be removed. Development of software on such platform is made possible by incorporating virtualization technologies. Each software is bundled as a virtual machine and is installed on a suitable computer. Virtual machines of the program run alongside the virtual machine of the host OS (which executes the initial network operations) on a fog computer hypervisor.

• *Cloud Layer*: Its layer forms the apex of hierarchical architecture, with virtual cloud servers being the offload points for computation. Theoretically, the cloud's unlimited scalability and high-end architecture allow computation to be performed and involve extensive computing and massive storage that cannot be achieved at the edge devices. In addition to the processing of application layers, the cloud layer contains baseband units that process data from RRHs and small cells to application servers via fronthaul and route processed data.

## 5.2   Application Architecture

A fog-ready program must be planned to exploit the full potential of the fog. Usually, as shown in Fig. 3 [13], an application ready for fog will have the following components:

- *Device component*: The part System is connected to the end devices. It executes operations at the system level, often handling resources, minimizing redundancy, etc. At periods where the end-user is not only a light client, it also houses application logic, which needs very low latency responses because this part is executed on the computer. But this package does not include heavy computing activities due to the underlying device's resource constraints.
- Fog component: An application's fog portion performs tasks that are crucial in terms of latency and requires computing power that end devices cannot provide.

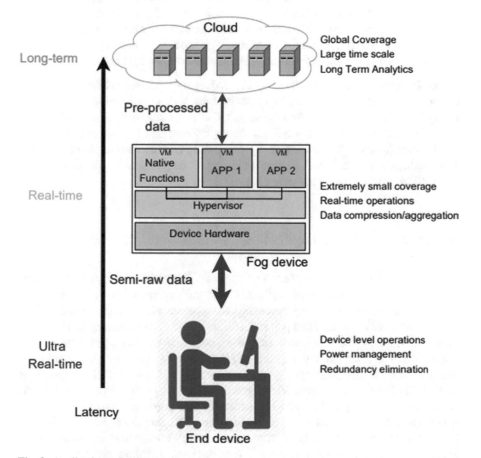

**Fig. 3** Application architecture

Additionally, because the fog feature is designed to operate on fog devices at the ground, this component's coverage isn't universal. Therefore, this portion should host logic that involves execution of only local state data. The part of fog is not attached to a particular unit. It is safe to exist between the edge (consisting of end devices) and the cloud in any form of the system. Mapping the fog components into devices relies on the offload points in the direction from the cloud to the edge. Based on the application's geographical scope and latency criteria, the fog portion could be hosted at either of these offload points. In addition, the positioning of fog components on suitable fog nodes forms a fascinating and significant research area.

- Cloud component: The cloud component of the core network is bounded to cloud servers. It contains logic for long-term analyses of the data obtained from the lower layers and for operations with no latency restrictions of any sort per sec. Application activities requiring tremendous computing power and storage are ideal to be put in the cloud portion, allowing them to leverage the cloud's unlimited resources. Furthermore, because the cloud layer is at the center of the network, it gathers input from all users and therefore has a broad understanding of the entire system. Therefore, application logic requiring knowledge of the global state of the system should be placed in the cloud component of the application.

Application output is calculated by the coding logic through different layers of a fog-ready program. Incorrect logic positioning will cripple an application and make it unable to leverage the advantages that fog computing has to offer. The following sections address multiple use cases, the criteria of which can be fulfilled by the specific level of service that fog computing offers when implemented over a 5G cellular network. We also present the correct mapping of application logic to application layers for each use-case [14].

## 6  Latency Critical Application: Mobile Gaming

Gambling is no longer just for fun; it is for profit. Entertainment joys have become an integral part of the corporate culture. Cloud gaming provides a chance to have more fun inside the game, by involving the players all over the world. Cloud gaming, also called on-demand gaming, is thus a new type of gaming environment made possible by the emergence of online computing, allowing geographically remote users to compete with one another. Cloud computing is an easy and cost-effective way to offer high-quality gaming content and has opened up numerous market possibilities as a result. Video games run on powerful game servers in the cloud on an online gaming network, while players communicate with the game using the Internet-connected thin client program. The thin client is an application that is lightweight and can be accessed on resource-constrained computers, such as handheld devices. Cloud gaming is omnipresent, enabling players to play a game from anywhere and at any time, whereas software creators may customize their games for a particular computer configuration. A cloud gaming framework effectively allows a game program on cloud storage, then broadcasts the gameplay scenes back to the player as a video series. A game player communicates with the game through a thin client, which is responsible for viewing the video obtained from the webserver and transmitting the player's experiences with the game to the web. Cloud gaming is one of the technologies requiring stringent assurances of latency, and the inability to have would have a negative effect on user experience. Additionally, cloud gamers are often particular about the consistency of the video that is made on their light servers. From this, we can infer that the design of a cloud gaming program must take into account the distribution of energy, scalability, and acceptance to faults as well as the needs of gamers. Traditional mobile gaming deployment includes storing both the computing and data in the cloud, thus making mobile gaming synonymous with online gaming. Nevertheless, connecting with the cloud with any request cannot always be the best idea, particularly where latency requirements are high. Through a large-scale empirical analysis, Choy et al. [15] have shown that contemporary cloud networks cannot satisfy strict latency criteria needed for reasonable gameplay for many end-users, thereby putting a cap on the number of potential users for an on-demand gaming service. Based on observations, they concluded that extending the cloud network with edge servers would substantially improve the viability of on-demand gaming or online gaming. Consequently, unloading any computing that is involved in the cloud-based game to the edge makes sense. They presented three approaches to computing: cloud-only, edge-only, and a mixed approach in [15]. Experiments show that the number of users supported in a hybrid environment that used both cloud and edge servers rose from 70% in a single-service implementation to 90% in multiple users.

These studies give an ample support to the fact that fog computing is an efficient platform for deploying on-demand games, and we can discuss the deployment of a cloud-game on fog infrastructure in the following sections.

- *Requirements*: Cloud gaming is a dynamically immersive technology with strict latency and video quality specifications and can have a significant effect on user experience. This addresses the usual criteria of on-demand game as follows:
- *Interaction Delay*: The authors [16] performed a categorical study of state-of-the-art cloud gaming systems and took the innovation to the fore in their system architecture. They also stressed contact latency and streaming efficiency as the two attributes of cloud gaming service specifications. Standard online games will render on the local computer, then later refresh the game state on the game server. Therefore, a typical online gaming player doesn't experience the impact of pause in interaction. However, the processing is off-loaded to the server in the case of online games, and a thin client does not have the ability to mask the user's interface delay. This makes cloud gaming tolerant to less delay than conventional online gaming systems. The average interaction delay for all cloud-based games will be at an average of 200 ms. Other games, especially action-based games like First-Person Shooter games, are likely to allow less than 100 ms interaction delay so that the level of experience of players is not affected.

- *Video Streaming and Encoding*: If a cloud-based game player sends an instruction, it needs to go across the internet to the cloud-based game server, interpreted by the gaming logic, made by the processing machine, compressed by the video encoder, and transmitted back to the player. The encoding/compression and delivery to end-users must take place very quickly in order to prevent loss of the Quality of Experience (QoE) of the consumers. Besides timeliness in encoding the consistency of the transmitted video is also an important element in deciding user experience [17].
- *Design Requirements*

  - Low-latency response: In case of high response time, the user interface would be disrupted, rendering low latency response a crucial feature of mobile gaming. To ensure low response time, the network should be sufficiently flexible to enable user inputs to access the game server when the game logic is being processed and to record, encrypt and transmit audio/video in a timely manner.
  - High bandwidth: Sharing of video streams forms much of the cloud-game data exchange. To transmit such an immense amount of data that requires a high bandwidth link between the game server and the device, even in real-time.
  - Global coverage: The online gaming platform has to be available from anywhere to be able to accommodate users from different geographical areas. Hence, ensuring a regional scope is imperative for such an operation.

# 7  5G Gaming

Both controllers ought to be replaced in the first light of the 5G as there will no doubt a need to play computer games. In addition to purchasing and installing the cards, there would likely be a major reduction in demand for a decent desktop machine, which usually costs about $1,500 on average. Figure 4 shows what sports would look like in 5G games.

**Fig. 4** Gaming under the 5G [16]

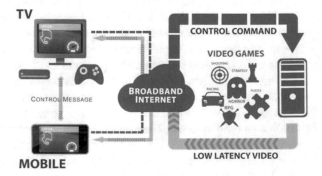

For 5G, all you need is a TV and 5G internet service, which makes gaming consoles seem meaningless. Although speed could be an enticing attribute that advertisers can use and communicate to promote 5G to customers, the guarantee of incredibly low latency may be one of the most promising aspects of 5G networks for gamers personally. Reducing latency is the foundation on which systems can be designed to stream games. With 5G offering latency in the sub-1ms region, any immersive networking technology that the cloud will arise over the next five years is likely to become the center of it.

The other advantage of extremely low latency is the developers' ability to offload some of the computing burdens from a computer. If data can be transmitted quickly between the source and a computer, a greater proportion of the processing can be performed remotely. It is particularly important for mobile devices which literally don't have the battery-operated or thermal power to make a 4K or 8K triple-A game without melting down entirely. The growth of edge computing—the development of smaller, nearby devices rather than a large data center located tens or hundreds of miles away—could interact with 5G networks to unlock experiences on mobile devices that were previously impossible [18].

## 8    5G with Bigdata

As in the near future, we will communicate with each other thousand times larger than we now have, and a huge flood of data will be added into the data flow. Big data is a term that refers to data obtained through the internet or from their own initiatives by any of the organizations. This analysis helps forecast issues such as conditions or health care. Unlike 4G/LTE, 5G would be more than a one-way cable, providing a purpose-built network developed and optimized to allow both mobile devices and automated systems. 5G would in several ways be a facilitator and catalyst in the new digital revolution, also called Technology 4.0. 5G promises to offer ultra-low latency (less than a millisecond delay) fast data speeds (in the Gbps range) for applications in Industrial Automation, Tactile Internet, Robotics and AR/VR applications, etc. [19].

### 8.1    Data Analytics

Data processing takes full advantage of the features of the 5the G network, such as high speed, low latency and mobile edge computing (MEC). The ability of 5G to enable vast networking through a range of devices (sensors/gateways/controllers), supported by the centralized network infrastructure, provides the potential to turn large-scale data-at-rest and data-in-motion into real-time information through powerful intelligence. In 5G Information Analytics should play a dual role. On the one hand, analytics will continue to help multiple enterprise applications/use cases over 5G networks, and on the other hand, analytics will play a vital role in 5G and network operations roll-out [20].

## 8.2   Application Intelligence

5G technology covers a very broad range of situations, from wearables, mobile phones, smart cities, electric vehicles and modernization of the industry. IoT and Industry 4.0 will be the big engines for 5G implementations. Context-Aware Engine (CAE) will become an important part of 5G and will make the network aware of the underlying meaning and become intelligent enough to have a smart interface and improved flow control decisions for an individual user on the network.

## 8.3   Network Intelligence

5G networks are highly dynamic with several levels of virtual services, software and physical RAN properties, spectrum use, remote networking nodes that are based on virtualization (NFV) principles for SDN as network features. In order to create a scalable 5G network, network management must become very important where roll-out and organizational complexities need to be streamlined. Network planning and optimization (NPO) determines whether to optimize particular network operations, and the serviced method should be based on a machine learning algorithm that analyzes the dynamics of network use and traffic data more closely. In short, Operations and Business Support Systems (OSS/BSS) will have analytics integrated and embedded into their toolset, unlike the traditional system where analytics has been an afterthought [20].

## 9   5G and Satellites

The new 5th generation of mobile networks (5G) is deliberately designed to offer by-design standards of extreme versatility to enable increasingly heterogeneous performance, scalability, and implementation scenarios for services and applications. To achieve these ambitious targets, proposed 5G specification should be called a "network of networks," because it will enable the introduction and combination (as required by the overlying applications) of different and alternative network stacks and communication technologies. The main cross-cutting enabler for 5G architecture is the "virtualization" approach. It would enter the 5G infrastructure on every layer and have "as-a-service" relevant services. Clear and practical manifestations of this phase are the technical structures Network Functions Virtualization (NFV), Software-Defined Networking (SDN), and Software Defined Radio (SDR), which together constitute the 5G architecture's "virtualization" engine [21]. The future role of satellite networking in these environments becomes apparent when referring to this model of slicing into which satellite services can be integrated. When viewed in their present implementation, they are integrated either as Physical Network Functions (PNFs), or of much greater significance, by integrating their virtualized operating elements as functional structures within the 5G architecture system. Satellite networks can play several roles in 5G, due to their inherent ubiquity and transmission capabilities. The satellite will serve as a main single backhaul link for rural areas, ships, boats, trains, or an alternate means of delivering opportunistic external communication and bandwidth resources—also improving quality of service—or as a mere transportation subnetwork. The incorporation and utilization of satellite technologies within the 5G network clearly raise additional criteria and challenges around infrastructure and operation. For instance, on one side, it is reasonable to assume that satellite subnetworks can be directly applied to those traffic flows (e.g., mission-critical data) that are associated with 3GPP 5G [22] Quality of Service (QoS) and Indicators (5QI) allows a delay of the order of 1-2 hundred milliseconds. On the other hand, satellite subnetworks can be implemented to promote and allow delivery and operation of other intermediate 5G subsystems more effective, such as edge computing nodes required to cope with tighter and more demanding 5QI rates, such as for Augmented Reality applications. Satellite interconnectivity can be used in the edge computing scenario for the simultaneous unicast/multicast/broadcast regional delivery of binaries of video, audio, and computer applications to a wide number of terminals.

Satellites will have unique opportunities for providing 5G services in rural areas. In addition, satellites will also support machine-type communications, paving the way for future technologies like smart production, environmental conservation, shipping, animal monitoring, etc. By 2020–2025, there will be more than 100 High Throughput Satellite (HTS) networks utilizing Geostationary Earth Orbit (GEO) spacecraft, but also Low Earth Orbit (LEO) mega-constellations, providing terabits per second of bandwidth worldwide. These developed satellite systems are expected to provide Radio Access Networks (RANs), dubbed satellite RANs, to be integrated into the 5G system along with other wireless technologies, including cellular systems, Wi-Fi, and so forth. A native feature of 5G would be the smooth transition between heterogeneous wireless access technologies, as well as continuous use of different radio access technologies to improve flexibility, efficiency, and capacity. Figure 5 portrays the deployment of 5G satellites [22].

## 10 An Introduction to Telecom Network Security

Today's telecommunication networks are generally separated into four logical parts: radio access network, core network, transport network and interconnect network. Every section of the network consists of three so-called planes, each of which is responsible for carrying some type of traffic, namely: the control plane carrying

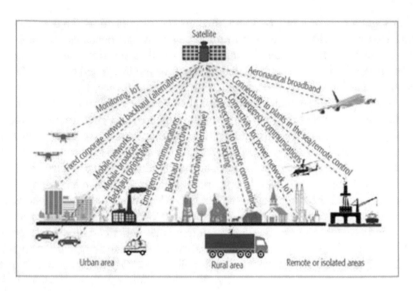

**Fig. 5** Integration of satellite with 5G [23]

the signaling traffic; the user plane carrying the payload (actual-) traffic; and the administration plane carrying the administrative traffic. As regards network defense, each of the three planes can be subjected to specific types of threats. There are also standard hazards, which can potentially impact all three planes. Telecommunication network security is defined by the following components:

- Standardization: a mechanism by which carriers, suppliers and other stakeholders set guidelines for how the world's networks should operate together. This also covers how best to protect networks and applications from hostile actors.
- Network architecture: Network suppliers plan, create and enforce negotiated specifications for usable network components and structures, which play an essential role in making the end-network product both efficient and secure.
- Network configuration: networks are designed to a specified level of protection during the implementation process, which is necessary for defining protection criteria and further improving network reliability and durability.
- Network deployment and operation: operating processes that allow networks to function and achieve targeted protection levels that are highly dependent on network deployment and operation [24].

## 11   Understanding Security in the Era of 5G

Telecommunications networks are increasingly expanding through a wide technical landscape including virtualization, IoT and 4.0 Industry. This is achieved by an ecosystem of cybersecurity that is as large and degrading. Advances in technology, along with wider network growth beyond 5G Runs, are projected to have a major effect on health, such as SDN, NFV, and edge computing. The 5G 3GPP protocol is agnostic in that it is sufficiently robust to require various forms of physical and virtual connectivity between, for example, the radio access network (RAN) and the core network, from a remote computer to the core network. The distinction between RAN and core roles raises concerns about competition and efficiency. From a technical, strategic and efficiency viewpoint, failure to make use of technological advances in the design and implementation of 5G commercial networks would eventually prove counterproductive in the realization of particular 5G use cases, such as essential machine-type communication or applications that belong to autonomous latency-sensitive systems [25].

With the introduction of 5G networks, the security issues facing service providers are expected to escalate. Many providers would then want to invest in these capabilities in an IaaS platform. If this is the case, then the operators will search at some main features to ensure optimum security. Service companies are facing a variety of cyber challenges never before seen. Architecture is changing to support 5G networks, and the bad guys will just be opening further doors. 5G would endorse specific use cases such as e-health and wired vehicles, as opposed to 4G and previous iterations. In these scenarios, the defense may be a matter of life and death.

Furthermore, network slicing involves new and complex network protection for each slice and for each individual device. Additionally, there would be an increasing DDoS threat from 5G applications on the RAN side which may have been hacked. What is the answer to all of this? The requisite costs and know-how can be huge. Here are the five key things that need to be based on: Vulnerability detection will mitigate certain underlying problems that already account for other security concerns. Consideration of firewalls to defend the network from external networks and access controls to reduce user harm. Tools for detecting and preventing intrusion may also aid by blocking essential risks to health.

We need to protect and patch sophisticated malware. To do this, one needs to go beyond signature-based methods to find the stuff intended to circumvent simple filters. Behavior-based inspections of endpoints are necessary, probably using sandboxing. If a threat has been detected, all instances of it on the network must be deleted and blocked. Detection of irregularities uses the analysis of packets, big data, and machine learning to identify risks not detected by simple filters. It is much more efficient when installing in network switches and routers because it transforms such tools into safety sensors.

DNS intelligence is critical because today, it is a major vector of attack. Tools that track DNS behavior and that guard against something harmful are of great benefit. Yet this in-house production is highly costly and resource intensive. But one needs to search for an experienced supplier who can support. Any successful 5G security policy should be focused on threat intelligence. Service providers continue to search for partners who can help interpret their activities as profile hackers. This proposed attempting to provide information from the largest possible variety of sources. Ensure that the company delivers only actionable information and is provided promptly [25].

## 12    5G Security and New Use Cases

The coming 5G networks have the ability to disrupt vertical markets, allowing a large variety of new technologies to be created—all of which would need new, differing protection standards. Automated vehicles will be the first such example. The threat of cyberattacks on automobiles will rise as autonomous vehicles become more common. To counter this, the National Highway Traffic Safety Agency takes a multi-layered approach to technology, as it supports innovations for driver assistance.

Among other advancements in the healthcare sector, 5G technologies can help with quicker delivery of big patient data, virtual surgery and virtual patient tracking through IoT devices. These advancements, however, are being offset by the need for ever-greater protection. Creating threats including misuse of patient identification, violation of health privacy and protection of medical records. Smart homes may need stronger verification mechanisms, such as biometric detection, used in Sensory-made apps that use speech and facial recognition, or fingerprint entry door locks in hardware stores. In general, IoT devices and sensors will demand more complex authentication to prevent unauthorized access [26].

## 13    5G Security and New Network Architectures

In expectation of 5G networks, modern cloud virtualization systems such as SDN and NFV are booming, but they still come with new security issues. Due to its open, flexible, programmable existence, SDN and NFV create a new avenue of threats to security. An SDN network feature, such as the management interfaces, may be used to target the SDN controller or management system and bring the device down. In addition to the standard, the security of 5G network infrastructure will evolve. Due to the capacity of 5G networks to be separated into separate slices, each virtual slice of the network will need new security requirements depending on the needs of specific use scenarios. Compromised RAN side 5G devices could also pose a greater DDoS threat [26] (Fig. 6).

In December 2018 the UK Government's Department for Internet, Arts, Media and Sport published a comprehensive report on 5G infrastructure and security. It outlined four security requirements that need to be fulfilled by 5G networks. Cross-layer protection will be the first security feature. A cohesive structure is required to organize various security approaches, such as software or the IoT, for each security layer.

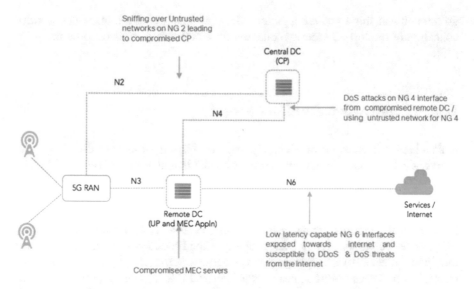

**Fig. 6** Vulnerabilities for a network with a distributed 5G core [26]

Secondly, the monitoring will be end-to-end. A safe link between the user and the core network should be provided for the communication paths. It is made difficult by the centralized existence of 5G network. Cross-domain authentication is a must-have. 5G networks generate a huge quantity of new usage cases and particular demands. Because the vertical market can only expand to accommodate these new use cases, the report calls for collaboration among those in the 5G network to incorporate comprehensive security solutions across domains. A final mechanism would be the concept of secure-by-design. As the network changes and evolves, security must be built into the design during development [26].

## 14  Security Concerns for Cloud-Based Services

According to the Cloud Security Alliance (CSA), more than 70% of the world's companies and markets are currently running on the cloud—at least in part. For advantages such as reduced operating costs, more accessibility, automated app upgrades, improved teamwork and the freedom to operate from anywhere, 70% is not an unprecedented figure. The cloud does have its share of security problems, though. The "Data Protection Focus Survey" recently found that "90 percent of organizations are extremely or mildly worried about the protection of the public cloud." These concerns run the gamut from vulnerability to hijacked accounts to malicious insiders to full-scale data breaches. Although cloud services have ushered in a new age of transmitting and storing data, many companies are still hesitant or make the move without a clear plan for security in place. Here, we are about to go through and show you a big picture view (we just giving a hints about security concerns) of the top 12 security concerns for cloud-based services to be aware of [27].

## 14.1 Security Concern: Data Breaches

A data breach is an event in which a person who is not allowed to do so exposes, shares, steals or uses important, secure or confidential information. A data loss may be the main purpose of a targeted attack or may merely be the product of human error, flaws in the code, or public safety practice. These may contain any sort of information not meant for public dissemination including, but not limited to, personal health records, financial information, Personally Identifiable Information (PII), trade secrets and intellectual property. Cloud-based data from an enterprise may be of interest to multiple parties for various purposes. Organized crime, for example, often uses political, health, and personal information to perform a series of illegal activities. Competitors and foreign nationals may be involved in the secrets of confidential knowledge, intellectual property, and trade. Activists may choose to reveal details that may cause shame or harm. Unauthorized sources become a big problem for companies to access data inside the cloud.

Cloud infrastructure is not special to the possibility of data loss but is regularly rated as a top concern for cloud customers. Cloud infrastructure is exposed to the same threats as a conventional corporate network, as well as new avenues of attack through shared services, cloud provider employees and their devices and cloud provider third-party partners. Highly open cloud services and the large amount of data they host makes them an enticing target [27].

## 14.2 Security Concern: Insufficient Identity, Credential and Access Management

Data compromises and facilitating attacks will occur due to lack of robust identity access management framework, inability to use multifactor authentication, poor use of passwords, and a lack of continuous automated recycling of cryptographic keys, passwords, and certificates. It is not necessary to embed passwords and cryptographic keys in the source code or to share them in shared databases such as GitHub because there is a substantial risk of detection and misuse. Keys need to be sufficiently maintained and there needs to be a well-secured Public Key Infrastructure (PKI) to ensure that key security operations are carried out. Identity applications need to expand and tackle lifecycle management for millions of users and the Content Protection Policy CSPs. Identity management programs must facilitate the automatic de-provision of access to services when changes of staffing occur, such as cessation in employment or change in position. Identity networks are increasingly integrated, and federating identity with a cloud service (e.g., Authentication Claim Markup Language SAML assertions) is becoming more common in order to ease the app management burden. Organizations preparing to federate their identities with a cloud provider need to consider the protection of the identities system for the cloud provider, including procedures, technology, user segmentation (in the case of a shared identity system), and cloud provider implementation. To cloud service users and operators, multifactor authentication schemes—to example, smartcard, one-time passwords (OTP), and phone authentication are required. Such a method of authentication helps tackle the stealing of passwords, where the stolen passwords require access to services without user consent. Code stealing can manifest lateral movement attacks in standard networks, such as "pass the hash." In situations where the legacy system involves the use of passwords alone, the authentication mechanism must endorse protocol enforcement such as assurance of secure password use as well as organizationally defined rotation cycle policies. Cryptographic keys, including Transport Layer Security (TLS) certificates, are used to secure user access as well as to encrypt user that must be refreshed regularly during rest times. Doing this can fix attacks where insufficient authorization keys are reached. When cryptographic keys are stolen, a lack of key rotation policy can significantly increase the time and complexity of a successful elapsed breach. Every centralized data hidden management system (e.g., passwords, private keys, sensitive client communication database) is a highly valuable option for attackers. The alternative of centralizing passwords and keys is a compromise where an enterprise must balance the efficiency trade-off with centralized key control against the danger presented by centralizing keys. Similar for any high-value item, a high priority will be the surveillance and security for identity and key management systems [27].

## 14.3   Security Concern: Insecure Interfaces and APIs

Cloud infrastructure platforms are providing a series of User Interfaces (UIs) or Application Interfaces (APIs) applications that users utilize to access and communicate with cloud resources. Such APIs are mostly used for provisioning, control, orchestration, and tracking. The reliability of these specific APIs depends on the stability and availability of general cloud services. These interfaces will be configured to defend against all unintended and intentional efforts to bypass the regulation, from authentication and access management to encryption and operation tracking. In addition, organizations and third parties can draw on these frameworks to give their clients value-added services. It raises the difficulty of modern hierarchical API; it also increases the risk as it can allow companies to give their certificates to third parties to enable their service. Generally speaking, APIs and UIs are the most vulnerable component of a program, and maybe the only commodity with an IP address accessible beyond the respected organizational cap.

Such properties assist the objective of times for intense attack, and the first line of protection and surveillance is appropriate controls to shield them from the Internet [27].

## 14.4  Security Concern: System Vulnerabilities

Software flaws are exploitable bugs in systems that can be exploited by attackers to penetrate a computer network to capture data, gain control of the device or disrupt service operations. Vulnerabilities within the operating system components— kernel, system libraries, and programming resources—posed a serious risk to the protection of all services and records. That sort of hazard isn't new. Bugs became an overlying concern after machine invention. Problems only grew further as networks are developed and are potentially exploitable. With the emergence of multitenancy in cloud computing, devices from multiple organizations are positioned in close proximity to each other, and granted access to common memory and services, providing a new surface of attack. While the harm done by attacks on network weaknesses can be substantial, these attacks can be mitigated by simple IT processes. Daily vulnerability testing, monitoring identified network risks and deploying protection fixes or updates go a long way towards closing security holes left open by bugs in the framework. Secure design and architecture can lessen the chances of an attacker taking full control of every part of an information system by limiting who has access to specific systems [27].

## 14.5  Security Concern: Account Hijacking

Hijacking of accounts or utilities is not recent. Methods of attack such as phishing, hacking, and manipulation of bugs in applications will also yield the expected results. Credentials and passwords are frequently reused, further exacerbating the effect of these attacks. Web technology introduces new landscape hazards. When an attacker has access to your passwords, they can search your activities and transactions, manipulate records, return falsified information and redirect your customers to unauthorized websites. The instances of your account or operation will become a new foundation for the attackers. From there they will exploit your reputation's strength to initiate subsequent attacks. Organizations should be mindful of these forms of attacks as well as specific defense-in-depth security techniques to mitigate the harm arising from a violation—and potential lawsuits. Organizations will aim to discourage customer and provider exchange of account credentials and use robust two-factor authentication mechanisms wherever possible. All records and actions, including service accounts, should be tracked and traceable to a human user [27].

## 14.6   Security Concern: Malicious Insiders

In the defense industry, danger posed by an unauthorized insider has been frequently addressed. Although the extent of vulnerability is left to question, it is not the case that an individual enemy is an insider vulnerability. CERN describes an insider threat as follows: "A malicious insider threat to an entity is an actual or former employee, contractor or another business partner who has or has allowed access to the network, program or data of an entity and who has intentionally violated or misused that access in a way that adversely affects the secrecy, credibility or quality of the organization's information."

## 14.7   Security Concern: Advanced Persistent Threats

Advanced Persistent Threats (APTs) are a type of cyberattack that infiltrates networks to create a foothold in target corporations' computing infrastructure from which they smuggle data and intellectual property. APTs follow their targets over long stretches of time, frequently responding to the protective mechanisms intended to protect themselves against them. Spear phishing, direct intrusion schemes, distribution of attack code through USB devices, infiltration through partner networks and the use of unsecured or third-party networks are typical entry points for APTs. When setting up, APTs will travel laterally through data center networks and blend in with regular network traffic to achieve their goals. This helps to educate these teams of the new sophisticated malware threats targeting businesses and public institutions. While APTs can be difficult to identify and remove, aggressive monitoring procedures can protect others. As an example, teaching users to understand and manage social manipulation tactics such as spear phishing, which are widely used for deploying APTs, is important. Regularly updated detection campaigns are one of the strongest protections against these forms of threats since all of these exploits need user interaction or action. Until opening an attachment or clicking on a connection, staff should be acquainted with thinking twice.

## 14.8   Security Concern: Data Loss

Any possibility of losing one's data forever is frightening for customers and enterprises alike. Data saved in the cloud can be destroyed, rather than for malicious attacks. An unintentional delete by the cloud service company or, worse, a physical catastrophe such as a fire or earthquake will result in a significant loss of customer data unless the vendor or cloud user takes appropriate steps to back up records, including best practices in business continuity and disaster recovery—as well as regular data retention and likely off-site storage. In fact, the risk of preventing data loss doesn't only lie on the shoulders of the vendor. When a company encrypts the data before transferring it to the cloud then it removes the encryption key, otherwise, the data will still be lost. Cloud users should check the rules on contracting data loss, inquire about the reliability of a provider's system to see which company is responsible for data loss and under what circumstances. Many services provide regional replication options, cloud data recovery, and premise-to-cloud backups. The cost of depending on a vendor to hold, back up and secure the data may be considered against performing the task in-house, and if data is extremely sensitive decision to do so, maybe possible.

## 14.9   Security Concern: Insufficient Due Diligence

As managers develop corporate plans, they need to understand digital technology and CSPs. This is important to build a clear plan and checklist for due diligence when reviewing innovations and CSPs with the best chance of performance. A company that attempts to accept cloud computing and pick CSPs without due consideration is exposed to a multitude of economic, political, technological, regulatory, and enforcement threats that endanger its performance. Whether the organization is contemplating moving to the cloud or combining with or replacing an organization that has migrated to or is considering moving to the cloud, this applies.

## 14.10   Security Concern: Abuse and Nefarious Use of Cloud Services

Poorly protected cloud server installations, free cloud server trials and fraudulent account sign-ups from payment instrument fraud reveal malicious attacks on cloud infrastructure models such as IaaS, PaaS and SaaS. Malicious actors may exploit tools from cloud infrastructure to threaten customers, organizations or other service providers. Examples of cloud service-based infrastructure abuse include launching DDoS attacks, email spam, and phishing campaigns; digital currency "mining;" large-scale automatic button fraud; compromised account database brute-force computing assaults; and hosting fraudulent or pirated materials. Cloud infrastructure protection entails CSP prevention of payment system theft and abuse of cloud services, providing cases of Denial-of-Service (DoS) attacks on inbound and outbound networks. A cloud service needs to provide an incident management system to resolve resource misuse, as well as a way of reporting misconduct from a cloud provider to clients. A service provider will have adequate controls to allow a client to manage the quality of their service workload [28].

## 14.11   Security Concern: Denial of Service

DoS attacks are threats designed to prohibit a service's customers from getting access to their data or their programs. Through causing the targeted cloud server to use disproportionate quantities of limited machine resources, such as processing capacity, memory, storage space, or network bandwidth, the attacker—or attackers, as is the case for DDoS attacks—causes an unacceptable device bottleneck and leaves all legitimate service customers puzzled and furious about why the infrastructure is not reacting. Although DDoS attacks appear to create fear and media attention—especially when perpetrators act out of a spirit of "hacktivism" in politics—they are by no means the only type of DoS assault. Application-level asymmetric DoS attacks take advantage of flaws in web servers, databases, or other cloud services, allowing a malicious attacker to carry out a single incredibly small attack payload application—in certain instances less than 100 bytes long. Other attacks can target similarly restricted resources: an economic DoS challenges the cash flow of a company, leveraging the complex complexity of cloud to overpower the capacity of a startup to pay. Similarly, an organization's intellectual resources can be caught up easily with a governmental DoS in a legal job and leaving a corporation similarly unable to deliver business [28].

## 14.12   Security Concern: Shared Technology Vulnerabilities

Cloud service providers are providing their services scalably by connecting networks, websites, or devices. Cloud infrastructure splits the "as a service" package by modifying hardware/software significantly from the shelf—sometimes at the cost of reliability. Underlying modules (e.g., CPU caches, GPUs, etc.) underlying the infrastructure that facilitates the delivery of cloud services could not have been configured to provide robust separation properties for multi-tenant architecture (IaaS), re-deployable systems (PaaS), or multi-customer applications (SaaS). It will lead to flaws in shared infrastructure and can theoretically, be abused in all distribution models. Robust defense policy is recommended and will include compliance and control of computing, data, network, device, and user protection, whether the business model is IaaS, PaaS, or SaaS. The point is that a single bug or misconfiguration will result in a cloud-wide failure across a whole network. Mitigations should be introduced to deter violations of common infrastructures, such as multi-factor authentication on all servers, the Host-based Intrusion Detection System (HIDS) and Network-based Intrusion Detection Systems (NIDS) on internal networks; applying concepts of networking least privilege and segmentation, and keeping shared resources patched [28].

## 15   The Risks to 5G IoT: Preparing for the Next Generation of Cybersecurity Threats

5G's bold new future also hopes to add unparalleled speed and size to IoT activities which also includes a new wave of security threats. 5G IoT networks can no longer blend seamlessly into conventional 4G/LTE cellular-optimized security frameworks as set out in the NSA white paper, 'Modern paradigms such as disconnected operation, small cell data connections, edge-focused communication and more will transform the central authority authentication model on its head.' Security analysts warn of risks to the 5G-IoT system, such as the increased risk of distributed denial of service (DDoS) attacks and Proximity Service (ProSe) intrusions [29]. The massive deployment of 5G IoT open, small-cell networks would challenge holding every device up-to-date and ready to deal with increasingly changing cyber-attacks.

Increased bandwidth would pose more possible security threats, even as the amount of wired device is exposed to unauthorized access is expected to explode. Botnets, attacks at the network level, and other disruptive operations pose a complex security image of advanced, mutable threats. Studies were undertaken by the University of Lorraine and Dundee University also found "important security holes" in 5G links, allowing for massive quantities of data theft [26].

## 16   IoT and Self-Driving Cars

Integrating the Internet into cars that drive themselves is contributing to the modern age of transport. Real-time coordination on a coordinated basis with different sensing media will change our understanding with autonomous driving, provided the following, the number of injuries is dramatically decreased due to the complete collaboration of the sensors. In turn, real-time traffic management, identifying the right route, and speed change would reduce travel costs and effects on the community. Continuous application with the new 5G system offers urgent emergency geolocation coverage. The use of a central network that collects data from various sensors installed on central highways and from the autonomous vehicle navigation system can be considered to be a revolutionary method for automatic driving. In turn, in the event of pollution or adverse weather situations, the device should be able to notify autonomous vehicles, thereby enhancing transport safety [30].

## 17   Virtual Reality Mobile Application with 5G

Virtual Reality (VR), also referred to as one of the next-generation computing technologies, is basically a computer-generated immersive interface. In addition, it takes place with not just audio and video inputs but also other sensory inputs in a virtual environment. The interactive VR interface is also provided via the VR headsets, consisting of head-mounted displays (HMDs). With the expectation of 5G rollout beginning at the end of 2020, fast adoption of fast-end smartphones, and low-cost mobile devices, VR HMD shipments and VR apps have risen steadily. Market research firms predict that, by 2023, the global VR industry will hit a size of over $34 billion. Despite these opportunities, the latest smartphone VR (apps) and platform technologies tend to face many technological difficulties, including high network latency to load and view high-quality interactive content and experiences with 3D and VR environments. Therefore, both lead to reduced user interaction (UX). Experts [31] also believe that a significant hurdle to the universal acceptance of VR applications is the current lack of 5G. The technology will leverage Multi-Access Edge Computing (MEC) and 5G technologies to address these problems. The Digiotouch VR app relies on cutting-edge video technology that requires high bandwidth. With the advent and evolution of modern video formats, particularly 360 degree, even greater bandwidth will be needed. LTE manages 360-degree video at 4K 30fps, but with running at 8K 90fps, 5G is expected to enhance the experience. Generally speaking, VR applications will quickly hit LTE's Gbps cap as devices grow and become more sophisticated. In [31] app will offer state-of-the-art user experience with 5G's expanded capability and a $10\times$ volume. Additionally, its physical location would limit the 5G powered VR applications less. Due to dramatically increased infrastructure distribution capabilities and substantially lower profiles of resource usage. Using MEC ensures that end-users get good communication efficiency as VR demands a reliable, steady signal [31].

There are primarily four kinds of VR HMD devices: the first type is PC VR attached to Desktop, such as Oculus Rift, HTC Vive, etc.; the second type is Console VR attached to a game system, such as PlayStation VR; the third type is handheld VR, untethered with PC/system but with a smartphone inside, such as Samsung Gear VR, Google Daydream, etc [32]

## 18 Security of 5G-V2X

Cellular-Vehicle to All (C-V2X) seeks to address problems related to the conventional accessibility of networking between Vehicle to Infrastructure (V2I) and Vehicle to Vehicle (V2V). Specifically, C-V2X decreases the number of organizations participating in vehicle interactions and allows V2X to be extended to provide wireless protection solutions. For this, the development of LTE-V2X is groundbreaking, but despite its security framework, it struggles to satisfy the demands of fast performance, ultra-high reliability, and ultra-low latency. To address this, 5G-V2X is considered an integral approach that not only addresses LTE-V2X problems but also offers a function-based network setup. Several reports were issued for 5G security but none of them focuses specifically on 5G-V2X security. Rising from Vehicle-to-Network (V2I) to Vehicle-to-Vehicle (V2V), V2X expands its reach by relying on a multitude of innovations, such as Dedicated Short-

Range Communications (DSRC), Wireless Networking in Vehicle Setting (WAVE), CellularV2X (C-V2X), which involves Long-Term Evolution V2X (LTEV2X), 5G Public-Private Network Alliance (5GPPPV2X) [33–35]. CV2X is also seen as a base for implementing technologies like Low Power Wide Area Network (LPWAN), IPv6-Low-Power Wireless Personal Area Network (6LoWPAN) and Long-Range Wide Area Network (LoRaWAN) where conservation of energy is the primary motive of the deployed technology [36].

However, this development poses other security issues including the risk of autonomous vehicles using real-time data and guidance from different sensors connecting to the cellular network. The real-time planning feedback maps can be obtained via C-V2X communications. C-V2X's protection feature helps avoid any intrusion by impersonation and replay that might misguide the vehicle and lead to interruptions and accidents. Continuously, driver authentications in supported cars and secure C-V2X operations can help verify drivers through third-party authentications.

## 19   Conclusion

To conclude, cloud computing is seen as a distribution mechanism that offers a convenient way to store consumer data and safe access to personal and company information. Users are offered Internet-based on-demand services. Fog computing doesn't substitute cloud computing. Fog computing is a major step towards a centralized cloud by managing data at all node locations, fog computing helps users to transform data centers into centralized cloud platforms. Fog is an extension that builds cloud services. Cloud-based data may be segregated and held close to consumers. Moreover, with all 5G network benefits that we explained in the previous sections that will be available for people, it will improve the cloud computing and fog computing work through increasing storage, speed and process and that can bring another benefit to users and costumers from many different aspects like Online games, self-drive and big data collect and control.

## References

1. Hill, S.: What's the difference between 4G and LTE ... and does it even matter? In: Digital Trends. https://www.digitaltrends.com/mobile/4g-vs-lte/ (2019). Accessed 2 Dec 2019
2. Resources: 4G LTE security for mobile network operators. https://www.csiac.org/journal-article/4g-lte-security-for-mobile-network-operators/2/ (2019). Accessed 12 Dec 2019
3. Techblog: Ericsson: MOU with SK Telecom for 5G SA core network; KT commercial contract for 5G roll-out in April. https://techblog.comsoc.org/2019/03/25/ericsson-mou-with-sk-telecom-for-5g-sa-core-network-kt-commercial-contract-for-5g-roll-out/ (2019). Accessed 12 Dec 2019

4. Gupta, H.G., Chakraborty, S.C., Ghosh, S.K.G., Buyya, R.B.: Fog computing in 5G networks: an application perspective. In: Cloud and Fog Computing in 5G Mobile Networks: Emerging Advances and Applications, pp. 23–56. Institution of Engineering and Technology, London (2017). https://doi.org/10.1049/pbte070e_ch2
5. Use of millimeter wave carrier frequencies in 5G. https://www.5gitaly.eu/2018/wp-content/uploads/2019/01/5G-Italy-White-eBook-Use-of-millimeter-waves.pdf. Accessed 30 May 2020
6. Li X, Bjornson E, Zhou S, Wang J.: Massive MIMO with multi-antenna users: when are additional user antennas beneficial? In: 23rd IEEE international conference on telecommunications (ICT), pp 1–6. IEEE, Thessaloniki, Greece, (2016)
7. Hosseini K, Hoydis J, Ten Brink S, Debbah M.: Massive MIMO and small cells: how to densify heterogeneous networks. In: 2013 IEEE international conference on communications (ICC), pp 5442–5447. IEEE, Budapest, Hungary (2013)
8. Jafari, A.H., Lopez-Perez, D., Song, H., Claussen, H., Ho, L., Zhang, J.: Small cell backhaul: challenges and prospective solutions. EURASIP J. Wirel. Commun. Netw. **2015**(1), 206 (2015)
9. 5G – Baicells Fixed Wireless LTE Small Cell Solutions. https://na.baicells.com/5g/. Accessed 31 May 2020
10. IEEE Press Series on Digital and Mobile Communication: Millimeter Wave Communication Systems, pp. 276–276. Wiley, Hoboken, NJ (2011). https://doi.org/10.1002/9780470889886.scard
11. Cranford, N.: The role of fog computing in 5G. https://www.rcrwireless.com/20180131/the-role-of-fog-computing-in-5g-tag27-99 (2018). Accessed 20 Mar 2020
12. Peng, M., Li, Y., Jiang, J., Li, J., Wang, C.: Heterogeneous cloud radio access networks: a new perspective for enhancing spectral and energy efficiencies. IEEE Wirel. Commun. **21**(6), 126–135 (2014). https://doi.org/10.1109/mwc.2014.7000980
13. Choy, S., Wong, B., Simon, G., Rosenberg, C.: (2012). The brewing storm in cloud gaming: a measurement study on cloud to end-user latency. 2012 11th Annual Workshop on Network and Systems Support for Games (NetGames), IEEE, Venice, Italy, 22–23 Nov 2012. doi:https://doi.org/10.1109/netgames.2012.6404024
14. Bessis, N., Dobre, C.: Big data and internet of things: a roadmap for smart environments. Springer, Cham (2014)
15. Choy, S., Wong, B., Simon, G., Rosenberg, C.: A hybrid edge-cloud architecture for reducing on-demand gaming latency. Multimedia Systems. **20**(5), 503–519 (2014). https://doi.org/10.1007/s00530-014-0367-z
16. Orland, K..: Ubisoft CEO. Cloud gaming will replace consoles after the next generation. https://arstechnica.com/gaming/2018/06/ubisoft-ceo-cloud-gaming-will-replace-consoles-after-the-next-generation/ (2018). Accessed 21 Dec 2019
17. Shea, R., Liu, J., Ngai, E.C.-H., Cui, Y.: Cloud gaming: architecture and performance. IEEE Network. **27**(4), 16–21 (2013). https://doi.org/10.1109/mnet.2013.6574660
18. Bradley, A.: What the advent of 5G means for gaming. https://www.gamesradar.com/what-does-5g-mean-for-gaming/ (2019). Accessed 20 Mar 2020
19. IEEE: Big data analytics in 5G. https://futurenetworks.ieee.org/images/files/pdf/applications/Data-Analytics-in-5G-Applications030518.pdf Accessed 22 Dec 2019
20. The role of big data and advanced analytics in SDN/NFV. https://accedian.com/wp-content/uploads/2015/06/BTE15_The-Role-of-Big-Data-and-Advanced-Analytics-in-SDN-NFV.pdf (2015). Accessed 12 Jan 2020
21. Ordonez-Lucena, J., Ameigeiras, P., Lopez, D., Ramos-Munoz, J.J., Lorca, J., Folgueira, J.: Network slicing for 5G with SDN/NFV: concepts, architectures, and challenges. IEEE Commun. Mag. **55**(5), 80–87 (2017). https://doi.org/10.1109/mcom.2017.1600935
22. 5G; System Architecture for the 5G System. https://www.etsi.org/deliver/etsi_ts/123500_123599/123501/15.02.00_60/ts_123501v150200p.pdf (2018). Accessed Jan 2020

23. Giambene, G., Kota, S. L., Pillai, P.: Figure 1 from Satellite-5G integration: a network per-
    spective: semantic scholar. https://www.semanticscholar.org/paper/Satellite-5G-Integration:-
    A-Network-Perspective-GiambeneKota/6d4222a4fc421d7d1c836947492b90e8b834feb6/
    figure/0 (2019). Accessed 12 Jan 2020
24. A guide to 5G network security insight report. https://www.ericsson.com/en/security/a-guide-
    to-5g-network-security (2019). Accessed 20 Jan 2020
25. 5 key requirements for a secure 5G network. https://www.cisco.com/c/m/en_us/network-
    intelligence/service-provider/digital-transformation/secure-5g-network.html (2019). Accessed
    12 Jan 2020
26. What Are the Top 5G Security Challenges?. https://www.sdxcentral.com/5g/definitions/top-
    5g-security-challenges/ (2017).
27. Ma, J.: Top 10 security concerns for cloud-based services. https://www.imperva.com/blog/top-
    10-cloud-security-concerns/ (2018). Accessed 12 Jan 2020
28. Cloud Security Alliance Releases 'The Treacherous Twelve' Cloud Computing Top Threats
    in 2016. https://cloudsecurityalliance.org/press-releases/2016/02/29/cloud-security-alliance-
    releases-the-treacherous-twelve-cloud-computing-top-threats-in-2016/ (2016). Accessed 20
    Mar 2020
29. 5G Security & Privacy - National Security Agency. https://www.nsa.gov/Portals/70/
    documents/resources/everyone/digital-media-center/publications/the-next-wave/TNW-21-
    4.pdf. Accessed 3 Feb 2020
30. Gazis, A., Ioannou, E., Katsiri, E.: Examining the sensors that enable self-driving vehicles.
    IEEE Potentials. **39**(1), 46–51 (2020). https://doi.org/10.1109/mpot.2019.2941243
31. Datta, S. K.: Virtual Reality Mobile Application Testing in a 5G Testbed. 2019 Eleventh
    International Conference on Ubiquitous and Future Networks (ICUFN), IEEE, Zagreb, Croatia,
    (2019). doi:https://doi.org/10.1109/icufn.2019.8806058
32. Hou, X., Lu, Y., Dey, S.: Wireless VR/AR with Edge/Cloud Computing. 2017 26th Interna-
    tional Conference on Computer Communication and Networks (ICCCN), IEEE, Vancouver,
    BC, Canada, (2017). doi:https://doi.org/10.1109/icccn.2017.8038375
33. Machardy, Z., Khan, A., Obana, K., Iwashina, S.: V2X access technologies: regulation,
    research, and remaining challenges. IEEE Commun. Surv. Tutorials. **20**(3), 1858–1877 (2018).
    https://doi.org/10.1109/comst.2018.2808444
34. Wang, P., Di, B., Zhang, H., Bian, K., Song, L.: Platoon Cooperation in Cellular V2X Networks
    for 5G and Beyond. IEEE Trans. Wirel. Commun. **18**(8), 3919–3932 (2019). https://doi.org/
    10.1109/twc.2019.2919602
35. WLAN,      radar,     IoT,     V2X     to     complement     5G     at     IMS.     https://
    wrcyww.evaluationengineering.com/special-reports/article/13015712/wlan-radar-iot-v2x-
    to-complement-5g-at-ims (2019). Accessed Jan 2020
36. Anwar, W., Franchi, N., Fettweis, G.: Physical layer evaluation of V2X Communications
    Technologies: 5G NR-V2X, LTE-V2X, IEEE 802.11bd, and IEEE 802.11p. 2019 IEEE 90th
    Vehicular Technology Conference (VTC2019-Fall), IEEE, Honolulu, HI (2019). doi:https://
    doi.org/10.1109/vtcfall.2019.8891313.

# An Overview of the Edge Computing in the Modern Digital Age

Reinaldo Padilha França (iD), Ana Carolina Borges Monteiro (iD),
Rangel Arthur (iD), and Yuzo Iano (iD)

## 1 Introduction

Cloud computing was seen as a solution with infinite capacity, however, when this technology appeared, along with the popularization of smartphones, many companies realized the potential of cloud computing and started to invest intellectual and financial capital to make it more applicable and with greater capacity to meet the needs of ordinary people and large institutions. What made data traffic and processing demands grow so much, the objective was to solve the growing demand for data access at any time, from anywhere and, increasingly, in large volumes. It is currently observed that there is data congestion on the network and that the processing of this information is already overloaded. Services are increasingly slower and, with the expansion of the use of the Internet of Things (IoT), it tends to worsen [1, 2].

With edge computing, current and future devices with great processing power can be part of the work that is performed today by cloud servers. At the same time, intermediate servers installed closer to these devices physically perform data processing and send to the cloud only those that should be stored or that require more processing. Edge Computing is a breakthrough in distributed computing technology, transporting application, or project data processing directly to where it is needed, with no need to traffic this information to the cloud. This computation

R. P. França (✉) · A. C. B. Monteiro · R. Arthur · Y. Iano
School of Electrical and Computer Engineering (FEEC), University of Campinas – UNICAMP, Campinas, SP, Brazil
e-mail: padilha@decom.fee.unicamp.br; reinaldopadilha@live.com; monteiro@decom.fee.unicamp.br; carol94monteiro@gmail.com; rangel@ft.unicamp.br; yuzo@decom.fee.unicamp.br

© Springer Nature Switzerland AG 2021
W. Chang, J. Wu (eds.), *Fog/Edge Computing For Security, Privacy, and Applications*, Advances in Information Security 83,
https://doi.org/10.1007/978-3-030-57328-7_2

runs directly on the nodes of distributed devices and drives processing closer to the user, while also improving the response speed [3].

Fog or Edge Computing growth is the iteration (or repetition, or duplication) of a well-known technology cycle that begins with centralized processing and then evolves into more distributed architectures. Just as the Internet itself went through this cycle, wherein the beginning, it was only a limited number of mainframes connected in government facilities and universities. The network only reached mass scale and affordability when terminals that interfaced with mainframes were replaced by desktops, which were able to render graphical pages of an emerging internet. Getting to Distributed Computing [4].

Edge computing denotes the concept of devices that have the ability to perform advanced processing and analysis and is gaining momentum with the development of more powerful processors and computer systems present in mobile devices. This technology provides new possibilities for the IoT (Internet of Things) concept, since devices that perform some kind of cognitive computing and rely on more powerful processing are benefiting from this kind of application for tasks such as face detection, processing natural language, and even recommendation systems [5].

In this context, one must think of the "edge" as the universe of devices connected by the internet, a counterpart to the cloud. Thus, this technological concept provides new possibilities in IoT applications, particularly those that rely on cognitive computing for tasks such as detection, face recognition, language processing, and obstacle prevention. In some scenarios where data volume and cost to operate cloud computing are acceptable, this type of application may be ideal considering that edge computing is one of the driving factors for IoT and Fog Computing, but it has a field application design for a distributed resource environment, addressing issues such as reliability. Similarly, the mobile revolution was greatly accelerated with the arrival of smartphones and other devices, just as gadgets no longer depended solely on gateways and other field devices. Thus, Edge computing will have a similar effect on IoT, fueling ecosystem growth as devices become more powerful and capable of running more complex applications [6].

Therefore, this chapter aims to provide an updated review and overview of Edge Computing, addressing its evolution and fundamental concepts, showing its relationship as well as approaching its success.

This survey carries out a bibliographic review of the main research of scientific articles related to the theme of Edge Computing, published in the last 5 years on renowned bases.

Thus, the present study is organized as follows: Sect. 1 presents the introduction and methodology used in this work; Sect. 2 covers Edge Computing Concepts; Sect. 3 refers to Evolution Edge Computing; Sect. 4 presents an overview on Edge Computing and Internet of Things (IoT); Sect. 5 points to Edge Computing Applications; Sect. 6 there is a discussion on the subject; Sect. 7 contains the Benefits and Challenges of Edge Computing and finally and in Sects. 8 and "References section" there is a discussion on the topic as well as the bibliographic references used to carry out this research.

## 2 Edge Computing Concepts

Edge computing is the technology in which processing takes place at or near the physical location of the user or data source. Provided that with the closest processing, users benefit from faster and more reliable services, while companies take advantage of the flexibility of hybrid cloud computing. The concept of edge computing begins with centralized processing that later evolves to distributed architectures, basically, it is a network of micro data centers for processing critical data locally, that is, at the "edge" of the network, instead of sending them to the cloud. This technology is one of the ways that a company can use and distribute a pool of resources across a large number of locations [7].

This technology emerged to deal with the demands for traffic and data processing, which have become increasingly voluminous and growing, taking into account those at the edges of the network, these elements are able to process urgent requests and select which data should be sent to the cloud. The use of augmented and virtual realities are often hampered by high latency and insufficient bandwidth, since when using these technologies, it is common to experience slowness or processing delays that hinder the immersive experience, considering to be solved with the use of edge computing. For this reason, it is composed, in general, by devices that perform advanced processing and analysis closer to the data source, performing a sorting of information to minimize the traffic sent to the central processing, helping to reduce the bandwidth necessary for communications between the network and devices [8].

Edge computing also allows the possibility of considering the implementation of these processes in the Internet of Things (IoT), as in smart cars, which are equipped with high-performance GPUs or CPUs, being effective "data center on wheels", consisting of an edge device with self-contained computing capabilities. Taking into account that without this local processing power, if it were totally dependent on the cloud, there would be issues such as latency, availability, and quality of the car's data transmission infrastructure [9].

Since in an autonomous car, which is packed with sensors that require data to be transmitted in real-time, as with all IoT devices, there can be no latency to process driving information, process data, and make decisions practically in time real. And that is why systems that cannot suffer from latency or availability difficulties benefit from edge computing. In the same sense that bandwidth per user is a bottleneck for those who need to transmit large volumes of data. The increase in the number of devices connected by IoT should worsen this scenario, because bandwidth congestion can be critical in certain situations [10].

Other benefits of edge computing include the ability to do large data analyzes and aggregations locally, enabling almost immediate decision making, reducing the risk of exposure to sensitive data because it stores processing power locally. In this context, companies have greater control over the spread of information, such as trade secrets, and are able to comply with regulatory policies, such as the GDPR (General Data Protection Regulation) [11].

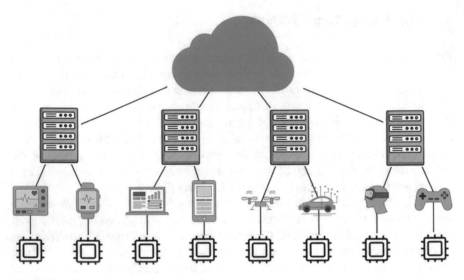

**Fig. 1** Edge computing

So, what is provided by edge computing is that, instead of processing in the cloud, it takes place at the edge of the network, making it possible that after this data is treated locally, the most frequently used information is stored in a nearby location and only those stored for long periods are sent to the cloud [12]. In short, more computing and analysis is done on the devices themselves, as illustrated in Fig. 1.

Businesses still benefit from the resilience and costs associated with edge computing. By maintaining computing power locally, regional facilities can continue to operate independently, even if there is an interruption in the operation of the main facility. The cost of bandwidth to move data between the main and regional facilities is also reduced when processing is kept close to the source. In this way, only the most relevant data crosses the network, making it easier for users to access their files, and thus overcoming connectivity and latency challenges by keeping content closer to the source [13].

Because latency is the response time of a request in this computational structure, and it is lower with edge computing technology precisely because of its proximity to the data source. When making a comparative parallel, a process using cloud computing, such information may have to travel long distances between sending and returning. Still considering that edge computing can be quite useful in places where the internet connection is not good. An oil platform in the middle of the ocean produces data that does not need to be sent to the network, as it is about the system itself. As a security measure, edge computing can be used to send daily reports to the cloud for long-term storage [7].

Then, the data obtained in the structure of edge computing can make business analysis and obtaining insights easier and more agile. Making sure companies should have the ability to move their operations from the cloud to the edge easily when needed [3].

## 3   Evolution Edge Computing

Cloud Computing is already known in corporate environments, since in modern terms, there is no way to think about developing a business without using archiving technology in the cloud. It means storing, accessing information and programs over the internet instead of using a traditional computer or server hard drive. This technology has enabled companies to rent online access to store files, replacing their own computing infrastructure or data centers. A significant reduction in costs and removing the need to expend staff to maintain traditional infrastructures [14].

The cloud supports a large number of functions including more everyday services such as e-mails or the backup of images from our cell phones. A device must send data about current conditions to a server that, in this case, could be in another state or even another country and wait for a response on how to proceed. In factories, homes and smart transport are examples that could not rely solely on cloud computing, especially given the growing number of sensors. In a factory environment, Cloud Computing is present when a factory's machinery or an autonomous car needs to obey commands from collected data, and small delays can lead to an accident. In this sense, Edge Computing helps to significantly reduce latency and improve the quality and security of services [15].

Edge Computing does not come to replace the cloud, as illustrated in Fig. 2, which will continue with the same importance, however, it will no longer monopolize all functions. Through the network architecture, the data will be collected and sent to a gateway, which is an intermediate device, and then forwarded to the cloud. "Edge", in this sense, corresponds to connected devices that produce data, in the so-called internet of things (IoT) concept. In the conventional model, the information is sent to the cloud structure, where calculations, analysis, and others are performed. But with the explosion in the number of connected devices, Edge Computing appears to streamline this flow and not overload the cloud structure [9–16].

It is possible to define Edge Computing as an open and distributed IT architecture that has decentralized processing power, allowing mobile computing and IoT technologies. Without Edge Computing, connected devices send all collected information to the cloud infrastructure. With this technology, it is possible to "shorten the path" and take consolidated information to the cloud. This increasingly performs processing on the edge and less in the cloud, for better use of the cloud [18].

Edge Computing's idea of making data obtained through an IoT network, for example, be processed close to the places of origin, will be able to solve latency impasses. Making the data read and processed by the local device, computer, or server, instead of all being transmitted to an online data center. By decentralizing the

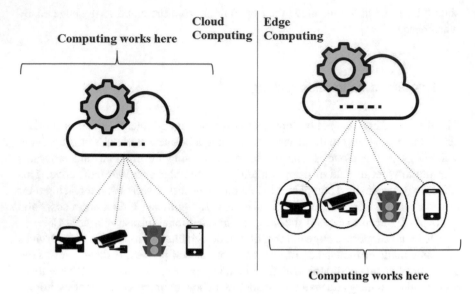

**Fig. 2** Cloud x Edge computing

processing and storage of data leaving them close to the location, edge computing will reduce traffic in the cloud, making response time faster, positively affecting latency and bandwidth [17].

The lack of security is another minimized challenge, since as most data will be on a gateway, from edge computing they will become more secure and less vulnerable to attack. As a result, fewer data in the cloud will make encryption easier. With edge computing the system allows the acceleration of data flow, including real-time processing without latency, i.e., smart devices and applications respond almost instantly, as they are created, eliminating waiting time. The processing is carried out on the equipment itself, before sending all data to the cloud. With specialized processors, the data can be processed when it is generated, filtered, and identified, is just stored in the cloud, instead of sending all information to the cloud to be processed, and then yes stored [19].

Edge Computing's growth will only be a dampener of the growth of the cloud, which will follow its exponential rise. The cloud will continue to be very important, but it will not be the only technology. Since it is related to the amount of data that is processed in modern times, Cloud Computing is not the best option for applications with latency intolerance and intensive use of computing. In this sense, Edge Computing arrives as a fast and flexible solution for more demanding demands [8, 9].

## 4   Edge Computing and Internet of Things (IoT)

The massive use of information technologies in conjunction with the internet has generated a large amount of data that is growing exponentially. Considering that in an increasingly connected world, the use of technologies such as IoT and Artificial Intelligence, make companies start to produce a lot of data and look for more agile ways to collect and analyze them. In this scenario, edge computing continually gains strength. The Internet of Things (IoT) is related to the current trend of common objects in people's daily lives to become "smart" due to the internet connection. These devices start to receive and send data through wireless networks, without requiring direct human intervention. In this context, there are drones, autonomous cars, smart houses, smartwatches, among many other IoT devices [20].

Through Edge Computing, IoT devices can transmit data to a nearby device, such as a gateway, capable of understanding and processing information and giving answers quickly, reducing the need to transfer data to the cloud, and then return the result. This is one of the cases in which Edge Computing is most needed, since these devices require real-time processing and transmission, as illustrated in Fig. 3, and cannot suffer from slow problems and connection failures, as in some cases [21].

This allows that IoT devices to have computational and processing capacity without relying so much on the cloud. The increase in these connected devices has led to an obvious and inevitable limitation related to the high volume of data that must be proportional to the computing resources, which can result in high business costs, high latency, excessive network usage, unavailability, and reduced reliability systems. In modern times when any and all electronic devices are connected to the internet, the current real trend in smart cars will have to make quick decisions and people will generate even more data. And speed and safety issues will become even more essential [19].

In the corporate environment, there is a high volume of data produced daily, both to power systems and to communicate between users as well as between machines and equipment. This dynamic is part of the IoT, characterized by the connection of different devices with the ability to communicate with each other and between

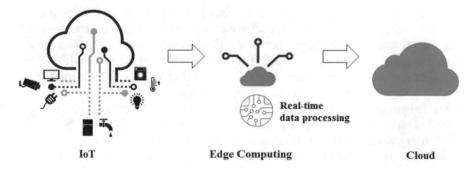

**Fig. 3**  Edge computing data processing

users in real-time. Where it has been using cloud computing to store data with the possibility of accessing it using any device connected to the internet. Although this technology is based on traditional models, the logic of edge computing, the move away from the cloud, and the rapprochement of computing towards the ends, it is essential to IoT. It deals with the decentralization of data storage and processing, allowing everything to happen closer to where the information is generated or used [22].

In reality, network devices collect information, sometimes in huge volumes, and send it to a data center or cloud for processing. What edge computing does is triage locally, which reduces traffic and prevents the information from having to travel to the data center of a centralized cloud. And that way, with less "travel" of the data, there is more possibility for the information to be secure, and your privacy becomes guaranteed, vacating networks and also the latency of the responses [23].

One of the main advantages of edge computing over IoT, and perhaps the main one, has to do with security. Since storage and processing are brought to the edge of the network, distributing to several points, substantially increasing the size of the surface exposed to attacks. And in that sense, there are not several protected entrance doors. Without a doubt, the cloud has made everything easier, flexible, and sustainable. However, this structure was not built for a large amount of data and the constant flow demanded by IoT seen in modern times. This fact triggers delays in communication and problems with the transmission capacity. And these problems leave the environment exposed to potentially dangerous flaws for the security of data and information, for companies that depend on important and sensitive information traveling on their network [24].

The move from the cloud to edge computing is a natural evolution of technology cycles. Which starts centralized to decentralize, just like with the internet itself and computers. With the increase in the volume of data and the need to generate algorithms, insights, and reports, edge computing complements the cloud model and brings more efficiency. Since the cloud-enabled several applications, the edge enables several pieces of information. The rise and need for edge computing are directly related to IoT in the sense of freeing up bandwidth, dealing with high volumes of data, and optimizing processing related to the era of connected devices [25].

With edge computing it is possible to obtain faster processing, with respect to the data being processed geographically closer to where they are generated, the response time is less. Obtain less network traffic, since networks to traffic so much data generated tend to be unoccupied if most actions are resolved without the need for constant communication with the cloud. Achieve lower costs since it is not necessary to occupy large storage spaces in outsourced companies, and in this sense reduce data management costs. Achieve greater efficiency with less latency, enabling more operations to be carried out more efficiently [26].

Achieve more security, as long as preventive measures are taken, the cloud is still fragile. In the event of a service crash, all of your customers also fall. In the same sense, greater privacy is obtained with data being processed locally, since less information is centralized and, as a result, they have less power over other people's data. And with the decentralization of edge computing, when a service goes down, the problem is individual. Still taking into account that the services can work offline, being useful in more remote places, where internet coverage is not ideal [27].

Looking back in time, Cloud Computing arrived at a time when there was a need to dispense with the use of hardware and increase the security levels of traditional computing. However, edge computing improves the system, becoming an intermediary for a more intelligent, effective, and secure use of the cloud. The potential of the technology is even greater when it is projected to be used in IoT initiatives, edge computing in industry 4.0, makes it possible to monitor and control production lines in real-time with closer servers and decentralized management capacity in micro data centers; in retail, it brings a better computing experience and benefits business verticals, with the use of sensors in stores creating more attractive interactions with customers, for example; in smart cities, where the infrastructure has to be dedicated, not shared, for compliance reasons. Edge Computing, therefore, can be considered both in the infrastructure of processes related to IoT, as well as in situations in which one cannot rely exclusively on the internet connection, or even in which it needs the most important data to be stored locally [28].

In the business environment, one of the great gains of Edge Computing is the non-dependence on a distant data center, making it possible to manage the periphery of the network, i.e., the points close to storage and processing. There are several application scenarios, which, while the world is increasingly connected and IoT becomes a present reality, edge computing promises to be a solution to handle the overhead and security of network data [13].

## 5  Edge Computing Applications

In an increasingly connected and intelligent world, it has become one of the main demands for the success of companies to find ways to obtain agile and accurate answers. For this reason, digital business initiatives have demanded new solutions "in real-time", which will facilitate the local actions of their operations, as illustrated in Fig. 4. In this scenario, Edge Computing emerges with ample prominence, directly meeting corporate needs for high localized computing power [10].

Video game developers face real and perceived delays between in-game actions in an online environment. What is solved by edge computing, applied in an online multiplayer game, meaning a much more fluid action and at the same timeless demand from the cloud, which in turn represents less cost. Today's players are tied to a single platform, physical hardware, and very slow software download. But gamers are demanding more flexibility, more mobility, and more freedom. In a cloud edge model, the gaming industry must design its edge computing networks to locate servers as close to gamers as possible. Adaptive Network's approach to cloud gaming, enhanced by edge computing, addresses the main issue of providing sufficient resources during peak usage periods [29].

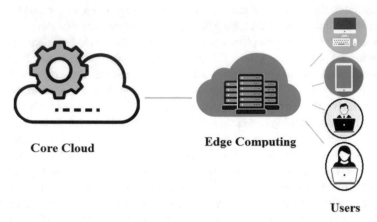

**Core Cloud**                     **Edge Computing**

                                                                    **Users**

**Fig. 4** Edge computing connection

The digital transformation of companies and industries was driven mainly by cloud computing and IoT, reaching the best levels of performance and productivity that allow for monitoring and optimized decision making. Edge Computing has accelerated the growth of the company's digital transformation, with IoT, data analysis has become the main mechanism for the complete understanding of industrial processes, along with Machine Learning and Artificial Intelligence techniques to extract information from data from sensors [5–30].

The most deployed architecture in the industry today consists of sensors that just collect data and send it to local servers or straight to the cloud. Data traffic has grown a lot and is likely to increase. The standard architecture of communication and information distribution on the factory floor in the current model, there is a chain of information that crosses the process in a linked way, passing from the machine (PLC–Programmable Logical Controller) to the management systems, until reaching an ERP (Enterprise Resource Planning) and/or CRM (Customer Relationship Management). Industries benefit by triggering the process, making the architecture decentralized, allowing better management of the supply chain and the industrial process [31, 32].

In telecommunications, the growth of IoT devices along with streaming services, has meant that cellular networks have reached high levels of data consumption. Still considering that many smartphone applications leave the computing part in the cloud, which demands greater data traffic. High latency and low network availability are inevitable. And to maximize the efficiency of network use, the decentralized architecture of Edge computing has the capacity to serve tens of billions of connected devices [33–36].

Shopfloor integration is more efficient with Edge Computing connecting devices and processes without the need to send data to the cloud. It allows the connection between the company's ERP and the shop floor, enabling the IT architecture to be more responsive and present data in real-time. Making the ERP system have more accurate numbers of the manufacturing process, allowing planning and the supply chain to be more efficient. There are still approaches that can make edge computing more efficient, depending on what will be served by a given application area [37].

*Human-Latency Sensitive* is related to the interaction between humans and the system, with tolerable response times or latencies. It includes cases in which services are optimized for human consumption, with reference to speed. This could mean that 15 milliseconds can be the limit between the customer completing a purchase on a given website or simply closing the window. A classic example involving IoT is that of speech recognition, which currently occurs in the cloud, but as the network of users grows, it tends to become impractical and will need to be placed closer to the user, with Edge Computing. Delay in data delivery negatively impacts a user's experience with the technology, reducing a retailer's sales and profitability. Application examples include small retail, website optimization, augmented reality, and natural language processing [38].

*Machine-to-Machine Latency* Sensitive consists of scenarios where systems consume data with each other, with a tolerable latency that is even lower. Because machines are able to process data much faster than humans, the consequences of slow delivery are greater than in the Human-Latency Sensitive archetype. It includes smart power grids, real-time analytics, smart security, low latency content delivery, and defense force simulation [39–42].

This is due to the amount of data that systems can consume and process, in these cases, each millisecond can represent a loss in revenue. Since delays in the trading of commodities and shares, in which prices fluctuate in fractions of a second, for example, representing potential business gains in losses. In the industrial context, subsystems of a production line can exchange data for manufacturing control in a synchronized way. A longer delay can cause loss of synchronism and system failure leading to a production stop. SmartGrid technologies applied in energy distribution to balance supply and consumption in an optimized way, are another example. Where everyone can be solved with the application of Edge Computing [43–48].

*Life Critical* encompasses cases that directly impact the health and safety of human beings. It occurs in cases where a system needs to obtain information, processes it, and makes decisions in a timely manner, otherwise, it poses a risk to people's lives and health. Consequently, speed and reliability are vital. They include smart transportation, connected/autonomous cars, digital health, autonomous robots, and drones [49].

Drones need up-to-date data to operate safely and be used for e-commerce delivery and correspondence. Ditto autonomous vehicles. Still considering that the latency tolerated between the decision making of an autonomous vehicle may be different from that tolerated in a cardiac monitoring device, for example. In both cases, edge computing is an excellent alternative [50–52].

Marketing and sales can also take advantage of edge computing. Related to digital marketing and its automation for e-commerce, networks with this edge computing technology can process payments and information more quickly than if they used exclusively cloud technology. Making the data can be stored and sent later to the central, in order to generate long-term reports, and the processes have already been executed at the edge. And so, edge and cloud computing work together to improve information processing and security [25, 27].

Finally, companies that need high data security can also benefit from edge computing, as long as there may be security breaches when data is processed exclusively in central or cloud computing platforms. They need data processed quickly, securely and with high performance, that benefits from edge computing technology, if they have devices that use IoT [24].


# 6  Discussion

Edge computing is linked to the evolution of IoT, reducing latency, decreasing cloud dependency, and enabling better data management. Also offering new possibilities in applications related to machine learning for facial recognition, object detection, language processing, prevention of obstacles, and derivatives.

In this context, it is the case of security cameras, since all images are sent to the cloud and a lot of space is needed to store them. With the association of edge computing and artificial intelligence, combined with machine learning, it is possible to save bandwidth and storage, with material filtering and sending to the cloud only what is really relevant.

What can be realized with edge computing technology is an apparent shift to the next computing cycle. Since long ago, computing was centralized in mainframes, with banks and other large companies relying on refrigerated offices to manage their business operations. Many of these mainframes have been deactivated to make room for the decentralized client-server era. Similarly, cloud computing technology is essentially the new mainframe, with the data center hosted by a vendor. And edge computing aiding the natural ebb and flow of that computing, and IoT accelerating the next step in distributed computing. Edge computing did not come to end cloud computing or replace it, but to act together with it and bring even more benefits to companies.

Edge computing is closely linked to the evolution of IoT, as illustrated in Fig. 5, since many of the connected devices take advantage of cloud computing, the technology also known as the next stage of IoT, related to the progress required for data processing and analysis increasingly complex carried out close to them. This method of doing more computing and analysis on the devices themselves generates less dependence on the cloud and allows better management of the amount of data generated by the IoT.

After all, with more devices connected, the greater the need for processing and the data traffic will intensify more and more. In this sense, edge computing optimizes the use of electronic devices that can be connected to the internet, since it seeks to bring computing closer to the source of the collected data. Thus, processing takes place as close as possible to the user's location or data source.

In IoT, the possibility to analyze information at the point where it is collected makes the operation more efficient, taking into account that data are the most

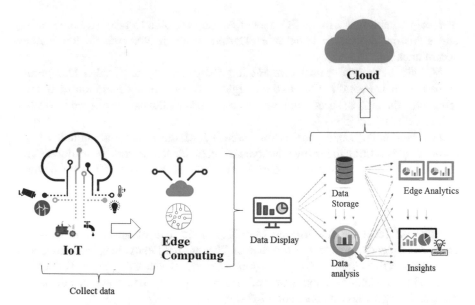

**Fig. 5** IoT in Edge computing

important assets of today, and being able to transform them into an immediate value is a great differential. The data collection is done through the devices, which gather what was collected, the amount depends on the equipment, and then they are stored and sent to a processing center, or cloud.

What edge computing does is instead of taking everything to these centers or the cloud, technology "classifies" this data locally and separates that which can be processed right there, thus decreasing data traffic and the need for sending them. Those that are processed at the ends of a network are just a part that is sent to central directories.

In summary, in this technological structure, cloud computing is above all and, below, there is another layer of infrastructure that is edge computing. It is then responsible for selecting the data that should be sent or received by the cloud. And everything else, that is, users and devices, is subordinate to edge computing.

Edge Computing is the next chapter in the cloud. Factories will be able to perform predictive maintenance on equipment that is about to fail, retailers will benefit from access to faster updates to consumer buying trends, and mobile operators will be able to support game applications, for example, for devices mobile and augmented reality [2, 8].

5G networks are already considered a mobile edge computing technology. It is a concept of network architecture that allows cloud computing resources, in an IT service environment to act at the edge of the cellular network, allowing the use of Edge in the hybrid cloud strategy. Edge Computing can also enable augmented reality applications, as long as the technology creates opportunities for speech and video analysis. It accelerates the Internet by improving the management of local content, and through 5G networks, they are able to take health, security, and industrial production to remote locations through ultra-reliable low-latency communications [53–55].

5G will become a critical element for companies to adopt Edge Computing as part of their hybrid multi-cloud strategies. With the new generation of mobile telephony, the highest speeds and the necessary bandwidth can be offered to further reduce data latency [56–59].

Fog computing stands out as an extension of the cloud and includes edge computing. It is through the mist, in layers of network access, that the resources and services for storage, processing, control, and communication closest to the devices will be shared. The advantage of this technology is that it is still within the cloud system, but it extends its capacity to the edges bringing it closer to the data source, as illustrated in Fig. 6 [60–62].

Edge AI (artificial intelligence of edge) is the use of artificial intelligence techniques incorporated in the Internet of Things (IoT) endpoints, gateways, and other devices that compute data at the point of use. A device using Edge AI does not need to be connected to function properly, as it can process data and make decisions independently, without a connection [63–64].

The smart edge is a continually expanding set of connected systems and devices that collect and analyze data, close to your users, data, or both. Users gain real-time insights and experiences from context-aware and responsive applications. Smart Edge is designed to help industries manage and get better information with the growing flow of data generated by warehouses, production facilities, retail stores, connected buildings, urban infrastructure, and other environments. They are intelligent applications focused on edge, based on systems ideal for external environments and compatible with open source software (open-source)

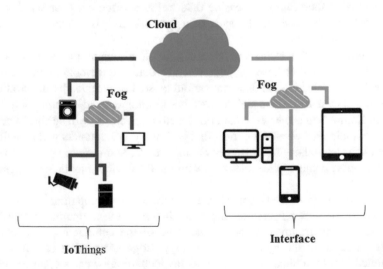

**Fig. 6** Fog computing

and unbundled hardware. Configured with a variety of memory and processor combinations so that customers can adapt their solutions to remote data center management for the smart edge [65].

Regarding Industry 4.0, Edge AI enables instant reactions, that is, it has the ability to process data in real-time and at the place where it is collected, which eliminates much of the reaction time of a traditional cloud system. To use this technology, only a gadget that includes a microprocessor and sensors is needed. In addition, by processing data on-site, possible problems with interrupting the transmission are avoided and storage of a lot of data in the cloud, for example. Still considering that this technology is capable of reducing energy consumption and, thus, improving battery life, a very important fact when thinking about portable devices [66, 67].

The trends of Edge Computing enhance the corporate network architecture in order to streamline processes and bring productivity with savings and security. Companies will again use their structure at the ends, as a strategic measure to make the best use of the benefits of the cloud and adapt their technology to the IoT [68, 69].

## 7 Benefits and Challenges of Edge Computing

The advantages of edge computing lie in the decentralization of computational processing that the cloud computing has traditionally aimed at in relation to services in some large data centers, as illustrated in Fig. 7, which is positive, since it is possible to scale and share resources more efficiently and maintain control and IT security in the enterprise.

Edge computing addresses use cases that cannot be resolved with the cloud computing centralization approach, due to network requirements or other restrictions in general. It focuses on several small computing locations, which reduces network costs, decreases transmission delays, avoids bandwidth limits, restricts service failures, and offers greater control over the movement of sensitive data. Loading time is reduced and with online services deployed closer to users it is possible to enable static and dynamic caching features.

The poor connectivity of IoT devices benefits edge computing, considering the need to send information over long distances. One of the advantages is to minimize the need to pass data over long distances between the device and the server, reducing latency and requiring less internet bandwidth. And with ever shorter response times, better for companies that are investing in this new form of technology.

The technology reduces the costs for data transmission, as the infrastructure for centralized points is bigger and more expensive. But the lower cost of edge computing infrastructure does not occur only on the network.

With better connectivity and better data transmission, this means a faster and more consistent experience for end-users, just as for companies and service

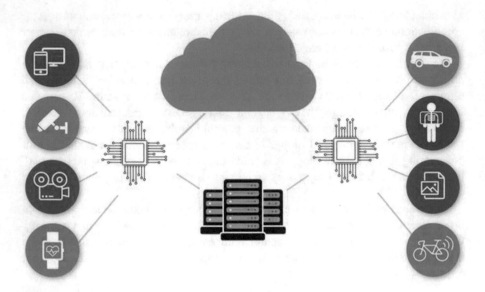

**Fig. 7** Edge computing decentralization

providers, it means high availability with real-time monitoring and low latency applications.

Edge computing enables greater integration between equipment by integrating devices from different generations on the same network, favoring a reduction in implementation costs, as equipment can be replaced on time, rather than completely.

Cloud computing platforms and central data processing centers receive high traffic all the time, which generally generates latency and, as a result, device downtime, which needs to wait for the data sent to continue operating. In this scenario, edge computing improves the response speed for IoT devices.

The existence of intelligence closer to the user, at the edges, ensures faster, cheaper, and more accurate decision making, as time and resources are saved by avoiding sending large amounts of data to be processed in the cloud. And when data is needed that's in the cloud, it can be brought to the edge.

The challenges of edge computing are related to the greater difficulty in the highly distributed scale, since it can be difficult for small businesses to manage the indirect costs of physical locations to scale the technology horizontally for many environments can be more complicated than adding the resources equivalent to a single primary data center.

It is necessary to have an infrastructure that can be repaired by unskilled employees and centrally managed by a small number of people with technical knowledge, because in general, edge computing sites are often remote and have few or no employees with specialized knowledge.

With regard to site management operations, they must be carried out to allow their reproduction at all points of edge computing in order to simplify management, facilitating problem-solving and preventing software from being implemented differently in different locations.

Although edge computing offers greater control over information flows by restricting data geographically, the physical security of these locations is often less, where these factors increase the risk of malicious actions or accidents.

# 8 Conclusions

In essence, edge computing optimizes cloud computing systems by executing data processing at the "edge" of the network, close to the data source. This method reduces the communication bandwidth needed between the sensors and the data center, performing analyzes, and generating knowledge at or near the data source.

Edge computing can be used mainly in places where latency, which is the time required for data to be captured and processed on a network, needs to be as short as possible. As is the case in IoT environments, this is because edge computing speeds up the processing of data collected on the IoT network, making it processed close to where it was collected, instead of having to go through a central location, such as on cloud computing.

In other words, it refers to the processing power at the edge of a network rather than retaining that processing power in a cloud or a data center. The factors that should be considered when making a decision on using edge computing should be energy, cost of deployment, size of infrastructure, internet bandwidth, processing capacity, and type of service for a given company.

So far, the role of edge computing has been used primarily to absorb, store, filter, and send data to cloud systems. However, in modern times, it is seen that these systems have more computing, storage, and analytical power to consume and act on the data on the machine's network.

Since the evolution of technologies and their presence in the most diverse everyday situations, there is a growing need for solutions for faster and more assertive processing. And this is where Edge Computing comes in to minimize the problems caused by traffic overload and the bandwidth necessary to send data to the cloud, acting on the processing and analysis of the data locally, in addition to compiling and filtering what should actually be sent to these systems online.

With this, it becomes possible to minimize the traffic and the necessary bandwidth, in addition to avoiding problems such as high response time (latency) and connectivity, since the technology allows the processing of critical data in a faster and more efficient way without difficulties.

# References

1. França, R.P., Iano, Y., Monteiro, A.C.B., Arthur, R.: Lower memory consumption for data transmission in smart cloud environments with CBEDE methodology. In: Smart Systems Design, Applications, and Challenges, pp. 216–237. IGI Global, Hershey (2020)
2. França, R.P., Iano, Y., Monteiro, A.C.B., Arthur, R.: Intelligent applications of WSN in the world: a technological and literary background. In: Handbook of Wireless Sensor Networks: Issues and Challenges in Current Scenario's, pp. 13–34. Springer, Cham (2020)
3. Ai, Y., Peng, M., Zhang, K.: Edge computing technologies for internet of things: a primer. Digit. Commun. Netw. **4**(2), 77–86 (2018)
4. Dolui, K., Datta, S.K.: Comparison of edge computing implementations: fog computing, cloudlet and mobile edge computing. In: 2017 Global Internet of Things Summit (GIoTS), pp. 1–6. IEEE, Piscataway (2017, June)
5. Li, H., Ota, K., Dong, M.: Learning IoT in edge: deep learning for the internet of things with edge computing. IEEE Netw. **32**(1), 96–101.7 (2018)
6. Dastjerdi, A.V., Buyya, R.: Fog computing: helping the internet of things realize its potential. Computer. **49**(8), 112–116 (2016)
7. Olaniyan, R., Fadahunsi, O., Maheswaran, M., Zhani, M.F.: Opportunistic edge computing: concepts, opportunities and research challenges. Futur. Gener. Comput. Syst. **89**, 633–645 (2018)
8. Shi, W., Dustdar, S.: The promise of edge computing. Computer. **49**(5), 78–81 (2016)
9. Satyanarayanan, M., Shi, W.: Overview of Edge Computing. IEEE, Piscataway (2018)
10. Satyanarayanan, M.: The emergence of edge computing. Computer. **50**(1), 30–39 (2017)
11. Wachter, S.: Data protection in the age of big data. Nat. Elect. **2**(1), 6–7 (2019)
12. Khan, W.Z., Ahmed, E., Hakak, S., Yaqoob, I., Ahmed, A.: Edge computing: a survey. Futur. Gener. Comput. Syst. **97**, 219–235 (2019)
13. Chen, B., Wan, J., Celesti, A., Li, D., Abbas, H., Zhang, Q.: Edge computing in IoT-based manufacturing. IEEE Commun. Mag. **56**(9), 103–109 (2018)
14. Zhang, K., Mao, Y., Leng, S., He, Y., Zhang, Y.: Mobile-edge computing for vehicular networks: a promising network paradigm with predictive off-loading. IEEE Veh. Technol. Mag. **12**(2), 36–44 (2017)
15. Jararweh, Y., Doulat, A., AlQudah, O., Ahmed, E., Al-Ayyoub, M., Benkhelifa, E.: The future of mobile cloud computing: integrating cloudlets and mobile edge computing. In: 2016 23rd International Conference on Telecommunications (ICT), pp. 1–5. IEEE, Piscataway (2016, May)
16. Pham, Quoc-Viet, et al. A survey of multi-access edge computing in 5G and beyond: Fundamentals, technology integration, and state-of-the-art. IEEE Access **8**, 116974–117017 (2020)
17. Abbas, N., Zhang, Y., Taherkordi, A., Skeie, T.: Mobile edge computing: a survey. IEEE Internet Things J. **5**(1), 450–465 (2017)
18. Li, H., et al.: Mobile edge computing: progress and challenges. In: 2016 4th IEEE International Conference on Mobile Cloud Computing, Services, and Engineering (MobileCloud), pp. 83–84. IEEE, Piscataway (2016)
19. Roman, R., Lopez, J., Mambo, M.: Mobile edge computing, fog et al.: a survey and analysis of security threats and challenges. Futur. Gener. Comput. Syst. **78**, 680–698 (2018)
20. Tran, C., Misra, S.: The technical foundations of IoT. IEEE Wirel. Commun. **26**(3), 8–8 (2019)
21. Sun, X., Ansari, N.: EdgeIoT: Mobile edge computing for the internet of things. IEEE Commun. Mag. **54**(12), 22–29 (2016)
22. Lyu, X., Tian, H., Jiang, L., Vinel, A., Maharjan, S., Gjessing, S., Zhang, Y.: Selective offloading in mobile edge computing for the green internet of things. IEEE Netw. **32**(1), 54–60 (2018)
23. Liu, X., Liu, Y., Song, H., Liu, A.: Big data orchestration as a service network. IEEE Commun. Mag. **55**(9), 94–101 (2017)

24. He, Y., Guo, J., Liu, L., Liu, H., Zhang, X., Zhao, Q., et al.: IoT for the power industry: recent advances and future directions with Pavatar. In: Proceedings of the 16th ACM Conference on Embedded Networked Sensor Systems, pp. 353–354 (2018, November)
25. Hassan, N., Gillani, S., Ahmed, E., Yaqoob, I., Imran, M.: The role of edge computing in internet of things. IEEE Commun. Mag. **56**(11), 110–115 (2018)
26. Park, J.H., Piuri, V., Chen, H.H., Pan, Y.: Guest editorial special issue on advanced computational technologies in mobile edge computing for the internet of things. IEEE Internet Things J. **6**(3), 4742–4743 (2019)
27. Xiao, Y., Jia, Y., Liu, C., Cheng, X., Yu, J., Lv, W.: Edge computing security: state of the art and challenges. Proc. IEEE. **107**(8), 1608–1631 (2019)
28. Xu, L.D., Xu, E.L., Li, L.: Industry 4.0: state of the art and future trends. Int. J. Prod. Res. **56**(8), 2941–2962 (2018)
29. Zhang, X., Chen, H., Zhao, Y., Ma, Z., Xu, Y., Huang, H., et al.: Improving cloud gaming experience through mobile edge computing. IEEE Wirel. Commun. **26**(4), 178–183 (2019)
30. Wang, S., Tuor, T., Salonidis, T., Leung, K.K., Makaya, C., He, T., Chan, K.: Adaptive federated learning in resource-constrained edge computing systems. IEEE J. Sel. Areas Commun. **37**(6), 1205–1221 (2019)
31. Zhang, H., Li, S., Yan, W., Jiang, Z., Wei, W.: A knowledge sharing framework for green supply chain management based on blockchain and edge computing. In: International Conference on Sustainable Design and Manufacturing, pp. 413–420. Springer, Singapore (2019, June)
32. Buttle, F., Maklan, S.: Customer Relationship Management: Concepts and Technologies. Routledge, New York (2019)
33. Padilha, R.F.: Proposta de um método complementar de compressão de dados por meio da metodologia de eventos discretos aplicada em um baixo nível de abstração= Proposal of a complementary method of data compression by discrete event methodology applied at a low level of abstraction. (2018)
34. Padilha, R., et al.: Computational performance of an model for wireless telecommunication systems with discrete events and multipath Rayleigh. In: Brazilian Technology Symposium. Springer, Cham (2017)
35. Padilha, Reinaldo, et al. "Proposal for improvement of information transmission in OFDM systems through the CBEDE methodology." Set Int. J. Broadcast Eng. 5 (2020): 9
36. França, R.P., et al.: Potential proposal to improve data transmission in healthcare systems. In: Deep Learning Techniques for Biomedical and Health Informatics, pp. 267–283. Academic Press, London (2020)
37. Soldatos, J., Lazaro, O., Cavadini, F.: The Digital Shopfloor: Industrial Automation in the Industry 4.0 Era. River Publishers, Gistrup (2019)
38. Patel, C., Doshi, N.: Internet of Things Security: Challenges, Advances, and Analytics. CRC Press, Boca Raton (2018)
39. Monteiro, A.C.B., et al.: Development of a laboratory medical algorithm for simultaneous detection and counting of erythrocytes and leukocytes in digital images of a blood smear. In: Deep Learning Techniques for Biomedical and Health Informatics, pp. 165–186. Academic, London (2020)
40. Wuest, T., et al.: Machine learning in manufacturing: advantages, challenges, and applications. Prod. Manufact. Res. **4**(1), 23–45 (2016)
41. Zhu, X., Goldberg, A.B.: Introduction to semi-supervised learning. In: Synthesis Lectures on Artificial Intelligence and Machine Learning, vol. 3, pp. 1–130. Morgan & Claypool Publishers, San Rafael (2009)
42. Chen, J., Ran, X.: Deep learning with edge computing: a review. Proc. IEEE. **107**(8), 1655–1674 (2019)
43. Ashraf, S.A., et al.: Ultra-reliable and low-latency communication for wireless factory automation: from LTE to 5G. In: 2016 IEEE 21st International Conference on Emerging Technologies and Factory Automation (ETFA). IEEE, Piscataway (2016)
44. Kabalci, Y.: A survey on smart metering and smart grid communication. Renew. Sust. Energ. Rev. **57**, 302–318 (2016)

45. Yoldaş, Y., et al.: Enhancing smart grid with microgrids: challenges and opportunities. Renew. Sust. Energ. Rev. **72**, 205–214 (2017)
46. Wang, K., et al.: Wireless big data computing in smart grid. IEEE Wirel. Commun. **24**(2), 58–64 (2017)
47. Dileep, G.: A survey on smart grid technologies and applications. Renew. Energy. **146**, 2589–2625 (2020)
48. Colak, I.: Introduction to smart grid. In: 2016 International Smart Grid Workshop and Certificate Program (ISGWCP). IEEE, Piscataway (2016)
49. Sendin, A., et al.: Telecommunication Networks for the Smart Grid. Artech House, Boston (2016)
50. Custers, B.: Drones Here, there and everywhere introduction and overview. In: The Future of Drone Use, pp. 3–20. TMC Asser Press, The Hague (2016)
51. Maurer, Kathrin, and Andreas Immanuel Graae. Introduction: Debating Drones: Politics, Media, and Aesthetics. Politik 20.1 (2017)
52. Hassanalian, M., Abdelkefi, A.: Classifications, applications, and design challenges of drones: a review. Prog. Aerosp. Sci. **91**, 99–131 (2017)
53. França, R.P., et al.: Improvement for channels with multipath fading (MF) through the methodology CBEDE. In: Fundamental and Supportive Technologies for 5G Mobile Networks, pp. 25–43. IGI Global, Hershey (2020)
54. Dragičević, T., Siano, P., Prabaharan, S.R.: Future generation 5G wireless networks for smart grid: a comprehensive review. Energies. **12**(11), 2140 (2019)
55. Ezhilarasan, E., Dinakaran, M.: A review on mobile technologies: 3G, 4G and 5G. In: 2017 Second International Conference on Recent Trends and Challenges in Computational Models (ICRTCCM). IEEE, Piscataway (2017)
56. Taleb, T., et al.: On multi-access edge computing: a survey of the emerging 5G network edge cloud architecture and orchestration. IEEE Commun. Surv. Tutorials. **19**(3), 1657–1681 (2017)
57. Tran, T.X., et al.: Collaborative mobile edge computing in 5G networks: new paradigms, scenarios, and challenges. IEEE Commun. Mag. **55**(4), 54–61 (2017)
58. Rimal, B.P., Van, D.P., Maier, M.: Mobile edge computing empowered fiber-wireless access networks in the 5G era. IEEE Commun. Mag. **55**(2), 192–200 (2017)
59. Kiani, A., Ansari, N.: Edge computing aware NOMA for 5G networks. IEEE Internet Things J. **5**(2), 1299–1306 (2018)
60. Dolui, K., Datta, S.K.: Comparison of edge computing implementations: fog computing, cloudlet and mobile edge computing. In: 2017 Global Internet of Things Summit (GIoTS). IEEE, Piscataway (2017)
61. Iorga, M., et al.: Fog computing conceptual model. No. Special Publication (NIST SP)-500-325. (2018)
62. Dubey, H., et al.: Fog computing in medical internet-of-things: architecture, implementation, and applications. In: Handbook of Large-Scale Distributed Computing in Smart Healthcare, pp. 281–321. Springer, Cham (2017)
63. Dai, Y., et al.: Artificial intelligence empowered edge computing and caching for internet of vehicles. IEEE Wirel. Commun. **26**(3), 12–18 (2019)
64. Deng, S., et al.: Edge intelligence: the confluence of edge computing and artificial intelligence. arXiv preprint arXiv:1909.00560 (2019)
65. Condry, M.W., Nelson, C.B.: Using smart edge IoT devices for safer, rapid response with industry IoT control operations. Proc. IEEE. **104**(5), 938–946 (2016)
66. Carvalho, A., et al.: At the edge of industry 4.0. Proc. Comput. Sci. **155**, 276–281 (2019)
67. Hasan, T.K., Sokolov, A., Tantawi, O.: Advances in industrial robotics: from industry 3.0 automation to industry 4.0 collaboration. In: 2019 4th Technology Innovation Management and Engineering Science International Conference (TIMES-iCON). IEEE, Piscataway (2019)
68. Bilal, K., et al.: Potentials, trends, and prospects in edge technologies: fog, cloudlet, mobile edge, and micro data centers. Comput. Netw. **130**, 94–120 (2018)
69. Baktir, A.C., Ozgovde, A., Ersoy, C.: How can edge computing benefit from software-defined networking: a survey, use cases, and future directions. IEEE Commun. Surv. Tutorials. **19**(4), 2359–2391 (2017)

# Part II
# Security in Fog/Edge Computing

# Secure Search and Storage Services in Cloud and Fog/Edge Computing

Qin Liu

## 1 Introduction

Cloud computing, which enables ubiquitous and convenient network access to a shared pool of configurable computing resources (e.g., networks, servers, storage, applications, and services), has become one of the most promising computing paradigms. According to the report from Research and Markets [1], cloud computing market in 2017 is US$25.171 billion, and it will reach US$92.488 billion by 2022 at a compound annual growth rate of 29.73% from 2017 to 2022.

In clouds, resources are delivered as services that can be subscribed and unsubscribed by customers on-demand. In other words, everything is a service (XaaS) [2] in clouds, where customers enjoy cloud services anytime and anywhere, using any kinds of devices connecting to the Internet. Meanwhile, cloud computing, as an evolved paradigm of distributed computing, parallel computing, grid computing, and utility computing, has a lot of merits like fast deployment, high availability, high scalability, rapid elasticity, low costs, and so on [3]. Cloud computing is a win-win business model. On the one side, cloud users can achieve cost savings and productivity enhancements, by using cloud-based services to manage projects, collaborate on documents and presentations, manage enterprisewide contacts and schedules, and the like. Especially for startups, they can businesses with reduced upfront investment and expected performance, so as to concentrate more on developing the core business without worrying about the underlying deployment details. On the other side, the cloud service provider (CSP) can take full advantage of idle resources by resource integration and optimizing configuration. Several

Q. Liu (✉)
College of Computer Science and Electronic Engineering, Hunan University, Changsha, P.R. China
e-mail: gracelq628@hnu.edu.cn

© Springer Nature Switzerland AG 2021
W. Chang, J. Wu (eds.), *Fog/Edge Computing For Security, Privacy, and Applications*, Advances in Information Security 83,
https://doi.org/10.1007/978-3-030-57328-7_3

companies (e.g., Amazon Web Services, Microsoft Azure, and Google) offer cloud services to users in a pay-as-you-go fashion.

Although cloud computing has overwhelming superiorities over traditional computing models, the application of clouds is still far from expected. Most of the enterprises only be willing to outsource the data and services that are unrelated to their business to the cloud. The main reason is that customers worry that their sensitive data may be deliberately or unintentionally leaked by the CSP. Actually, the concerns about cloud security are not unnecessary. State-of-art CSPs experience noteworthy outages and security breaches from time to time due to attacks, malfunctions or misconfigurations. For example, Gmail's mass email deletions in 2006, Microsoft Azure had an outage lasting 22 h in 2008, and the recent news about Apple iCloud leaking out celebrities' sensitive photos [4], Dropbox password leak [5] and medical data leak on Amazon [6]. Therefore, a natural way to keep sensitive data confidential against an untrusted CSP is to apply cryptographic approaches, by disclosing decryption keys only to authorized users. In this way, only the authorized entities can decrypt the data with appropriate keys. The unauthorized entities, even if the CSPs, cannot know data contents. Actually, the state-of-art CSPs already adopt cryptographic techniques to preserve data confidentiality. For example, Amazon EC2 encrypt users' data by default, and Amazon Simple Storage Service(S3) allows users to encrypt their data before outsourcing.

This book chapter researches the problem of secure search and storage services in cloud computing. We will first introduce related work on preserving storage security and search privacy in Sect. 2, before discussing the system model and threat model in Sect. 3. Then, we will describe cryptographic primitives in Sect. 4 before proposing possibly feasible approaches to provide secure cloud storage and search services in Sects. 5 and 6, respectively. Finally, we will discuss how the proposed schemes can be applied to the fog/edge computing environment in Sect. 7 before concluding this book chapter in Sect. 8.

# 2 Related Work

## 2.1 Fine-Grained Access Control on Encrypted Data

For data encryption, symmetric key and public key cryptosystems are widely used tools. In symmetric key cryptosystem [7], a shared key between the sender and the recipient is used as an encryption key and a decryption key. It is quick and efficient, but its security relies on the length of the shared key. Moreover, different messages require different shared keys for confidentiality, and thus the number of keys maintained by each user grows linearly with the number of messages. In the public key cryptosystem [8], each user has a public/private key pair. Messages are encrypted with the recipient's public key, and can only be decrypted with the corresponding private key. No keys need to be shared between the sender and the recipient before communication. It ensures high security, but it is much slower than a symmetric key cryptosystem.

In cloud computing environments, data is generally shared by many data users of different roles and attributes, thus how to achieve fine-grained access controls on ciphertexts becomes a burning question. A lot of researches have been conducted on achieving fine-grained access controls based on symmetric and public key cryptosystems, all of which face various challenges. For example, Kallahalla et al. [9] classified data with similar Access Control Lists (ACLs) into a file group, and then encrypted each file group with a symmetric key. This key will be distributed to the users in the ACL, so that only they can access this group of files. The main drawback of this approach is that the number of symmetric keys managed by each user grows linearly with the number of file groups he/she is authorized to access. Goh et al. [10] encrypted the data with a symmetric key, which was in turn encrypted with the public keys of the users in the ACL. Therefore, only the users in the ACL can use their secret keys to recover the symmetric key, and then use the symmetric key to recover the data. The main drawback of this approach is that the encryption cost grows linearly with the number of users in the ACL. Liu et al. [11] constructed an efficient data sharing scheme in cloud computing by using the *one-to-many* property of the Hierarchical Identity-Based Encryption (HIBE) [12], where a ciphertext can be decrypted by the recipient and all his ancestors with their own private keys. Though each data needs to be encrypted only once, the length of ciphertext is affected by the number of recipients.

Attribute-based encryption (ABE) is an effective cryptosystem for ensuring fine-grained access controls on ciphertexts. The predecessor of ABE is fuzzy identity-based encryption [13], which is first designed to achieve fault tolerance in biometric identities. Now, ABE has developed to two branches, key-policy ABE (KP-ABE) [14] and ciphertext-policy ABE (CP-ABE) [15, 16]. The main difference between KP-ABE and CP-ABE lies in the location of the access structure, which is in key and in ciphertext, respectively. CP-ABE allows the data owner to take more initiative on specifying access structure for the ciphertext, and thus is more suitable for a data sharing environment.

The original ABE systems only support monotone access policy and assume the existence of a single private key generator (PKG). A lot of research has been done to achieve more expressive access policy [17–20], and distributed key management [21–23]. To achieve a flexible access control in cloud computing, our previous work [24–26] proposed a Hierarchical Attribute-Based Encryption (HABE) scheme, by combining HIBE and ABE systems. The HABE scheme supports both ID-based and attribute-based access policies, with hierarchical key generation properties. On the basis of the ABE scheme, Zhu et al. [27] proposed a Comparison-based encryption (CBE) scheme by making use of forward and backward derivation functions and applied CBE to the cloud environment. However, the encryption cost of the CBE scheme grows linearly with the number of attributes in the access policy. To solve this problem, our previous work proposed a Hierarchical CBE scheme by incorporating the attribute hierarchy into the CBE scheme [28]. Wu et al. [29] proposed a Multi-message Ciphertext Policy Attribute-Based Encryption

(MCP-ABE) scheme to achieve scalable media data sharing in clouds. Li et al. [30] leveraged ABE techniques to achieve fine-grained and scalable data access control on Personal Health Records (PHRs) in cloud environments. ABE inherently exposes the access structure. To achieve both payload hiding and attribute hiding, Predicate Encryption (PE) [31, 32] was proposed as a stronger security notion of ABE. In this book chapter, we will introduce the HABE scheme, which simultaneously achieves a fine-grained access control, high performance, practicability, and scalability, so as to be more applicable in cloud computing.

## 2.2   User Revocation

User revocation is a well studied, but non-trivial task. The key problem is that the revoked users still retain the keys issued earlier, and thus can still decrypt ciphertexts. Therefore, whenever a user is revoked, the re-keying and re-encryption operations need to be executed by the data owner to prevent the revoked user from accessing the future data. For example, when ABE is adopted to encrypt data, the work in reference [33] proposed to require the data owner to periodically re-encrypt the data, and re-distribute new keys to authorized users. This approach is very inefficient due to the heavy workload introduced on the data owner.

A better solution is to let the data owner delegate a third party to execute some computational intensive tasks, e.g., re-encryption, while leaking the least information. Proxy re-encryption [34, 35] is a good choice, where a semi-trusted proxy is able to convert a ciphertext that can be decrypted by Alice into another ciphertext that can be decrypted by Bob, without knowing the underlying data and user secret keys. For example, the work in reference [36] is the first to combine KP-ABE and PRE to delegate most of the computation tasks involved in user revocation to the CSP. Our previous work [25] is the first to combine PRE and a CP-ABE system (HABE) to achieve a scalable revocation mechanism in cloud computing. The work in reference [37] that supports attribute revocation may be applicable to a cloud environment. This approach requires that once a user is revoked from a system, the data owner should send PRE keys to the CSP, with which the CSP can be delegated to execute re-encryption. The main problem of this approach is that the data owner should be online in order to send the PRE keys to the CSP in a timely fashion, to prevent the revoked user from accessing the data. The delay of issuing PRE keys may cause potential security risks.

Shi et al. [38] proposed a dubbed directly revocable key-policy ABE with verifiable ciphertext delegation (drvuKPABE) scheme to achieve direct revocation and verifiable ciphertext delegation. Liu et al. [39] proposed a time-based proxy re-encryption (TimePRE) scheme, which allowed the cloud to automatically re-encrypt the ciphertexts based on time. Yang et al. [40] presented a conditional proxy re-encryption scheme to achieve cloud-enabled user revocation. Yang et al. [41] proposed a novel scheme that enables efficient access control with dynamic policy updating in cloud computing. They designed policy updating algorithms for different types of access policies so as to simultaneously achieve correctness and completeness and meet security requirements. Inspired by the work in [41], our previous work [28] developed the DPU scheme to efficiently achieve dynamic policy updating in cloud-based PHR environments.

In this book chapter, we will introduce the TimePRE scheme, which extends the HABE [24–26] by incorporating the concept of time to perform automatic proxy re-encryption. The main difference from prior work is that the TimePRE scheme enables each user's access right to be effective in a pre-determined time, and enables the CSP to re-encrypt ciphertexts automatically based on its own time. Thus, the data owner can be offline in the process of user revocations.

## 2.3 Searchable Encryption

The first searchable encryption scheme was first proposed by Song et al. [42], where both the user query as well as the data is encrypted under a symmetric key setting. Depending on the selection of *index key*, *pk*, and *search key*, *sk*, existing SE solutions can be classified into symmetric-key settings ($sk = pk$) and public-key settings ($sk \neq pk$). In their scheme, each word in the file is encrypted independently with a symmetric key, and is encapsulated with a two-layer structure. On receiving a trapdoor from the user, the server can get rid of the outer layer and determine whether the file matches user query or not. The main drawback of this approach is that the server has to scan the whole file collection while conducting searches, and thus the searching cost grows linearly with the number of files in the collection. Since then, there has been much work conducted in this field, for example, both Goh [43] and Chang et al. [44] developed secure searchable index schemes to solve the problem of linear searching cost. Furthermore, to allows the users to verify the integrity of the search results returned by an untrusted platform, Kurosawa et al. [45] proposed a verifiable searchable symmetric encryption scheme. However, their scheme only supports verification of static data. Kamara et al. [46] proposed a dynamic verifiable searchable symmetric encryption scheme to achieve integrity verification of search results from dynamic data sets.

The work in [47] proposed the first public key-based searchable encryption protocol, where anyone with the public key can encrypt data, but only users with the private key can generate queries. The typical application of their work is to employ a gateway to test whether certain keywords are contained in an email without learning any information about the email and keywords. The above work supports only OR semantic in the keywords. As an attempt to enrich search predicates, searchable encryption schemes that support conjunctive keyword search [48], subset query [50], and range query [51], have also been proposed. Popa et al. [52] proposed a CryptDB system that allows users to execute SQL queries over encrypted data, by using adjustable searchable encryption incorporating with an SQL-aware encryption strategy and onion encryption.

Recently, ranked SE and fuzzy keyword-based SE were proposed to optimize the query results. Ranked SE enables users to retrieve the most matched files from the untrusted server. For example, Wang et al. [53] encrypted files and queries with Order Preserving Symmetric Encryption [54] and utilized keyword frequency to rank results; Cao et al. [55] proposed a multi-keyword ranked searchable encryption scheme by applying the secure KNN technique [56] to rank results based on inner products. Fuzzy keyword-based SE aims to improve the matching ratio in the case that the users are not sure about the accurate keywords. For example, Li et al. [57] proposed fuzzy keyword search over encrypted data in cloud computing, which would return the matching files even if users' searching inputs does not exactly match the predefined keywords; Wang et al. [58] extended fuzzy keyword search to multi-keyword environments by using Bloom filter [59] and locality-sensitive hashing [60]. To enrich the search patterns, Fu et al. [61] designed a content-based symmetric SE scheme to enable efficient semantic searches; Wang et al. [62] proposed a scheme for a generalized pattern-matching string-search. To improve the security, Ding et al. [63] proposed a random traversal algorithm, which produced different visiting paths on the index for the identical queries with different keys. Boldyreva et al. [64] improved the security of existing fuzzy search schemes based on closeness graphs. To improve search efficiency, Moataz et al. [65] employed letter orthogonalization to allow testing of string membership by computing inner products; Hahn et al. [66] transformed the problem of secure substring search into range queries for fast execution time.

In terms of fine-grained search authorization, Bao et al. [67] proposed an authorized searchable encryption in a multi-users setting, which allowed the data owner to enforce an access policy by distributing some secret keys to authorized users; Li et al. [68] constructed an authorized private keyword search scheme based on hierarchical predicate encryption (HPE) [32]. Zheng et al. [69] proposed an attribute-based keyword search (ABKS) scheme based on ABE, which allowed the data owner to control the keyword search operation over the outsourced encrypted data. This book chapter will introduce the dynamic attribute-based keyword search (DABKS) scheme [70, 71] that incorporate proxy re-encryption (PRE) and a secret sharing scheme (SSS) into ABKS to achieve achieve policy updates in an efficient way.

# 3 Problem Formulation

## 3.1 System Model

As shown in Fig. 1, our system model consists of three types of entities: data owner, data user, and cloud server.

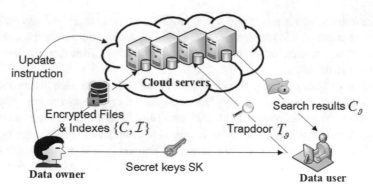

**Fig. 1** System model

- *Data owner* possesses a large-scale collection of files $F$ and decides to outsource them in the encrypted forms $C$ for reduced cost and convenient access. After outsourcing all the above information, she can perform updates (add/delete) on ciphertexts on demand by sending an update instruction to the cloud server. To enable efficient searches, she builds a searchable index from file collection $F$ and a universal keyword set $W$ and uploads file ciphertexts and an encrypted index, $\{C, \mathcal{I}\}$, to the cloud server. For access authorization, she is responsible for the secure distribution of key information to qualified data users.
- *Data user* sends the cloud server a trapdoor, $T_\vartheta$, to retrieve files matching query $\vartheta$ after obtaining a warrant from the data owner. Upon receiving search results $C_\vartheta$ from the cloud server, he performs decryption locally to recover file contents.
- *Cloud server* maintained by the CSP centralizes abundant resources to provide data storage and query services to data owners and data users, respectively. Upon receiving the store request from the data owner, it stores the encrypted files and index $\{C, \mathcal{I}\}$ to appropriate locations. Given a trapdoor $T_\vartheta$ sent by the data user, it evaluates $T_\vartheta$ on the encrypted index $\mathcal{I}$ and returns all matched ciphertexts $C_\vartheta$ as search results. Besides, it also follows the data owner's command to perform updates on $\{C, \mathcal{I}\}$ appropriately.

## 3.2 Threat Model

We assume that the data owner and the authorized data users are fully trusted. Furthermore, we assume that communication channels are secured under existing security protocols such as SSL. According to the attack ways initiated by the CSP, we mainly consider the Honest but curious model: A honest but curious CSP would correctly execute the prespecified protocol, but still attempt to learn extra information about the stored data and the received message.

Therefore are two kinds of privacy against the adversary: *file privacy* and *query privacy*. In terms of file privacy, the file should be accessed by only the authorized users. That is, the attacker cannot deduce the contents of files from the encrypted data stored in the cloud.

In terms of query privacy, the ideal case is that the server should learn nothing during the query process. However, this perfect level of privacy requires expensive primitives, such as oblivious RAM [72] and fully homomorphic encryption [73]. To be practical, existing work trades leakage for efficiency. This leakage typically includes *search pattern* and *access pattern*. As defined in [74], *access pattern* refers to the outcome of search results, i.e., which documents have been returned; *search pattern* refers to whether two searches have been performed for the same keyword.

## 4  Preliminaries

### 4.1  Definitions and Assumptions

**Definition 1 (Bilinear Map)**  Let $\mathbb{G}_1$ and $\mathbb{G}_2$ be two cyclic groups of some large prime order $q$, where $\mathbb{G}_1$ is an additive group and $\mathbb{G}_2$ is a multiplicative group. A bilinear map, $\hat{e}$: $\mathbb{G}_1 \times \mathbb{G}_1 \rightarrow \mathbb{G}_2$, satisfies the following properties: (1) Computable: There is a polynomial time algorithm to compute $\hat{e}(P, Q) \in \mathbb{G}_2$, for any $P, Q \in \mathbb{G}_1$. (2) Bilinear: $\hat{e}(\alpha_0 P, \alpha_1 Q) = \hat{e}(P, Q)^{\alpha_0 \alpha_1}$ for all $P, Q \in \mathbb{G}_1$ and all $\alpha_0, \alpha_1 \in \mathbb{Z}_q^*$. (3) Non-degenerate: The map does not send all pairs in $\mathbb{G}_1 \times \mathbb{G}_1$ to the identity in $\mathbb{G}_2$.

**Definition 2 (BDH Parameter Generator)**  A randomized algorithm $\mathcal{IG}$ is called a BDH parameter generator if $\mathcal{IG}$ takes a sufficiently large security parameter $K$ as input, runs in polynomial time in $K$, and outputs a prime number $q$, the description of two groups $\mathbb{G}_1$ and $\mathbb{G}_2$ of order $q$, and the description of a bilinear map $\hat{e}$ : $\mathbb{G}_1 \times \mathbb{G}_1 \rightarrow \mathbb{G}_2$.

**Definition 3 (BDH Problem)**  Given a random element $P \in \mathbb{G}_1$, as well as $\alpha_0 P$, $\alpha_1 P$, and $\alpha_2 P$, for some $\alpha_0, \alpha_1, \alpha_2 \in \mathbb{Z}_q^*$, compute $\hat{e}(P, P)^{\alpha_0 \alpha_1 \alpha_2} \in \mathbb{G}_2$.

**Definition 4 (BDH Assumption)**  If $\mathcal{IG}$ is a BDH parameter generator, the advantage $Adv_{\mathcal{IG}}(\mathcal{B})$ that an algorithm $\mathcal{B}$ has in solving the BDH problem is defined to be the probability that $\mathcal{B}$ outputs $\hat{e}(P, P)^{\alpha_0 \alpha_1 \alpha_2}$ on inputs $q$, $\mathbb{G}_1$, $\mathbb{G}_2$, $\hat{e}$, $P$, $\alpha_0 P$, $\alpha_1 P$, $\alpha_2 P$, where $(q, \mathbb{G}_1, \mathbb{G}_2, \hat{e})$ are the outputs of $\mathcal{IG}$ for a sufficiently large security parameter $K$, $P$ is a random element $\in \mathbb{G}_1$, and $\alpha_0, \alpha_1, \alpha_2$ are random elements of $\mathbb{Z}_q^*$. The BDH assumption is that $Adv_{\mathcal{IG}}(\mathcal{B})$ is negligible for any efficient algorithm $\mathcal{B}$.

## 4.2  Proxy Re-encryption

Let us illustrate the motivation of the PRE scheme [34] by the following example: Alice receives emails encrypted under her public key $PK_A$ via a semi-trusted mail server. When she leaves for vacation, she wants to delegate her email to Bob whose public key is $PK_B$, but does not want to share her secret key $SK_A$ with him. The PRE scheme allows Alice to provide a PRE key $RK_{A \to B}$ to the mail server, with which the mail server can convert a ciphertext that is encrypted under Alice's public key $PK_A$ into another ciphertext that can be decrypted by Bob's secret key $SK_B$, without seeing the underlying plaintext, $SK_A$, and $SK_B$.

Let $\mathbb{G}$ be a multiplicative group of prime order $q$, and $g$ be a random generator of $\mathbb{G}$. The PRE scheme is consisted of the following algorithms:

**Key Generation**  Alice can choose a random element $a \in \mathbb{Z}_q^*$ as her secret key $SK_A$, and her public key $PK_A$ is $g^a \in \mathbb{G}$. In the same way, Bob's public/secret key pair $(SK_B, PK_B)$ are $(b, g^b)$. The PRE key $RK_{A \to B} = b/a(\mod q)$ is used to transfer a ciphertext that is encrypted under $PK_A$ to the ciphertext that can be decrypted with $SK_B$, and vice versa.

**Encryption**  To encrypt a message $m \in \mathbb{G}$ to Alice, the sender randomly chooses $r \in \mathbb{Z}_q^*$, and generates ciphertext $C_A = (C_{A1}, C_{A2}) = (g^r m, g^{ar})$.

**Decryption**  Given the ciphertext $C_A = (C_{A1}, C_{A2})$, Alice can recover message $m$ with her secret key $a$ by calculating $C_{A1}/(C_{A2})^{1/a}$.

**Re-encryption**  Given $RK_{A \to B}$, the mail server can convert $C_A$ to $C_B$ that can be decrypted by Bob as follows: $C_{B1} = C_{A1}$ and $C_{B2} = (C_{A2})^{RK_{A \to B}}$. Given the ciphertext $(C_{B1}, C_{B2})$, Bob can recover message $m$ with his secret key $b$ by calculating $C_{B1}/(C_{B2})^{1/b}$.

Note that although the data is encrypted twice, first encrypted with Alice's public key, and then re-encrypted with a PRE key, Bob only needs to execute decryption once to recover data. The PRE scheme is based on ElGamal encryption [75], and thus the ciphertext is semantically secure, and given the PRE key, the mail server cannot guess the secret keys $a$ nor $b$. Please refer to [34] for more details.

## 5  Secure Storage Services

Since the cloud service provider (CSP) is outside the users' trusted domain, existing research suggests encrypting data before outsourcing. In cloud computing environments, data is generally shared by many data users of different roles and attributes, thus how to achieve fine-grained access controls on ciphertexts becomes a burning question.

In CP-ABE, users are identified by a set of *attributes* rather than an exact identity. The data is encrypted with an *attribute-based access structure*, such that only the

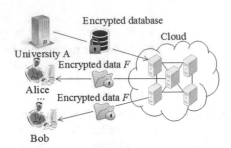

| User information | | |
|---|---|---|
| **User ID** | **Attributes** | **Effective time** |
| Alice | Staff, CIS | 01/01/2020-12/31/2020 |
| Bob | Student, CIS | 05/01/2020-06/30/2020 |
| ... | ... | ... |

**Fig. 2** Company A outsources an encrypted database to the cloud

users whose attributes satisfy the access structure can decrypt the ciphertext using their private keys. For example, for data which is encrypted with the access structure {(Student ∧ CIS) ∨ Staff}, either users with attributes Student and CIS, or users with attribute Staff, can recover data. The original ABE systems only support monotone access policy and assume the existence of a single private key generator (PKG). A lot of research has been done to achieve more expressive access policy [17, 18], and distributed key management [22, 23].

However, new problems, such as fine-grained access control on the encrypted data and scalable user revocation, emerge for ABE schemes. To illustrate, let us consider the following application scenario, as shown in Fig. 2. Suppose that University A outsources the electronic library database to a cloud for easy access by its staff and students. For the protection of copyright, each piece of data is encrypted before outsourcing. In this application, the staff and students are users, and University A is the data owner who will specify the access structure for each data, and will distribute decryption keys to users. Once joining University A, each user will first be assigned an access right with certain validity for accessing the outsourced database. Once the period of validity passes, this user should request an extension for his access right from University A. In Fig. 2, data $F$'s access structure stipulates that only Alice or the students in computer information science (CIS) department have the right to access it. In this access structure, the data owner describes the intended recipients using not only their IDs but also descriptive attributes, such as Staff, Student, and CIS. Therefore, the adopted encryption scheme should have the ability to efficiently implement a fine-grained access control over ID and attributes simultaneously.

Furthermore, from the above application scenario, we observe that each user's access right is only effective in a predetermined period of time. For example, the effective time of Alice's access right is from 01/01/2020 to 12/31/2020 and she can access the database in year 2020, but the effective time of Bob's access right is from 05/01/2020 to 06/30/2020 and thus he cannot access the database after June. Therefore, the adopted encryption scheme should support a scalable revocation mechanism to efficiently achieve a dynamic set of users.

To achieve a flexible access control in cloud computing, our previous work [24–26] proposed a Hierarchical Attribute-Based Encryption (HABE) scheme, by combining HIBE and ABE systems. The main merits of the HABE scheme is as follows: (1) It supports both ID-based and attribute-based access policies, with hierarchical key generation properties. (2) It requires only a constant number of bilinear map operations during decryption, is tailored for best serving the needs of accessing data anytime and anywhere.

Our subsequent work proposed a time-based proxy re-encryption (TimePRE) scheme, by incorporating the concept of time into a combination of HABE and proxy-re-encryption (PRE). The TimePRE take full advantage of abundant resources in a cloud by delegating the CSP to execute computationally intensive tasks in user revocations, while leaking the least information to the CSP. The main merits of the TimePRE scheme is as follows: (1) It enables the CSP to automatically re-encrypt data without receiving any PRE keys from the data owner. Therefore, our scheme can avoid the security risks caused by the delay of issuing PRE keys. (2) It allows a user's access right to automatically expire after a predetermined period of time. Therefore, the data owner, who can be offline in the process of user revocations, has much less workload. In the next subsections, we will expatiate the HABE scheme and the TimePRE scheme. The most relevant notations used in HABE and TimePRE are shown in Table 1.

## 5.1 Hierarchical Attribute-Based Encryption

The access structure in HABE is expressed as disjunctive normal form (DNF). For example, access structure $\mathbb{A} = \overset{N}{\underset{i=1}{\vee}} (CC_i) = \overset{N}{\underset{i=1}{\vee}} (\overset{n_i}{\underset{j=1}{\wedge}} a_{ij})$ consists of $N$ conjunctive clauses, $CC_1, \ldots, CC_N$, where the $i$-th conjunctive clause $CC_i$ is expressed as $a_{i1} \wedge a_{i2} \wedge \ldots \wedge a_{in_i}$. The users that possess all of the attributes in $CC_i$ can decrypt the ciphertext. The original HABE allows a delegation mechanism in the generation of decryption keys. For ease of illustration, we simplify the HABE scheme to a one-layer structure and provide a modified version as follows:

$Setup(K, \mathbb{UA}) \rightarrow (PK, MK)$: takes the security parameter $K$ and the universal attribute $\mathbb{UA}$ as input, and outputs system public key $PK$ and system master key $MK$ as follows:

$$PK = (\{PK_a\}_{a \in \mathbb{UA}}, q, \mathbb{G}_1, \mathbb{G}_2, Q_0, \hat{e}, P_0, P_1)$$
$$MK = (\{sk_a\}_{a \in \mathbb{UA}}, mk_0, mk_1, SK_1)$$

where $(q, \mathbb{G}_1, \mathbb{G}_2, \hat{e})$ are the outputs of a BDH parameter generator $\mathcal{IG}$, $P_0$ is a random generator of $\mathbb{G}_1$, and $P_1$ is a random element in $\mathbb{G}_1$; $sk_a \in \mathbb{Z}_q^*$ is the secret key of attribute $a$ and $PK_a = sk_a P_0 \in \mathbb{G}_1$ is the public key of attribute $a$; $mk_0$ and $mk_1$ are random elements in $\mathbb{Z}_q^*$, $Q_0 = mk_0 P_0 \in \mathbb{G}_1$, and $SK_1 = mk_0 P_1 \in \mathbb{G}_1$.

$GenKey(PK, MK, ID_u, a) \rightarrow$: takes system public key $PK$, system master key $MK$, user identity $ID_u$, and attribute $a$ as inputs to generate a decryption key $dk_{u,a} = (SK_u, SK_{u,a})$ as follows:

**Table 1** Summary of Notations in HABE

| Notation | Description |
|---|---|
| $K$ | Security parameter |
| $\mathbb{UA}$ | Universal attributes |
| $PK$ | System public key |
| $MK$ | System master key |
| $s$ | Root secret key |
| $PK_a$ | Initial public key of attribute $a$ |
| $sk_a$ | Initial secret key of attribute $a$ |
| $T$ | A specific day $(y, m, d)$, month $(y, m)$, or year $(y)$ |
| $PK_a^T$ | Time-based public key of attribute $a^a$ |
| $\mathbb{A}$ | Data access structure |
| $t$ | Data access time[b] |
| $T_u$ | An effective time period[c] |
| $PK_u$ | User public key |
| $SK_u$ | User identity secret key (UIK) |
| $SK_{u,a}^{T_u}$ | Time-based user attribute secret key (UAK)[d] |
| $\subseteq$ | Satisfying a condition |

[a] If $T$ is a particular day $(y, m, d)$, $PK_a^T = PK_a^{(y,m,d)}$; If $T$ is a particular month $(y, m)$, $PK_a^T = PK_a^{(y,m)}$; If $T$ is a particular year $(y)$, $PK_a^T = PK_a^{(y)}$

[b] Data access time is a particular day $(y, m, d)$

[c] An effective time period is a particular day $(y, m, d)$, month $(y, m)$, or year $(y)$

[d] If $T_u$ is a particular day $(y, m, d)$, $SK_{u,a}^{T_u} = SK_{u,a}^{(y,m,d)}$; If $T$ is a particular month $(y, m)$, $SK_{u,a}^{T_u} = SK_{u,a}^{(y,m)}$; If $T$ is a particular year $(y)$, $SK_{u,a}^{T_u} = SK_{u,a}^{(y)}$

$$SK_u = mk_1 mk_u P_0 \in \mathbb{G}_1$$
$$SK_{u,a} = SK_1 + mk_1 mk_u PK_a \in \mathbb{G}_1$$

where $mk_u = H_1(PK_u) \in \mathbb{Z}_q^*$ and $H_1 \colon \mathbb{G}_1 \to \mathbb{Z}_q^*$ is a hash function which can be modeled as random oracle.

$Encrypt(PK, \mathbb{A}, M) \to (C_\mathbb{A})$ : takes system public key $PK$, access structure $\mathbb{A}$, and message $M$ as inputs to generate a ciphertext $C_\mathbb{A}$.

$$\mathbb{A} = \bigvee_{i=1}^{N} (CC_i) = \bigvee_{i=1}^{N} (\bigwedge_{j=1}^{n_i} a_{ij}),$$
$$U_0 = r P_0,$$
$$\{U_i = r \sum_{a \in CC_i} PK_a\}_{1 \leq i \leq N},$$
$$V = F \cdot \hat{e}(Q_0, r n_\mathbb{A} P_1)$$

where $N \in \mathbb{Z}^+$ is the number of conjunctive clauses in $\mathbb{A}$, $n_i \in \mathbb{Z}^+$ is the number of attributes in the *i-th* conjunctive clause $CC_i$, and $a_{ij}$ is the *j-th* attribute in $CC_i$; $r$ is a random element in $\mathbb{Z}_q^*$, and $n_\mathbb{A}$ is the lowest common multiple (LCM) of $n_1, \ldots,$ $n_N$.

$Decrypt(PK, C_\mathbb{A}, \{dk_{u,a}\}_{a \preceq \mathbb{A}}) \rightarrow (M)$ : takes system public key $PK$, ciphertext $C_\mathbb{A}$, and user decryption keys $\{dk_{u,a}\}_{a \preceq \mathbb{A}}$ as inputs, and outputs message $M$, where $a \preceq \mathbb{A}$ denotes $u$'s attributes satisfy the access structure $\mathbb{A}$.

$$F = V / \left( \frac{\hat{e}(U_0, \frac{n_\mathbb{A}}{n_i} \sum\limits_{a \in CC_i} SK_{u,a})}{\hat{e}(SK_u, \frac{n_\mathbb{A}}{n_i} U_i)} \right)$$

From the above construction, we know that The *Encrypt* and *Decrypt* algorithms require only a constant number of bilinear map operations, and the length of the ciphertext is related to the number of conjunctive clauses $(O(N))$, instead of the number of attributes $(O(m))$, in the access structure. To encrypt a data, we need to execute one bilinear map and $O(N)$ number of point multiplication operations to output a ciphertext of $O(N)$ length, where $N$ is the number of conjunctive clauses in the access structure; To recover a data, we only need to execute $O(1)$ bilinear map operations. The HABE scheme is proven to be semantically secure under the random oracle model and the BDH assumption. In Table 2, we briefly compare our scheme with the work by Bethencourt et al. [15] and the work by Muller et al. [16]. We believe that the most expensive computation is bilinear map operation, abbreviated as *map*, the next is the exponentiation operation, abbreviated as *exp*. In Table 2, $n$, $S$, $N$, $T$, and $P$ denote the number of attributes associated with a user, the number of attributes in an access structure, the number of conjunctive clauses in an access structure, and the number of attributes in an access structure that is matched by attributes in a user's private key, respectively. More details can be found in [25, 26].

## 5.2   Overview of TimePRE

While a user exits the system, his/her access right should be revoked. A typical application is that a company authorizes its staffs to access corporate data; while

**Table 2** Comparisons of CP-ABE schemes

| Properties | Reference [15] | Reference [16] | HABE |
|---|---|---|---|
| User key size | $O(2n)$ | $O(n)$ | $O(n)$ |
| Ciphertext | $O(2S)$ | $O(3N)$ | $O(N)$ |
| Encryption (exp) | $O(2N)$ | $O(3N)$ | $O(N)$ |
| Decryption (map) | $O(2P)$ | $O(1)$ | $O(1)$ |
| Full delegation | No | No | Yes |
| Multiple AAs | No | Yes | Yes |
| Access structure | Monotone | DNF | DNF |

a staff left the company, his/her access right should be revoked. In the case of accessing encrypted data, the revoked users still retain the keys issued earlier, and thus can still decrypt ciphertexts. Therefore, whenever a user is revoked, the re-keying and re-encryption operations need to be executed by the data owner to prevent the revoked user from accessing the future data.

User revocation is a well studied, but non-trivial task. In an ABE encryption system, it is more tricky to achieve effective user revocation, since each attribute is conceivably possessed by multiple different users so that multiple different users might match the same decryption policy [76]. The usual solution is to require each attribute to contain a time frame within which it is valid, and the latest version of attributes and user information will be periodically distributed. As discussed in Bethencourt et al. [33], this type of solution, which demands the users to maintain a large amount of private key storage, lacks flexibility and scalability.

Other than requiring the data owner to do all of work on user revocation, a better solution is to let the data owner delegate a third party (e.g., the service provider) to execute some computational intensive tasks, e.g., re-encryption. Since the third party is not fully trusted, delegation should leak the least information. Proxy Re-Encryption (PRE) [34, 77] is a good choice, where a semi-trusted proxy is able to convert a ciphertext that can be decrypted by Alice into another ciphertext that can be decrypted by Bob, without knowing the underlying data and user secret keys.

In recent work, Yu et al. [36] and our previous work [24–26] applied proxy re-encryption (PRE) [34] into KP-ABE and CP-ABE, respectively, to achieve a scalable revocation mechanism in cloud computing. Once a user is revoked from a system, this kind of approaches, also called instruction-based PRE, require that the data owner should distribute new keys to unrevoked users, and send PRE keys to the CSPs, which can be delegated to execute re-encryption in a *lazy* way [9]. Although the revocation can be performed on demand, the data owner needs to send the PRE keys to the servers in a timely fashion, to prevent the revoked user from accessing the data. The delay of issuing PRE keys may cause potential security risks. To solve this problem, Liu et al. [39] proposed a TimePRE scheme, which allows the servers to perform re-encryption automatically. In their work, each user's access right is effective within a pre-determined time, and enables the servers to re-encrypt ciphertexts automatically based on its own time. Thus, the data owner can be offline in the process of user revocations. The differences between the instruction-based PRE and the time-based PRE are shown in Fig. 3. Suppose that $T1 < T2 < \ldots < T6$. In the instruction-based PRE, data user A who is revoked at $T2$ can access data from the clouds before the CSP re-encrypts the data at $T4$.

The main idea of the TimePRE scheme is to incorporate the concept of time into the combination of HABE and PRE. Intuitively, each user is identified by a set of *attributes* and a set of *effective time periods* that denotes how long the user is eligible for these attributes, i.e., the period of validity of the user's access right. The data accessed by the users is associated with an *attribute-based access structure* and *an access time*. The access structure is specified by the data owner, but the access time is updated by the CSP with the time of receiving an access request. The data

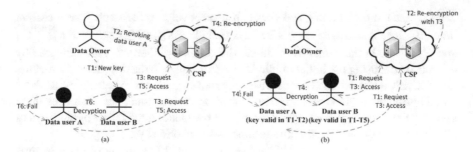

**Fig. 3** Proxy re-encryption in clouds. (**a**) Instruction-based PRE. (**b**) Time-based PRE

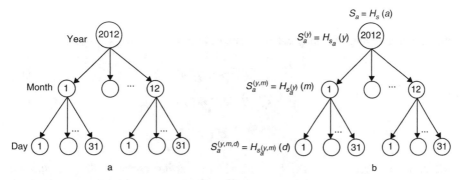

**Fig. 4** Three-level time tree and attribute $a$'s PRE key tree. (**a**) Sample time tree. (**b**) Sample PRE key tree for attribute $a$

can be recovered by only the users whose attributes satisfies the access structure and whose access rights are effective in the access time.

To enable the CSP to update the access time automatically, we first express *actual time* as a *time tree*. The height of the time tree can be changed as required. For ease of presentation, in this paper we only consider a three-layer time tree as shown in Fig. 4a, where time is accurate to the day, and the time tree is classified into three layers in order: year, month, and day. We use $(y, m, d)$, $(y, m)$, and $(y)$ to denote a particular day, month, and year, respectively. For example, $(2020, 4, 5)$ denotes April 5, 2020. The access time associated with a data corresponds to a leaf node in the time tree, and the effective time periods associated with a user correspond to a set of nodes in the time tree. If there is a node corresponding to an effective time period that is an ancestor of (or the same as) the node corresponding to the access time, then the user's access right is effective in the access time.

Then, we allow the data owner and the CSP to share a root secret key $s$ in advance, with which the CSP can calculate required PRE keys based on its own time and re-encrypt corresponding ciphertext automatically. Specifically, at any time, each attribute $a$ is associated with one initial public key $PK_a$, and three time-based public keys: day-based public key $PK_a^{(y,m,d)}$, month-based public key $PK_a^{(y,m)}$, and year-based public key $PK_a^{(y)}$, each of which denotes $a$'s public key in a particular day $(y, m, d)$, month $(y, m)$, and year $(y)$, respectively. For example, given current time $(2020, 4, 5)$, attribute $a$'s public keys include $PK_a$, $PK_a^{(2020,4,5)}$, $PK_a^{(2020,4)}$, and $PK_a^{(2020)}$. In the TimePRE scheme, the original ciphertexts are encrypted by the data owner using the initial public keys of attributes in the data access structure. On receiving a request, the CSP first uses the root secret key $s$ to calculate PRE keys on all attributes in the access structure based on its own time, and then uses these PRE keys to re-encrypt the original ciphertext by updating the initial public keys of all attributes in the access structure to time-based public keys.

We use $s_a^{(y)}$, $s_a^{(y,m)}$, and $s_a^{(y,m,d)}$ to denote the PRE keys on attribute $a$ in time $(y)$, $(y, m)$, and $(y, m, d)$, which can be used to update attribute $a$'s initial public key $PK_a$ to time-based public keys $PK_a^{(y)}$, $PK_a^{(y,m)}$, and $PK_a^{(y,m,d)}$, respectively. As shown in Fig. 4b, for each attribute $a$, the CSP can use the root secret key $s$ and the time tree to hierarchically calculate the time-based PRE keys with Eq. (1):

$$s_a^{(y)} = H_{s_a}(y) \tag{1a}$$

$$s_a^{(y,m)} = H_{s_a^{(y)}}(m) \tag{1b}$$

$$s_a^{(y,m,d)} = H_{s_a^{(y,m)}}(d) \tag{1c}$$

where $s_a = H_s(a)$, $a, y, m, d \in \{0, 1\}^*$ is a string corresponding to specific attribute, year, month, and day; and $H_s, H_{s_a}, H_{s_a^{(y)}}, H_{s_a^{(y,m)}} : \{0, 1\}^* \rightarrow \mathbb{Z}_q^*$ are hash functions with indexes $s$, $s_a$, $s_a^{(y)}$, and $s_a^{(y,m)}$, respectively.

Furthermore, to incorporate the concept of time to HABE, each user is granted with a set of *time-based user attribute secret keys* (UAK). Each time-based UAK is associated with a user, an attribute, and an effective time period. If user $u$ is eligible for attribute $a$ in day $(y, m, d)$, the data owner first uses the root secret key $s$ to obtain day-based attribute public key $PK_a^{(y,m,d)}$ from initial attribute public key $PK_a$, and then uses $PK_a^{(y,m,d)}$ to generate a day-based UAK $SK_{u,a}^{(y,m,d)}$ for user $u$. The same situation holds for the case that user $u$ is eligible for attribute $a$ in a month $(y, m)$ or a year $(y)$.

Return to the application in Sect. 1, Alice is authorized to possess attributes Staff and CIS, and her effective time period is $(2020)$, she will be issued time-based UAK as shown in Table 3; Bob is authorized to possess attributes Student and CIS, and his effective time periods are $(2020, 5)$ and $(2020, 6)$, he will be issued time-based UAK as shown in Table 4. Given an access time $(2020, 7, 1)$ and data $F$ with access structure $\mathbb{A} = \{(\text{Student} \wedge \text{CIS}) \vee \text{Staff}\}$, the CSP will use the root secret key $s$ to calculate the PRE keys in $(2020)$, $(2020, 7)$, and $(2020, 7, 1)$ for all attributes in $\mathbb{A}$, say $\{s_{\text{Student}}^{(2020)}\}$, $\{s_{\text{Student}}^{(2020,7)}\}$, $\{s_{\text{Student}}^{(2020,7,1)}\}$, $\{s_{\text{CIS}}^{(2020)}\}$, $\{s_{\text{CIS}}^{(2020,7)}\}$, $\{s_{\text{CIS}}^{(2020,7,1)}\}$, $\{s_{\text{Staff}}^{(2020)}\}$, $\{s_{\text{Staff}}^{(2020,7)}\}$, $\{s_{\text{Staff}}^{(2020,7,1)}\}$. Then, it will use these PRE keys to re-encrypt original ciphertext by updating initial public keys $\{PK_{\text{Student}}, PK_{\text{CIS}}, PK_{\text{Staff}}\}$ to year-based attribute public keys $\{PK_{\text{Student}}^{(2020)}, PK_{\text{CIS}}^{(2020)}, PK_{\text{Staff}}^{(2020)}\}$, month-based attribute public

**Table 3** Alice's time-based user attribute secret keys

| Key | Description |
|---|---|
| $SK_{(\text{Alice, Staff})}^{(2020)}$ | UAK on attribute `Staff` effective in 2020 |
| $SK_{(\text{Alice, CIS})}^{(2020)}$ | UAK on attribute `CIS` effective in 2020 |

**Table 4** Bob's time-based user attribute secret keys

| Key | Description |
|---|---|
| $SK_{(\text{Bob, Student})}^{(2020,5)}$ | UAK on attribute `Student` effective in $(2020, 5)$ |
| $SK_{(\text{Bob, Student})}^{(2020,6)}$ | UAK on attribute `Student` effective in $(2020, 6)$ |
| $SK_{(\text{Bob, CIS})}^{(2020,5)}$ | UAK on attribute `CIS` effective in $(2020, 5)$ |
| $SK_{(\text{Bob, CIS})}^{(2020,6)}$ | UAK on attribute `CIS` effective in $(2020, 6)$ |

keys $\{PK_{\text{Student}}^{(2020,7)}, PK_{\text{CIS}}^{(2020,7)}, PK_{\text{Staff}}^{(2020,7)}\}$, and day-based attribute public keys $\{PK_{\text{Student}}^{(2020,7,1)}, PK_{\text{CIS}}^{(2020,7,1)}, PK_{\text{Staff}}^{(2020,7,1)}\}$. Given the re-encrypted ciphertext, only the users who possess $\{SK_{u,\text{Student}}^{(2020)}, SK_{u,\text{CIS}}^{(2020)}\}$ (or $\{SK_{u,\text{Staff}}^{(2020)}\}$, or $\{SK_{u,\text{Student}}^{(2020,7)}, SK_{u,\text{CIS}}^{(2020,7)}\}$, or $\{SK_{u,\text{Staff}}^{(2020,7)}\}$, or $\{SK_{u,\text{Student}}^{(2020,7,1)}, SK_{u,\text{CIS}}^{(2020,7,1)}\}$, or $\{SK_{u,\text{Staff}}^{(2020,7,1)}\}$) can recover data $F$. Therefore, Alice, who possesses year-based UAK $\{SK_{\text{Alice,Staff}}^{(2020)}\}$ can recover data $F$, but Bob, whose effective time periods are overdue in $(2020, 7, 1)$ cannot recover data $F$ any more.

## 5.3 Construction of TimePRE

1. $Setup(K, \mathbb{UA}) \rightarrow (PK, MK, s)$ : The data owner takes a security parameter $K$ and the universal attribute $\mathbb{UA}$ as inputs, and outputs the system public key $PK$, the system master key $MK$, and a root secret key $s \in \mathbb{Z}_q^*$ as follows:

$$PK = (\{PK_a\}_{a \in \mathbb{UA}}, q, \mathbb{G}_1, \mathbb{G}_2, Q_0, \hat{e}, P_0, P_1)$$
$$MK = (\{sk_a\}_{a \in \mathbb{UA}}, mk_0, mk_1, SK_1)$$

where $(q, \mathbb{G}_1, \mathbb{G}_2, \hat{e})$ are the outputs of a BDH parameter generator [49] $\mathcal{IG}$, $P_0$ is a random generator of $\mathbb{G}_1$, $P_1$ is a random element in $\mathbb{G}_1$; $sk_a \in \mathbb{Z}_q^*$ is the initial secret key of attribute $a$ and $PK_a = sk_a P_0 \in \mathbb{G}_1$ is the initial public key of attribute $a$; $mk_0$ and $mk_1$ are random elements in $\mathbb{Z}_q^*$, $Q_0 = mk_0 P_0 \in \mathbb{G}_1$, and $SK_1 = mk_0 P_1 \in \mathbb{G}_1$. $PK$ will be published, $MK$ will be kept secret, and $s$ will be sent to the CSP.

2. $GenKey(PK, MK, s, PK_u, a, T_u) \rightarrow (SK_u, SK_{u,a}^{T_u})$ : After authenticating user $u$ is eligible for attribute $a$ and his access right is effective in time period $T_u$, the data owner takes the system public key $PK$, the system master key $MK$,

user public key $PK_u$, and the root secret key $s$ as inputs, and generates a UIK $SK_u$ and a time-based UAK $SK_{u,a}^{T_u}$, as follows:

$$SK_u = mk_1 mk_u P_0$$
$$SK_{u,a}^{T_u} = SK_1 + mk_1 mk_u PK_a^{T_u}$$

where $mk_u = H_1(PK_u) \in \mathbb{Z}_q^*$, $H_1 \colon \mathbb{G}_1 \to \mathbb{Z}_q^*$ is a hash function which can be modeled as random oracle, and $PK_a^{T_u,a}$ is the time-based public key of attribute $a$ in time $T_u$. Specifically, we have the following three cases: (1) $T_u$ is a particular day $(y, m, d)$ and $SK_{u,a}^{(y,m,d)} = SK_1 + mk_1 mk_u PK_a^{(y,m,d)}$; (2) $T_u$ is a particular month $(y, m)$ and $SK_{u,a}^{(y,m)} = SK_1 + mk_1 mk_u PK_a^{(y,m)}$; (3) $T_u$ is a particular year $(y)$ and $SK_{u,a}^{(y)} = SK_1 + mk_1 mk_u PK_a^{(y)}$.

Here, time-based attribute public keys can be calculated with Eq. (2):

$$PK_a^{(y)} = PK_a + s_a^{(y)} P_0 \tag{2a}$$

$$PK_a^{(y,m)} = PK_a + s_a^{(y,m)} P_0 \tag{2b}$$

$$PK_a^{(y,m,d)} = PK_a + s_a^{(y,m,d)} P_0 \tag{2c}$$

where PRE keys $s_a^{(y)}$, $s_a^{(y,m)}$, $s_a^{(y,m,d)}$ can be calculated with Eq. (1).

3. $Encrypt(PK, \mathbb{A}, F) \to (C_\mathbb{A})$ : This algorithm is the same as the *Encryption* algorithm in HABE. The data owner encrypts data $F \in \mathbb{G}_2$ with access structure $\mathbb{A} = \overset{N}{\underset{i=1}{\vee}} (CC_i) = \overset{N}{\underset{i=1}{\vee}} (\overset{n_i}{\underset{j=1}{\wedge}} a_{ij})$ as follows: It first picks a random element $r \in \mathbb{Z}_q^*$, and then sets $n_\mathbb{A}$ to be the lowest common multiple (LCM) of $n_1, \ldots, n_N$. Finally, it calculates Eq. (3) to produce the ciphertext:

$$U_0 = r P_0, \tag{3a}$$

$$\{U_i = r \sum_{a \in CC_i} PK_a\}_{1 \le i \le N}, \tag{3b}$$

$$V = F \cdot \hat{e}(Q_0, r n_\mathbb{A} P_1) \tag{3c}$$

The original ciphertext is set to $C_\mathbb{A} = (\mathbb{A}, U_0, \{U_i\}_{1 \le i \le N}, V)$.

4. $ReEncrypt(C_\mathbb{A}, s, t) \to (C_\mathbb{A}^t)$ : On receiving a user's request for data $F$, the CSP first determines current time, say $t = (y, m, d)$. Then, it uses the root secret key $s$ and access time $t$ to re-encrypt the original ciphertext $C_\mathbb{A}$ with Eq. (4):

$$U_0^t = U_0 + r' P_0, \tag{4a}$$

$$U_{(y)i}^t = \sum_{a \in CC_i} (U_i + r' PK_a + s_a^{(y)} U_0^t), \tag{4b}$$

$$U^t_{(y,m)i} = \sum_{a \in CC_i} (U_i + r'PK_a + s_a^{(y,m)} U^t_0), \qquad (4c)$$

$$U^t_{(y,m,d)i} = \sum_{a \in CC_i} (U_i + r'PK_a + s_a^{(y,m,d)} U^t_0), \qquad (4d)$$

$$V^t = V \cdot \hat{e}(Q_0, r'n_{\mathbb{A}} P_1) \qquad (4e)$$

where $r'$ is randomly chosen from $\mathbb{Z}^*_q$ and the PRE keys $s_a^y$, $s_a^{(y,m)}$, $s_a^{(y,m,d)}$ can be calculated with Eq. (1). The ciphertext that is re-encrypted in time $t$ is set to $C^t_{\mathbb{A}} = (\mathbb{A}, t, U^t_0, \{U^t_{(y)i}\}_{1 \le i \le N}, \{U^t_{(y,m)i}\}_{1 \le i \le N}, \{U^t_{(y,m,d)i}\}_{1 \le i \le N}, V^t)$.

5. $Decrypt(PK, C^t_{\mathbb{A}}, SK_u, \{SK^{T_u}_{u,a}\}_{a \subseteq \mathbb{A}, T_u \subseteq t}) \rightarrow (F)$ : Given ciphertext $C^t_{\mathbb{A}}$, user $u$, whose attributes satisfy the $i\text{-}th$ conjunctive clause $CC_i$ and whose effective time period $T_u$ satisfies the access time $t$, uses his UIK $SK_u$ and UAKs $\{SK^{T_u}_{u,a}\}_{a \subseteq \mathbb{A}, T_u \subseteq t}$ to recover data $F$ with Eq. (5):

$$F = V^t / \frac{\hat{e}(U^t_0, \frac{n_{\mathbb{A}}}{n_i} \sum_{a \in CC_i} SK^{T_u}_{u,a})}{\hat{e}(SK_u, \frac{n_{\mathbb{A}}}{n_i} U^t_{T_u i})} \qquad (5)$$

Specifically, there are three cases:

(1) $T_u$ is a particular day $(y, m, d)$ and Eq. (5) is equivalent to:

$$F = V^t / \frac{\hat{e}(U^t_0, \frac{n_{\mathbb{A}}}{n_i} \sum_{a \in CC_i} SK^{(y,m,d)}_{u,a})}{\hat{e}(SK_u, \frac{n_{\mathbb{A}}}{n_i} U^t_{(y,m,d)i})}$$

(2) $T_u$ is a particular month $(y, m)$ and Eq. (5) is equivalent to:

$$F = V^t / \frac{\hat{e}(U^t_0, \frac{n_{\mathbb{A}}}{n_i} \sum_{a \in CC_i} SK^{(y,m)}_{u,a})}{\hat{e}(SK_u, \frac{n_{\mathbb{A}}}{n_i} U^t_{(y,m)i})}$$

(3) $T_u$ is a particular year $(y)$ and Eq. (5) is equivalent to:

$$F = V^t / \frac{\hat{e}(U^t_0, \frac{n_{\mathbb{A}}}{n_i} \sum_{a \in CC_i} SK^{(y)}_{u,a})}{\hat{e}(SK_u, \frac{n_{\mathbb{A}}}{n_i} U^t_{(y)i})}$$

The key technique of the TimePRE scheme is that the root secret key $s$ is simultaneously used by the data owner to generate time-based UAKs, and by the CSP to generate PRE keys. Note that each time-based UAK is generated with time-based attribute public key, which is in turn generated by $s$ and an effective time period $T_u$; each data is first encrypted with initial attribute public keys, and will be updated by the CSP to day-based attribute public keys, which are in turn generated by $s$ and the access time of receiving a request $t$. Therefore, even if $s$ is only shared between the data owner and the CSP, the users still can decrypt the ciphertext when their attributes satisfy the access structure and their access rights are effective in the access time. Furthermore, the *GenKey* algorithm should take the system master key $MK$ as inputs, which is kept secret by the data owner. Thus, given the root secret key that has nothing to do with the system master key, the CSP cannot know any information about the UAKs.

**Security Analysis of TimePRE** The *Encrypt* algorithm in the TimePRE scheme is the same as the *Encryption* algorithm in HABE, which has been proven to be semantically secure in [25]. Therefore, we consider that the TimePRE scheme is secure if the following propositions hold:

- **Proposition 1.** The keys produced by the *GenKey* algorithm are secure.
- **Proposition 2.** The ciphertext produced by the *ReEncrypt* algorithm is semantically secure.
- **Proposition 3.** Given the root secret key and the original ciphertext, the CSP cannot know neither the underlying data, nor UAKs while executing re-encryption.

For Proposition 1, we prove that the *GenKey* algorithm is as secure as the *Key Generation* algorithm in HABE. First, the way to generate UIK is the same in both algorithms. Then, given the system public key $PK$, the system master key $MK$, user public key $PK_u$, and attribute $a$, if the data owner takes the time-based attribute public key $PK_a^{(T_u)}$ as inputs of the *Key Generation* algorithm in HABE, then the produced UAK is the same as that of the *GenKey* algorithm that takes time $T_u$, the initial attribute public key $PK_a$, and the root secret key $s$ as inputs. As proven in [25], due to the BDH assumption, the malicious users cannot obtain $MK$, even if all of them collude. Therefore, Proposition 1 is correct.

For Proposition 2, we prove that the *ReEncrypt* algorithm is as secure as the *Encryption* algorithm in HABE. Given system public key $PK$ and data $F$ with access structure $\mathbb{A}$, if the data owner takes the time-based attribute public keys $\{PK_a^{(y)}\}_{a \in \mathbb{A}}$, $\{PK_a^{(y,m)}\}_{a \in \mathbb{A}}$, $\{PK_a^{(y,m,d)}\}_{a \in \mathbb{A}}$, and a random number $r'' = r + r'$ as inputs of the *Encryption* algorithm in HABE, then the produced ciphertext is the same as that of the the *ReEncrypt* algorithm that takes time $t = (y, m, d)$, the original ciphertext $C_{\mathbb{A}}$, and root secret key $s$ as inputs. Therefore, Proposition 2 is correct.

For completeness, we provide an intuitive security proof for the *ReEncrypt* algorithm as follows:

Recall that data $F$ is re-encrypted to $V^t = F \cdot \hat{e}(Q_0, (r + r')n_{\mathbb{A}}P_1)$ in time $t = (y, m, d)$. Therefore, an adversary $\mathcal{A}$ needs to construct $\hat{e}(Q_0, (r+r')n_{\mathbb{A}}P_1) = \hat{e}(U_0^t, SK_1)^{n_{\mathbb{A}}}$ to recover $F$. From the *GenKey* algorithm, we know that the only occurrence of $SK_1$ is in the UAKs. In our security model, we assume that the CSP will not collude with the malicious users, who possess UAKs. Therefore, we only consider the case that malicious users work independently, or collude to compromise data security.

We consider that the TimePRE scheme is insecure if one of the following cases happens: (1) Adversary $\mathcal{A}$, whose effective time period satisfies the access time, but whose attributes do not satisfy the access control, can recover data $F$. (2) Adversary $\mathcal{A}$, whose attributes satisfy the access control, but whose effective time does not satisfy the access time, can recover data $F$.

For case (1), we have the following assumptions for ease of presentation: Adversary $\mathcal{A}$ has requested UAKs on all but one of the attributes $a_{i1}, \ldots, a_{i(k-1)}, a_{i(k+1)}, \ldots, a_{in_i}$ in $CC_i$ for user $u$, and has requested a UAK on the missing attribute $a_{ik}$ for user $u'$. Both users' effective time periods $T_u$ and $T_{u'}$ satisfy the access time $t = (y, m, d)$. Based on Proposition 1, we know that the adversary cannot generate fake keys. The only occurrence of $SK_1$ is in the UAKs, so the adversary has to use UAKs requested for user $u$ and $u'$ for bilinear map, yielding for some $\alpha$:

$$\hat{e}(U_0^t, \frac{n_{\mathbb{A}}}{n_i} \sum_{j=1, j \neq k}^{n_i} SK_{u, a_{ij}}^{T_u} + \frac{n_{\mathbb{A}}}{n_i} SK_{u', a_{ik}}^{T_{u'}} + \alpha)$$

$$= \hat{e}(U_0^t, SK_1)^{n_{\mathbb{A}}} \hat{e}(r'' P_0, \alpha) \hat{e}(SK_{u'}, r'' PK_{a_{ik}}^{T_{u'}})^{\frac{n_{\mathbb{A}}}{n_i}} \hat{e}(SK_u, r'' \sum_{j=1, j \neq k}^{n_i} PK_{a_{ij}}^{T_u})^{\frac{n_{\mathbb{A}}}{n_i}}$$

where $r'' = r + r'$. To obtain $\hat{e}(U_0, SK_1)^{n_{\mathbb{A}}}$, the last three elements have to be eliminated. Note that $SK_{u'}$ and $SK_u$ are known to adversary $\mathcal{A}$, but $r$ is randomly chosen by the data owner for the original ciphertext $C_{\mathbb{A}}$ and $r'$ is randomly chosen by the CSP for the re-encrypted ciphertext $C_{\mathbb{A}}^t$. The adversary cannot know $r'' PK_{a_{ik}}^{T_{u'}}$ or $r'' \sum_{j=1, j \neq k}^{n_i} PK_{a_{ij}}^{T_u}$, even if he knows $U_{(T_u)i}^t$ and $U_{(T_{u'})i}^t$ due to the BDH assumption. Therefore, adversary $\mathcal{A}$ cannot recover the data from $V^t$.

For case (2), we have the following assumptions for ease of presentation: Adversary $\mathcal{A}$ has requested UAKs on all attributes in $CC_i$ for user $u$. Any effective time period $T_u$ of this user does not satisfy the access time $t = (y, m, d)$. Based on Proposition 1, we know that the adversary cannot generate fake keys. The only occurrence of $SK_1$ is in the UAKs, so the adversary has to use UAKs requested for user $u$ for bilinear map, yielding for some $\alpha$:

$$\hat{e}(U_0^t, \frac{n_{\mathbb{A}}}{n_i} \sum_{j=1}^{n_i} SK_{u, a_{ij}}^{T_u} + \alpha)$$

$$= \hat{e}(U_0^t, SK_1)^{n_{\mathbb{A}}} \hat{e}(r'' P_0, \alpha) \hat{e}(SK_u, r'' \sum_{j=1}^{n_i} PK_{a_{ij}}^{T_u})^{\frac{n_{\mathbb{A}}}{n_i}}$$

where $r'' = r + r'$. To obtain $\hat{e}(U_0, SK_1)^{n_{\mathbb{A}}}$, the last two elements have to be eliminated. Note that the $SK_u$ is known to adversary $\mathcal{A}$, but $r$ is randomly chosen by the data owner for the original ciphertext $C_{\mathbb{A}}$ and $r'$ is randomly chosen by the

CSP for the re-encrypted ciphertext $C_{\mathbb{A}}^t$. The adversary cannot know $r'' \sum\limits_{j=1, j \neq k}^{n_i} P_{a_{ij}}^{T_u}$, even if he knows $U_i^t$ and $U_i^{T_u}$ due to the BDH assumption. Therefore, adversary $\mathcal{A}$ cannot recover data from $V^t$.

For Proposition 3, we first prove that the CSP cannot derive the system master key $MK$ and UAKs from the root secret key $s$. As compared to HABE, the TimePRE scheme discloses an additional root secret key to the CSP, which is randomly chosen by the data owner and has nothing to do with the system master key. Therefore, the CSP cannot derive the system master key from the root secret key. Based on Proposition 1, the CSP cannot obtain UAKs without the system master key.

Then, we prove that the CSP cannot compromise data security given the original ciphertext. Note that the original ciphertext is encrypted with the *Encrypt* algorithm, which is semantically secure. Therefore, the ciphertext can be decrypted by only the entity who possesses UAKs on the initial attribute public keys. In the TimePRE scheme, a users' UAKs are generated on the time-based attribute public key, rather than the initial attribute public key. Therefore, only the data owner with the initial attribute secret keys can recover the data from the original ciphertext. Neither the users, nor the CSP can decrypt the original ciphertext.

## 6 Secure Search Services

Since the cloud service provider (CSP) may leak users' private data consciously or unconsciously, existing research suggests encrypting data before outsourcing to preserve user privacy. However, data encryption would make searching over ciphertexts a very challenging task. The simple solution that downloads the whole data set from the cloud will incur extensive communication costs. Therefore, searchable encryption (SE) [42, 47] was proposed to enable a user to retrieve data of interest while keeping user privacy from the CSP.

In previous SE solutions, the search permission is granted in a coarse-grained way. That is, the data user has the ability to generate search tokens for all the keywords by using the search key. In many situations, such a kind of search authorization will cause a potential risk of privacy disclosure. To illustrate, let us consider the following scenario: Company A outsources the file management system to Amazon S3 for easy access by its staff. Suppose that the collaboration agreement, F, described with keywords "Company B" and "Project X" can be accessed only by the manager of Company A. If attacker Alice is allowed to first search with keyword "Project X", and then with keyword "Company B", she can infer that Company A is cooperating with Company B on Project X from the search results returned, even if she cannot recover file content.

To alleviate this problem, the work in [69] proposed the attribute-based keyword search (ABKS) scheme, which utilizes attribute-based encryption (ABE) [14, 15] to achieve *fine-grained search authorization* for public-key settings. In ABKS, each keyword $w_i$ is associated with an access policy $AP$, and each search token is associated with a keyword $w_j$ and a set of attributes $S$. The data user can search the file only when her attributes satisfy the access policy, denoted as $S \preceq AP$, as well as $w_j = w_i$. However, ABKS never considered the problem of a dynamic access policy for keywords. If $AP$ is changed to $AP'$, it requires the data owner to re-encrypt the relevant keywords with $AP'$, so that only the users whose attributes satisfy $AP'$ have the search permission. For frequent updates on a large number of files, the workload on the data owner is heavy.

To achieve a fine-grained search authorization in an efficient way, this book chapter proposes a dynamic attribute-based keyword search (DABKS) scheme, by incorporating proxy re-encryption (PRE) and a secret sharing scheme (SSS) into ABKS. In DABKS, the CSP can be delegated to update the access policy for keywords without compromising user privacy. Specifically, DABKS expresses the access policy $AP$ as an access tree, and transforms the problem of updating an AND/OR gate in $AP$ to that of updating a threshold gate. For example, the AND gate is transformed to $(t, t)$ gate, and the OR gate is transformed to $(1, t)$ gate. Therefore, the updating of the AND gate can be treated as updating $(t, t)$ gate to $(t', t')$ gate, and the updating of the OR gate can be treated as updating $(1, t)$ gate to $(1, t')$ gate, where $t' = t + 1$ for adding an attribute to the AND/OR gate and $t' = t - 1$ for removing an attributes from the AND/OR gate.

## 6.1  Preliminaries

For ease of illustration, bilinear map is written in the form of $e(g^a, g^b) = e(g^b, g^a) = e(g, g)^{ab}$ for any $a, b \in \mathbb{Z}_p$. Let $\mathbb{A} = \{A_1, \ldots, A_M\}$ denote the universal attributes in the system. The cloud user $u$ is described by a set of attributes $S_u \subseteq \mathbb{A}$. The data owner holds a collection of files $\Omega = \{F_1, \ldots, F_n\}$, where each file $F_i$ can be described by a set of distinct keywords $W_i$. Before uploading file $F_i$ to the cloud, the data owner will first encrypt $F_i$ with ABE under access policy $AP_F$, and then encrypts each keyword in $W_i$ with DABKS under access policy $AP_K$. The ciphertext for keyword $w$ and file $F_i$ is denoted as $cph_w$ and $C_{F_i}$, respectively. It is worth noticing that $AP_F$ stipulates which entities can decrypt $F_i$ and $AP_K$ stipulates which entities can search over $W_i$ might be different. To retrieve files containing keyword $w$, the data user $u$ will issue a search token $Tok_w$ to the CSP, which will return $\{\{cph_w\}_{w \in W_i}, C_{F_i}\}$ only when $S_u \preceq AP_K$ and $w \in W_i$. To update $AP_K$ to $AP_K'$, the data owner will send an update instruction $\Gamma$ to the CSP, which will update related keyword ciphertexts to preserve the correctness of the access policy for keywords $W_i$.

**Access Tree**  As the work in [15] suggests, the access policy $AP_K$ can be depicted as an access tree $\mathcal{T}$, where each interior node denotes a gate, and each leave node is depicted as an attribute. For example, given $AP_K = (A_1 \vee A_2) \wedge A_3$, its access tree $\mathcal{T}$ is as shown in Fig. 5a. In $\mathcal{T}$, each node $x$ is associated with a threshold value

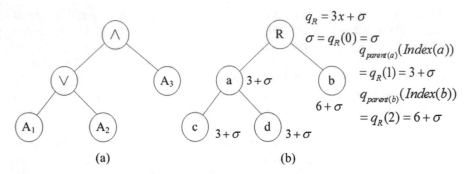

**Fig. 5** Access tree. (**a**) Sample access tree. (**b**) Sample tree implementation

$k_x$. For the interior node $x$ with $N_x$ children, $k_x = 1$ when $x$ is an OR gate, and $k_x = N_x$ when $x$ is an AND gate. For all leave nodes, the threshold value is 1.

Let $lev(\mathcal{T})$ denote leave nodes in $\mathcal{T}$. If $x \in lev(\mathcal{T})$, $att(x)$ is used to denote the attribute associated with node $x$. Furthermore, $\mathcal{T}$ defines an ordering between the children of each node, and $parent(x)$ and $index(x)$ returns the parent and the order number of children node $x$, respectively.

Let $\mathcal{T}$ with root $R$ correspond to the access tree of access policy $AP_K$. To check whether a set of attributes $S_u$ satisfies $AP_K$, denoted as $S_u \preceq AP_K$, we compute $\mathcal{T}_R(S_u)$ recursively as follows: Suppose that $\mathcal{T}_x$ denotes the subtree of $\mathcal{T}$ rooted at the node $x$. If $x$ is a non-leaf node, we evaluate $\mathcal{T}_b(S_u)$ for each child $b$ of node $x$. $\mathcal{T}_x(S_u)$ returns 1 if and only if at least $k_x$ children return 1. If $x$ is a leaf node, then $\mathcal{T}_x(S_u)$ returns 1 if and only if $att(x) \in S_u$.

**Secret Sharing Scheme (SSS)** To share the secret $\sigma$ in $\mathcal{T}$, the secret sharing scheme (SSS) generates $\Delta = \{q_x(0)\}_{x \in \mathcal{T}}$ as follows:

$SSS(\sigma, \mathcal{T}) \rightarrow \Delta$: A random polynomial $q_R$ of degree $k_R - 1$ is chosen for $q_R(0) = \sigma$. The rest of points in $q_R$ are randomly chosen. For each node $x \in \mathcal{T}$, a random polynomial $q_x$ of degree $k_x - 1$ is chosen for $q_x(0) = q_{parent(x)}(index(x))$. The rest of points in $q_x$ are chosen randomly. For example, Fig. 5b shows the tree implementation process. To recover the secret $\sigma$, the users with sufficient secret shares can perform Lagrange interpolation recursively. Please refer to [15] for more details.

## 6.2  Scheme Definition

The most used notations are shown in Table 5. The DABKS scheme consists of the following algorithms:

**Table 5** Summary of Notations in DABKS, respectively

| Notation | Description |
|---|---|
| $PK, MK$ | System public/master key |
| $S_u$ | Attribute set associated with cloud user $u$ |
| $sk_u$ | Search key for cloud user $u$ |
| $\mathcal{T}_w, \mathcal{T}_w'$ | Original/new access policy for keyword $w$ |
| $cph_w, cph_w'$ | Original/new ciphertext for keyword $w$ |
| $Tok_w$ | Search token for keyword $w$ |
| $\Delta$ | A set of secret shares for nodes in $\mathcal{T}_w$ |
| $UK$ | Update key for updating $\mathcal{T}_w$ to $\mathcal{T}_w'$ |
| $\Phi$ | Auxiliary information for Att2AND/Att2OR gate |

- $Init(\lambda) \rightarrow (PK, MK)$: The TTP takes a security parameter $\lambda$ as input to initialize the system, and outputs system public key $PK$ and system master key $MK$.
- $KeyGen(PK, MK, S_u) \rightarrow sk_u$: The TTP takes system public key $PK$, system master key $MK$, and a set of attributes $S_u$ associated with data user $u$ as input, and outputs search key $sk_u$ for $u$.
- $EncKW(PK, w, \mathcal{T}_w) \rightarrow (cph_w, \Delta)$: Given keyword $w$ associated with access tree $\mathcal{T}_w$, the data owner takes system public key $PK$, $\mathcal{T}_w$ and $w$ as inputs, and outputs the keyword ciphertext $cph_w$ and the secret shares $\Delta$ for nodes in $\mathcal{T}_w$.
- $TokenGen(sk_u, w) \rightarrow Tok_w$: The data user $u$ takes a keyword $w$ and the search key $sk_u$ as inputs, and outputs a search token $Tok_w$ for $w$.
- $Search(Tok_w, cph_w) \rightarrow \{0, 1\}$: On receiving the search request from data user $u$, the CSP evaluates the search token $Tok_w$ on keyword ciphertext $cph_w$, and outputs 1 if $S_u \preceq \mathcal{T}_w$. Otherwise, it outputs 0.
- $GenUpd(\Delta, \mathcal{T}_w') \rightarrow (UK, \Phi)$: The data owner takes the new access tree $\mathcal{T}_w'$ and the secret shares $\Delta$ for nodes in the original access tree $\mathcal{T}_w$ as inputs, and outputs an update key $UK$ and some auxiliary information $\Phi$.
- $ExeUpd(UK, \Phi, cph_w) \rightarrow cph_w'$: The CSP utilizes the update key $UK$ and the auxiliary information $\Phi$ to update the original ciphertext $cph_w$ to the new ciphertext $cph_w'$.

Given $(PK, MK)$ generated by the $Init$ algorithm, $sk_u$ generated by the $KeyGen$ algorithm, $cph_w$ generated by $EncKW$ algorithm, and $Tok_w$ generated by the $TokenGen$ algorithm, the DABKS scheme is correct if the following holds: (1) the $Search(Tok_w, cph_w)$ algorithm returns 1 if and only if $S_u \preceq \mathcal{T}_w$. (2) After running algorithms $GenUpd$ and $ExeUpd$, the $Search(Tok_w, cph_w')$ algorithm returns 1 if and only if $S_u \preceq \mathcal{T}_w'$.

Let $\Omega = \{F_1, \ldots, F_n\}$ be a set of files created by data owner $v$, where each file $F_i$ is described by a set of keywords $W_i$. For ease of illustration, we assume that the access policies for $W_i$ and $F_i$ are the same, i.e., $AP_K = AP_F$. Since our work focuses on preserving keyword privacy and query privacy, we omit the construction of these algorithms in this paper. From the systematic point of view, our scheme works as follows:

1. **System setup.** The TTP runs the *Init* algorithm to generate the system public key $PK$ and the system master key $MK$. It then sends $PK$ to the CSP.

2. **Data creation.** The data owner processes the data before outsourcing as follows: (1) it classifies $\Omega$ into several file groups, where a group of files share the same access policy. (2) Suppose that file $F_i$ is associated with access policy $AP_K$, which will be expressed as access tree $\mathcal{T}_w$. Given $\mathcal{T}_w$, it runs the *EncKW* algorithm to encrypt each keyword $w \in W_i$, and runs the *EncFile* algorithm to encrypt $F_i$. (3) It sends the keyword ciphertexts and file ciphertext, denoted as $\{\{cph_w\}_{w \in W_i}, C_{F_i}\}_{F_i \in \Omega}$, to the CSP.

3. **User grant.** When a new cloud user $u$ joins the system, the TTP first determines the attribute set $S_u$, which stipulates the search permission and data access right, for $u$. It then runs algorithms *GenKey* and *KeyGen* to generate decryption key $SK_u$ and search key $sk_u$ to $u$, respectively. Finally, the TTP sends the generated keys $(SK_u, sk_u)$ along with system public key $PK$ to $u$.

4. **Policy update.** If the access policy for file $F_i$ is changed to $AP_K'$, the data owner first builds a new access tree $\mathcal{T}_w'$ for $AP_K'$, and then runs the *GenUpd* algorithm to generate the update key $UK$ and some auxiliary information $\Phi$. The update instruction sent to the CSP is set to $\Gamma = \{Fid, UO, UK, \Phi\}$, where $Fid$ is the ID of file $F_i$ and $UO$ is the specific update operation. On receiving the policy update request, the CSP first locates $F_i$, and then runs the *ExeUpd* algorithm to generate new ciphertexts $\{cph_w'\}_{w \in W_i}$ for $W_i$ based on $\Gamma$.

5. **Data access.** If data user $u$ wants to retrieve files containing keyword $w$, it runs the *TokenGen* algorithm to generate a search token $Tok_w$, which will be sent to the CSP. On receiving the data access request, the CSP runs the *Search* algorithm by evaluating the search token on the whole keyword ciphertext set, and returns the searching results, denoted as $\{\{cph_w\}_{w \in W_i}, C_{F_i}\}_{Search(Tok_w, cph_w)=1}$. If $u$'s attributes satisfy the access policy of $F_i$, it can run the *Decrypt* algorithm to recover file content.

Taking Fig. 6 as an example, $\Omega = \{F_1, \ldots, F_7\}$ are classified into two groups, denoted as $FG_1$ and $FG_2$. Let $AP_{K,1}$ and $AP_{K,2}$ denote the access policy associated with file group $FG_1$ and $FG_2$, respectively. The data owner will encrypt $F_1, \ldots, F_4$ with $AP_{K,1}$ and $F_5, F_6, F_7$ with $AP_{K,2}$. Therefore, data user $u$ with attribute set $S_u = \{A_1, A_2\}$ has permission to search files only in $FG_1$. That is, $u$ can generate valid search tokens for keywords $a, b, c, d$ only. If $AP_{K,2}$ is updated to $AP_{K,2}' = \{A_1 \wedge A_2\}$, $u$ can search all files in $\Omega$.

## 6.3   Policy Updating

Efficient policy updating is still a challenging problem. A simple scheme is that the data owner retrieves the relevant keywords and re-encrypted them with the new access policy before sending them back to the cloud. However, for frequent updates

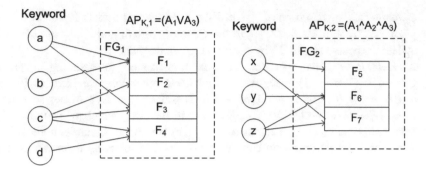

**Fig. 6** Example of keyword search

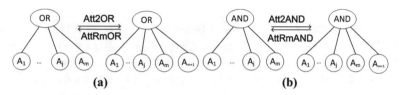

**Fig. 7** Policy updating operations. (**a**) Att2OR and AttRmOR. (**b**) Att2AND and AttRmAND

on a large number of files, the workload on the data owner is heavy. In this paper, we propose a DABKS scheme that achieving an efficient update of access policy by delegating the policy updating operations to the cloud.

Inspired by the work in [41], we consider four basic operations involved in policy updating (Fig. 7): $Att2OR$ denotes adding an attribute to an OR gate, $Att2AND$ denotes adding an attribute to an AND gate, $AttRmOR$ denotes removing an attribute from an OR gate, and $AttRmAND$ denotes removing an attribute from an AND gate. Given access tree $\mathcal{T}$ over access policy $AP_K$, the data owner needs to preserve $\Delta = \{q_x(0)\}_{x \in \mathcal{T}}$ generated by SSS, where $q_x(0)$ is the sharer of secret $\sigma$ for node $x$.

Let node $y$ be the AND/OR gate that will be updated, where $A_1, \ldots, A_m$ are the original attributes under $y$. Let $q_y(0)$ and $\{q_{x_1}(0), \ldots, q_{x_m}(0)\}$ denote the secret shares for node $y$ and $y$'s children nodes $x_1, \ldots, x_m$, where $att(x_i) = A_i$ for $i \in [1, m]$. Given an access tree $\mathcal{T}_w$, the DABKS scheme will produce a ciphertext for each leave node $x$ based on share $q_x(0)$. The original and new ciphertext for node $x_i$ is denoted as $C_i$ and $C'_i$, respectively.

- $Att2OR$ : This operation can be transformed to updating a $(1, m)$ gate to a $(1, m + 1)$ gate. Given $q_y(0)$ and the new access policy $\mathcal{T}'$, the data owner runs SSS to generate new shares $\{q'_{x_1}(0), \ldots, q'_{x_{m+1}}(0)\}$ for attributes $A_1, \ldots, A_{m+1}$. Since $q_{x_i}(0) = q'_{x_i}(0) = q_y(0)$ for $i \in [1, m + 1]$, the ciphertexts for the original attributes will not be changed, i.e., $C'_i = C_i$ for $i \in [1, m]$. For the newly added attribute $A_{m+1}$, the data owner needs to generate a new ciphertext $C'_{m+1}$ based on $q'_{x_{m+1}}(0)$. Finally, it sets the update instruction $\Gamma = \{Fid, Att2OR, NULL, C'_{m+1}\}$ to the CSP, which will add $C'_{m+1}$ to $cph_w$ and update the access tree to $\mathcal{T}'_w$ by adding $A_{m+1}$ under node $y$.

- $AttRmOR$ : This operation can be transformed to updating a $(1, m)$ gate to a $(1, m - 1)$ gate. As the $Att2OR$ operation, we have $q_{x_i}(0) = q'_{x_i}(0) = q_y(0)$ for $i \in [1, m - 1]$. Therefore, the data owner will send $\Gamma = \{Fid, AttRmOR, NULL, NULL\}$ to the CSP, which will remove $C_m$ from $cph_w$ and update the access tree to $\mathcal{T}'_w$ by removing $A_m$ under node $y$.

- $Att2AND$ : This operation can be transformed to updating a $(m, m)$ gate to a $(m + 1, m + 1)$ gate. Given $q_y(0)$ and the new access policy $\mathcal{T}'$, the data owner runs SSS to generate new shares $\{q'_{x_1}(0), \ldots, q'_{x_{m+1}}(0)\}$ for attributes $A_1, \ldots, A_{m+1}$. Next, it executes the $GenUpd$ algorithm to generate the update key $UK$ for attributes $A_1, \ldots, A_m$. Moreover, it generates a ciphertext $C'_{m+1}$ for the newly added attribute $A_{m+1}$ based on $q'_{x_{m+1}}(0)$. Finally, it sends $\Gamma = \{Fid, Att2AND, UK, C'_{m+1}\}$ to the CSP, which will execute the $ExeUpd$ algorithm to update the ciphertext $C_i$ to $C'_i$ for $i \in [1, m]$, add the new ciphertext $C'_{m+1}$ to the $cph_w$, and update the access tree to $\mathcal{T}'_w$ by adding $A_{m+1}$ under node $y$.

- $AttRmAND$ : This operation can be transformed to updating a $(m, m)$ gate to a $(m-1, m-1)$ gate. Given $q_y(0)$ and the new access policy $\mathcal{T}'$, the data owner runs SSS to generate new shares $\{q'_{x_1}(0), \ldots, q'_{x_{m-1}}(0)\}$ for attributes $A_1, \ldots, A_{m-1}$. Next, it executes the $GenUpd$ algorithm to generate the update key $UK$ for $C_1 \ldots, C_{m-1}$. Finally, it sends $\Gamma = \{Fid, AttRmAND, UK, NULL$ to the CSP, which will execute the $ExeUpd$ algorithm to update the ciphertext $C_i$ to $C'_i$ for $i \in [1, m - 1]$, remove $C_m$ from $cph_w$, and update the access tree to $\mathcal{T}'_w$ by removing $A_m$ under node $y$.

## 6.4   Construction

$Init(\lambda)$: Let $e : \mathbf{G}_0 \times \mathbf{G}_0 \to \mathbf{G}_1$ be the bilinear group, where $\mathbf{G}_0$ and $\mathbf{G}_1$ are cyclic groups of prime order $p$, the TTP takes security parameter $\lambda$ as input, and sets the system public key $PK$ and the system master key $MK$ as follows:

$$PK = \{H_1, H_2, e, g, p, g^a, g^b, g^c\}, \quad MK = (a, b, c),$$

where $H_1 : \{0, 1\}^* \to \mathbf{G}_0$ is a hash function modeled as a random oracle, $H_2 : \{0, 1\}^* \to \mathbb{Z}_p$ is a one-way hash function, $g \in \mathbf{G}_0$ is the random generator of $\mathbf{G}_0$, and $a, b, c$ are randomly chosen from $\mathbb{Z}_p$.

$KeyGen(PK, MK, S_u)$: For data user $u$ associated with attribute set $S_u$, the TTP generates search key $sk_u$ for $u$ as follows: it first randomly selects $r \in \mathbb{Z}_p$, and computes $D = g^{(ac-r)/b}$. Then, it selects a random $r_j \in \mathbb{Z}_p$ for each attribute

$A_j \in S_u$, and computes $B_j = g^r H_1(A_j)^{r_j}$ and $\bar{B}_j = g^{r_j}$. The search key is set as follows:

$$sk_u = (S_u, D, \{(B_j, \bar{B}_j)\}_{A_j \in S_u})$$

$EncKW(PK, w, \mathcal{T}_w)$: To encrypt keyword $w$ under access tree $\mathcal{T}_w$, the data owner first randomly selects $r_1, \sigma \in \mathbb{Z}_p$, and computes $K_1 = g^{cr_1}$, $K_2 = g^{a(r_1+\sigma)} g^{bH_2(w)r_1}$, and $K_3 = g^{b\sigma}$. Then, it computes $C_{x_i} = g^{q_{x_i}(0)}$ and $\bar{C}_{x_i} = H_1(att(x_i))^{q_{x_i}(0)}$, where $q_{x_i}(0)$ is the share of secret $\sigma$ for leave node $x_i$ in $\mathcal{T}_w$ generated by SSS. The ciphertext for keyword $w$ is set as:

$$cph_w = (\mathcal{T}_w, K_1, K_2, K_3, \{C_i = (C_{x_i}, \bar{C}_{x_i})\}_{x_i \in lev(\mathcal{T}_w)})$$

$TokenGen(sk_u, w)$: To retrieve files containing keyword $w$, data user $u$ associated with attribute set $S_u$ chooses a random $s \in \mathbb{Z}_p$, and computes $tk_1 = (g^a g^{bH_2(w)})^s$, $tk_2 = g^{cs}$, and $tk_3 = D^s = g^{(ac-r)s/b}$. In addition, for each $A_j \in S_u$, it computes $B'_j = B^s_j$ and $\bar{B}'_j = \bar{B}^s_j$. The search token for keyword $w$ is set as:

$$Tok_w = (S_u, tk_1, tk_2, tk_3, \{(B'_j, \bar{B}'_j)\}_{A_j \in S_u})$$

$Search(Tok_w, cph_w)$: On receiving the search token $Tok_w$ from data user $u$, the CSP first constructs a set $S \in S_u$ that satisfies the access tree $\mathcal{T}_w$ specified in $cph_w$, and then it computes $E_{x_i} = e(B'_i, C_{x_i})/e(\bar{B}'_i, \bar{C}_{x_i}) = e(g, g)^{rsq_{x_i}(0)}$ for each attribute $A_i \in S$, where $A_i = att(x_i)$ for $x_i \in lev(\mathcal{T})$. Next, it executes the Lagrange interpolation to recover $E_R = e(g, g)^{rs\sigma}$. Finally, it tests whether Eq. (6) holds.

$$e(K_2, tk_2) = e(K_1, tk_1)e(tk_3, K_3)E_R \tag{6}$$

If so, it outputs 1, and 0 otherwise.

$GenUpd(\Delta, \mathcal{T}'_w)$: Given the new access tree $\mathcal{T}'_w$ for keyword $w$, the data owner first locates the gate node that will be modified. Let node $y$ denote the AND/OR gate being updated, where $A_1, \ldots, A_m$ and $A_1, \ldots, A_{m'}$ are original and new attributes under node $y$, respectively. Given $q_y(0)$ the share associated with node $y$, it first takes $q_x(0)$ and $\mathcal{T}'_w$ as inputs of the SSS algorithm, and obtains the new secret shares for nodes $x_1, \ldots, x_{m'}$ in $\mathcal{T}'_w$, denoted as $\{q'_{x_i}(0)\}_{i \in [1, m']}$. Then, it generates update key for attributes $A_1, \ldots, A_m$ as follows:

$$UK = \{(UK_{1,i}, UK_{2,i})\}_{i \in [1,m]} \tag{7}$$

where $UK_{1,i} = g^\sigma$, $UK_{2,i} = H_1(att(x_i))^\sigma$, and $\sigma = q'_{x_i}(0) - q_{x_i}(0)$. Furthermore, for adding an attribute $A_{m+1}$ under gate node $y$, the data owner generates the new ciphertext $C'_{m+1}$ for $A_{m+1}$ as follows: it computes $C'_{x_{m+1}} = g^{q'_{x_{m+1}}(0)}$ and $\bar{C}'_{x_{m+1}} = H_1(att(x_{m+1}))^{q'_{x_{m+1}}(0)}$. The new ciphertext is set as $C'_{m+1} = (C'_{x_{m+1}}, \bar{C}'_{x_{m+1}})$.

$ExeUpd(UK, \Phi, cph_w)$: After receiving the policy updating request from the data owner, the CSP utilizes the update key $UK$ to update the original keyword ciphertext $cph_w$ as follows: For each leave node $x_i$ under the changing gate node $y$, it computes $C'_{x_i} = C_{x_i} \cdot UK_{1,i} = g^{q'_{x_i}(0)}$ and $\bar{C}'_{x_i} = \bar{C}_{x_i} \cdot UK_{2,i} = H_1(att(x_i))^{q'_{x_i}(0)}$. Then, the new ciphertext for keyword $w$ is set as:

$$cph'_w = (\mathcal{T}'_w, K_1, K_2, K_3, \{C'_i = (C'_{x_i}, \bar{C}'_{x_i})\}_{x_i \in lev(\mathcal{T}'_w)})$$

## 6.5   Correctness Proof

In the *Search* algorithm, for each attribute $A_j \in S$, we have:

$$\begin{aligned} E_{x_i} &= \frac{e(B'_j, C_{x_i})}{e(\bar{B}'_j, \bar{C}_{x_i})} = \frac{e(g^{rs} H_1(A_j)^{rjs}, g^{q_{x_i}(0)})}{e(g^{rjs}, H_1(att(x_i))^{q_{x_i}(0)})} \\ &= e(g, g)^{rsq_{x_i}(0)} \end{aligned} \tag{8}$$

If $S_u \preceq \mathcal{T}_w$, we can recover $E_R = e(g, g)^{rsq_R(0)} = e(g, g)^{rs\sigma}$ by executing the Lagrange interpolation. Therefore, the right side of Eq. (6) will evolve as follows:

$$\begin{aligned} &e(K_1, tk_1)e(tk_3, K_3)E_R \\ &= e(g^{cr_1}, g^{as}g^{bsH_2(w)})e(g^{(acs-rs)/b}, g^{b\sigma})e(g, g)^{rs\sigma} \\ &= e(g, g)^{acs(r_1+\sigma)}e(g, g)^{bcsH_2(w)r_1} \end{aligned} \tag{9}$$

The left side of Eq. (6) will evolve as follows:

$$\begin{aligned} e(K_2, tk_2) &= e(g^{a(r_1+\sigma)}g^{bH_2(w)r_1}, g^{cs}) \\ &= e(g, g)^{acs(r_1+\sigma)}e(g, g)^{bcsH_2(w)r_1} \end{aligned} \tag{10}$$

Therefore, Eq. (6) holds only when $S_u \preceq \mathcal{T}_w$.

Then, we prove that the output of the *Search* algorithm is still correct after the *GenUpd* and *ExeUpd* algorithms. Due to the limited space, we only provide the correct proof for the *Att2AND* operation in this paper. Let $q'_{x_1}, \ldots, q'_{x_m}, q'_{x_{m+1}}$ be the secret shares of $\sigma$ for the leave nodes in $\mathcal{T}'_w$. For the original attribute $A_i$ where $i \in [1, m]$, $C'_i = (C'_{x_i}, \bar{C}'_{x_i})$ will be constructed as follows:

$$\begin{aligned} C'_{x_i} &= C_{x_i} \cdot UK_{1,i} \\ &= g^{q_{x_i}(0)} g^{q'_{x_i}(0)-q_{x_i}(0)} = g^{q'_{x_i}(0)} \end{aligned}$$

$$\bar{C}'_{x_i} = \bar{C}_{x_i} \cdot UK_{2,i}$$

$$= H_1(att(x_i))^{q_{x_i}(0)} H_1(att(x_i))^{q'_{x_i}(0)-q_{x_i}(0)}$$

$$= H_1(att(x_i))^{q'_{x_i}(0)} \tag{11}$$

The new ciphertext $C'_{m+1}$ corresponding to $A_{m+1}$ is constructed as follows:

$$C'_{m+1} = (C'_{x_{m+1}}, \bar{C}'_{x_{m+1}})$$
$$= (g^{q'_{x_{m+1}}(0)}, H_1(att(x_{m+1}))^{q'_{x_{m+1}}(0)}) \tag{12}$$

Given the updated ciphertexts $C'_i$ for $i \in [1, m+1]$, $E_R = e(g,g)^{rsq_R(0)} = e(g,g)^{rs\sigma}$ can be recovered only when $S_u \preceq \mathcal{T}_w$. Therefore, the DABKS scheme is correct. ∎

## 6.6  Security Sketch

The work in [69] has proven that, given the one-way hash function $H_2$, the ABKS scheme is selectively secure against chosen-keyword attacks in the generic bilinear group model and achieves keyword secrecy in the random oracle model. The correctness proof has proven that the DABKS scheme carries out a correct search control after the update of access policy $AP_K$. Therefore, the security of our scheme can be derived from that of the ABKS scheme. ∎

## 6.7  Performance Analysis

We will analyze the performance of the DABKS scheme in terms of computational and communication complexity. For ease of understanding, we provide the following notations to denote the running time for various operations in our scheme: $H_1$ is used to denote the operation of mapping a bit-string to an element of $\mathbf{G}_0$, $e$ is used to denote the pairing operation, $E_0$ and $E_1$ are used to denote the exponentiation operation in $\mathbf{G}_0$ and $\mathbf{G}_1$, respectively. We neglect the multiplication in $\mathbf{G}_0$ and $\mathbf{G}_1$ and hash operations, since they are much less expensive compared to the above operations.

Table 6 shows the asymptotic complexity of the DABKS scheme, where $S$ denotes the number of attributes associated with a data user, $N$ denotes the number of attributes in a data owner's access policy, $m$ denotes the number of attributes under an AND/OR gate node, which will be updated, and $|\mathbf{G}_0|$ denotes the length of elements in $\mathbf{G}_0$. The $Init$ algorithm will be run in system initialization phase, and can be done once for all. Therefore, the TTP spends most of its time for generating

**Table 6** Performance analysis of DABKS

|          | Computation                     | Communication      |
|----------|---------------------------------|--------------------|
| KeyGen   | $(2S + 2)E_0 + SH_1$            | $(2S + 1)\|\mathbf{G}_0\|$ |
| EncKW    | $(2N + 4)E_0 + NH_1$            | $(2N + 3)\|\mathbf{G}_0\|$ |
| TokenGen | $(2S + 4)E_0$                   | $(2S + 3)\|\mathbf{G}_0\|$ |
| Search   | $(2N + 3)e + NE_1$              |                    |
| GenUpd   | $(2m + 2)E_0 + (m + 1)H_1$      | $(2m + 2)\|\mathbf{G}_0\|$ |
| ExeUpd   | $(2E_0 + H_1)m$                 |                    |

search key $sk$ for the data users, the complexity of which relates to $S$, the number of attributes associated with a user. For the data owner, the *EncKW* algorithm is mainly impacted by $N$, the number of attributes in the specified access policy, and the *GenUpd* algorithm is mainly impacted by $m$, the number of attributes under the updating gate node. For the data user, the cost of the *TokenGen* algorithm will grow linearly with $S$. For the CSP, the *ExeUpd* algorithm is also impacted by $m$.

# 7 Security in Fog/Edge Computing

Rapid development of sensing, communication, and micro-controller technologies are making everything connected and smarter, termed the Internet of Things (IoT). The cloud usually has almost unlimited resources, but it is located far away from IoT end devices. Fog computing, a.k.a edge computing, is proposed to enable computing directly at the edge of networks by delivering new applications and services for billions of connected devices. Fog devices are usually access points, set-top-boxes, road side units,

Our secure storage/secure schemes can be applied to fog/edge computing environment. In this way, many computation-heavy and resource-demanding tasks can be delegated to the edge layer, so as to alleviate the resource constraints at the user side. The general idea is to distribute multiple copies of each ciphertext to the fog devices that close to the user. The user will interact with the edge layer instead the central cloud to save communication cost. The system model is composed of three tiers: user tier, edge tier, and cloud tier. The user tier is composed of users' thin-clients that are assumed to be trusted. The user tier uploads ciphertexts to the cloud tier and interacts with the edge tier to retrieve appropriate data. The edge tier located between the user and the cloud is comprised of fog devices such as router, gateway, switch, and Access Points (APs). These fog devices can collaboratively share storage and computing facilities. We assume that the edge tier is a computing facility on the users premises (organization), hence, considered to be trusted. The user's search token is sent to the edge tier. On retrieving the searching request, the fog devices first check whether corresponding ciphertexts locate in the edge tier or not. If so, the fog devices evaluate the search query on the ciphertexts and return search results to user; Otherwise, the fog devices contact the cloud tier to retrieve corresponding ciphertexts before performing searches. The cloud tier centralizes sufficient storage and computing resources, providing storage services. The cloud servers are considered to be honest but curious. Therefore, all data will be encrypted before being uploading to the cloud.

In the tiered architecture, the main problem is which ciphertexts should be distributed to the fog devices. A naive solution is to let the fog devices store all the ciphertexts, so that the user can retrieve all the required data at the edge layer. However, this simple solution requires that the storage space of a fog device is as large as that of the central cloud. Therefore, a novel cache algorithm is required to locate appropriate ciphertexts to fog devices.

# 8  Conclusion

Despite a bit of hype, cloud computing is undeniably a fundamental trend of IT technologies. More and more IT companies are diving into launching cloud products, such as Amazon's EC2, Google's AppEng, Microsoft's Azure, etc. However, security is the main obstacle hindering the wide adoption of cloud computing. In this book chapter, we research the security and privacy problems in cloud storage and search services. The research results will play an important role in for providing secure services in cloud computing.

# References

1. Research and Markets: Cloud storage market - forecasts from 2017 to 2022. https://www.researchandmarkets.com/research/lf8wbx/cloud_storage/
2. Rimal, B.P., Choi, E., Lumb, I.: A taxonomy and survey of cloud computing systems. In: Proc. of the 5th International Joint Conference on INC, IMS and IDC (NCM 2009), pp. 44–51 (2009)
3. Zhou, M., Zhang, R., Zeng, D., Qian, W.: Services in the cloud computing era: a survey. In: Proc. of the 4th International Conference on Universal Communication Symposium (IUCS 2010), pp. 40–46 (2010)
4. Wikipedia: icloud leaks of celebrity photos. https://en.wikipedia.org/wiki/ICloud_leaks_of_celebrity_photos
5. Khandelwal, S.: Download: 68 million hacked dropbox accounts are just a click away!. https://thehackernews.com/2016/10/dropbox-password-hack.html
6. McGee, M.K.: Blood test results exposed in cloud repository. https://www.databreachtoday.com/blood-test-results-exposed-in-cloud-repository-a-10382
7. Ferguson, N., Kelsey, J., Lucks, S., Schneier, B., Stay, M., Wagner, D., Whiting, D.: Improved cryptanalysis of Rijndael. In: Fast Software Encryption, pp. 213–230. Springer, Berlin (2001)

8. Kleinjung, T., Aoki, K., Franke, J., Lenstra, A.K., Thomé, E., Bos, J.W., Gaudry, P., Kruppa, A., Montgomery, P.L., Osvik, D.A., et al.: Factorization of a 768-bit RSA modulus. In: Advances in Cryptology–CRYPTO 2010, pp. 333–350. Springer, Berlin (2010)
9. Kallahalla, M., Riedel, E., Swaminathan, R., Wang, Q., Fu, K.: Plutus: scalable secure file sharing on untrusted storage. In: Proc. of the 2nd USENIX Conference on File and Storage Technologies (FAST 2003), pp. 29–42 (2003)
10. Goh, E., Shacham, H., Modadugu, N., Boneh, D.: Sirius: securing remote untrusted storage. In: Proc. of the 10th Network and Distributed Systems Security Symposium (NDSS 2003), pp. 131–145 (2003)
11. Liu, Q., Wang, G., Wu, J.: Efficient sharing of secure cloud storage services. In: Proc. of the 10th IEEE 10th International Conference on Computer and Information Technology (CIT 2010), pp. 922–929 (2010)
12. Gentry, C., Silverberg, A.: Hierarchical ID-based cryptography. In: Proc. of the 8th International Conference on the Theory and Application of Cryptology and Information Security (ASIACRYPT 2002), pp. 149–155 (2002)
13. Sahai, A., Waters, B.: Fuzzy identity-based encryption. In: Proc. of the 24th International Conference on the Theory and Application of Cryptographic Techniques (EUROCRYPT 2005), pp. 557–557 (2005)
14. Goyal, V., Pandey, O., Sahai, A., Waters, B.: Attribute-based encryption for fine-grained access control of encrypted data. In: Proc. of the 13th ACM Conference on Computer and Communications Security (CCS 2006), pp. 89–98 (2006)
15. Bethencourt, J., Sahai, A., Waters, B.: Ciphertext-policy attribute-based encryption. In: Proc. of the 27th IEEE Symposium on Security and Privacy (SP 2007), pp. 321–334 (2007)
16. Müller, S., Katzenbeisser, S., Eckert, C.: Distributed attribute-based encryption. In: Proc. of the 7th International Conference on Information Security and Cryptology (ICISC 2009), pp. 20–36 (2009)
17. Waters, B.: Ciphertext-policy attribute-based encryption: an expressive, efficient, and provably secure realization. In: Public Key Cryptography–PKC 2011, pp. 53–70. Springer, Berlin (2011)
18. Lewko, A., Okamoto, T., Sahai, A., Takashima, K., Waters, B.: Fully secure functional encryption: attribute-based encryption and (hierarchical) inner product encryption. In: Advances in Cryptology–EUROCRYPT 2010, pp. 62–91. Springer, Berlin (2010)
19. Liang, K., Au, M.H., Liu, J.K., Susilo, W., Wong, D.S., Yang, G., Yu, Y., Yang, A.: A secure and efficient ciphertext-policy attribute-based proxy re-encryption for cloud data sharing. Future Gener. Comput. Syst. 28, 95–108 (2015)
20. Jung, T., Li, X.Y., Wan, Z., Wan, M.: Control cloud data access privilege and anonymity with fully anonymous attribute-based encryption. IEEE Trans. Inf. Forensics Secur. 10, 190–199 (2015)
21. Han, J., Susilo, W., Mu, Y., Zhou, J.: Improving privacy and security in decentralized ciphertext-policy attribute-based encryption. IEEE Trans. Inf. Forensics Secur. 10, 665–678 (2015)
22. Chase, M., Chow, S.S.: Improving privacy and security in multi-authority attribute-based encryption. In: Proceedings of the 16th ACM Conference on Computer and Communications Security, pp. 121–130. ACM, New York (2009)
23. Lewko, A., Waters, B.: Decentralizing attribute-based encryption. In: Advances in Cryptology–EUROCRYPT 2011, pp. 568–588. Springer, Berlin (2011)
24. Wang, G., Liu, Q., Wu, J.: Achieving fine-grained access control for secure data sharing on cloud servers. Concurrency Comput. Pract. Exp. 23(12), 1443–1464 (2011)
25. Wang, G., Liu, Q., Wu, J., Guo, M.: Hierarchical attribute-based encryption and scalable user revocation for sharing data in cloud servers. Comput. Secur. 30(5), 320–331 (2011)
26. Wang, G., Liu, Q., Wu, J.: Hierarchical attribute-based encryption for fine-grained access control in cloud storage services. In: Proceedings of the ACM Conference on Computer and Communications Security (CCS), pp. 735–737 (2010)
27. Zhu, Y., Hu, H., Ahn, G., et al.: Comparison-based encryption for fine-grained access control in clouds. In: Proceedings of ACM CODASPY, pp. 105–116 (2012)

28. Liu, X., Liu, Q., Peng, T., Wu, J.: Dynamic access policy in cloud-based personal health record (PHR) systems. Inf. Sci. **8**(7), 1332–1346 (2015)
29. Wu, Y., Wei, Z., Deng, H., et al.: Attribute-based access to scalable media in cloud-assisted content sharing. IEEE Trans. Multimed. **15**(4), 778–788 (2013)
30. Li, M., Yu, S., Zheng, Y., Ren, K., Lou, W.: Scalable and secure sharing of personal health records in cloud computing using attribute-based encryption. IEEE Trans. Parallel Distrib. Syst. **24**(1), 131–143 (2013)
31. Katz, J., Sahai, A., Waters, B.: Predicate encryption supporting disjunctions, polynomial equations, and inner products. In: Advances in Cryptology–EUROCRYPT 2008, pp. 146–162. Springer, Berlin (2008)
32. Okamoto, T., Takashima, K.: Hierarchical predicate encryption for inner-products. In: Advances in Cryptology–ASIACRYPT 2009, pp. 214–231. Springer, Berlin (2009)
33. Pirretti, M., Traynor, P., McDaniel, P., Waters, B.: Secure attribute-based systems. J. Comput. Secur. **18**(5), 799–837 (2010)
34. Blaze, M., Bleumer, G., Strauss, M.: Divertible protocols and atomic proxy cryptography. In: Proc. of the 17th International Conference on the Theory and Application of Cryptographic Techniques (EUROCRYPT 1998), pp. 127–144 (1998)
35. Green, M., Ateniese, G.: Identity-based proxy re-encryption. In: Proceedings of the International Conference on Applied Cryptography and Network Security (ACNS), pp. 288–306 (2007)
36. Yu, S., Wang, C., Ren, K., Lou, W.: Achieving secure, scalable, and fine-grained data access control in cloud computing. In: Proc. of the 29th IEEE International Conference on Computer Communications (INFOCOM 2010), pp. 534–542 (2010)
37. Yu, S., Wang, C., Ren, K., Lou, W.: Attribute based data sharing with attribute revocation. In: Proceedings of the ACM Symposium on Information, Computer and Communications Security (ASIACCS), pp. 261–270 (2010)
38. Shi, Y., Zheng, Q., et al.: Directly revocable key-policy attribute-based encryption with verifiable ciphertext delegation. Inf. Sci. **295**, 221–231 (2015)
39. Liu, Q., Wang, G., Wu, J.: Time-based proxy re-encryption scheme for secure data sharing in a cloud environment. Inf. Sci. **258**, 355–370 (2014)
40. Yang, Y., Zhu, H., et al.: Cloud based data sharing with fine-grained proxy re-encryption. Pervasive Mob. Comput. (2015). http://dx.doi.org/10.1016/j.pmcj.2015.06.017
41. Yang, K., Jia, X., Ren, K., et al.: Enabling efficient access control with dynamic policy updating for big data in the cloud. In: Proceedings of IEEE INFOCOM, pp. 2013–2021 (2014)
42. Song, D.X., Wagner, D., Perrig, A.: Practical techniques for searches on encrypted data. In: Proc. of the 2000 IEEE Symposium on Security and Privacy (SP 2000), pp. 44–55 (2000)
43. Goh, E.-J.: Secure indexes. Cryptology ePrint Archive, Report 2003/216, Tech. Rep. (2003)
44. Chang, Y.-C., Mitzenmacher, M.: Privacy preserving keyword searches on remote encrypted data. In: Applied Cryptography and Network Security, pp. 442–455. Springer, Berlin (2005)
45. Kurosawa, K., Ohtaki, Y.: UC-secure searchable symmetric encryption. In: Financial Cryptography and Data Security, pp. 285–298. Springer, Berlin (2012)
46. Kamara, S., Papamanthou, C.: Parallel and dynamic searchable symmetric encryption. In: Financial Cryptography and Data Security, pp. 258–274. Springer, Berlin (2013)
47. Boneh, D., Di Crescenzo, G., Ostrovsky, R., Persiano, G.: Public key encryption with keyword search. In: Advances in Cryptology-Eurocrypt 2004, pp. 506–522. Springer, Berlin (2004)
48. Golle, P., Staddon, J., Waters, B.: Secure conjunctive keyword search over encrypted data. In: Applied Cryptography and Network Security, pp. 31–45. Springer, Berlin (2004)
49. Boneh, D., Franklin, M.: Identity-Based Encryption from the Weil Pairing. In: Proceedings of CRYPTO 2001, *LNCS*, 2139, 213–229 (2001).
50. Boneh, D., Waters, B.: Conjunctive, subset, and range queries on encrypted data. In: Theory of Cryptography, pp. 535–554. Springer, Berlin (2007)
51. Shi, E., Bethencourt, J., Chan, T.-H., Song, D., Perrig, A.: Multi-dimensional range query over encrypted data. In: Proc. of the 2007 IEEE Symposium on Security and Privacy (SP 2007), pp. 350–364 (2007)

52. Popa, R.A., Redfield, C., Zeldovich, N., Balakrishnan, H.: Cryptdb: protecting confidentiality with encrypted query processing. In: Proc. of the 23rd ACM Symposium on Operating Systems Principles (SOSP 2011), pp. 85–100 (2011)
53. Wang, C., Cao, N., Li, J., Ren, K., Lou, W.: Secure ranked keyword search over encrypted cloud data. In: Proc. of the 30th IEEE International Conference on Distributed Computing Systems (ICDCS 2010), pp. 253–262 (2010)
54. Boldyreva, A., Chenette, N., Lee, Y., Oneill, A.: Order-preserving symmetric encryption. In: Advances in Cryptology-EUROCRYPT 2009, pp. 224–241. Springer, Berlin (2009)
55. Cao, N., Wang, C., Li, M., Ren, K., Lou, W.: Privacy-preserving multi-keyword ranked search over encrypted cloud data. IEEE Trans. Parallel Distrib. Syst. **25**(1), 222–233 (2014)
56. Wong, W.K., Cheung, D.W.-l., Kao, B., Mamoulis, N.: Secure KNN computation on encrypted databases. In: Proc. of the 2009 ACM SIGMOD International Conference on Management of Data (SIGMOD 2009), pp. 139–152 (2009)
57. Li, J., Wang, Q., Wang, C., Cao, N., Ren, K., Lou, W.: Fuzzy keyword search over encrypted data in cloud computing. In: Proc. of the 29th IEEE International Conference on Computer Communications (INFOCOM 2010), pp. 1–5 (2010)
58. Wang, B., Yu, S., Lou, W., Hou, Y.T.: Privacy-preserving multi-keyword fuzzy search over encrypted data in the cloud. In: Proc. of IEEE INFOCOM (2014)
59. Guo, D., Wu, J., Chen, H., Luo, X., et al.: Theory and network applications of dynamic bloom filters. In: Proc. of INFOCOM (2006)
60. Indyk, P., Motwani, R.: Approximate nearest neighbors: towards removing the curse of dimensionality. In: Proc. of ACM STOC (1998)
61. Fu, Z., Xia, L., Sun, X., Liu, A.X., Xie, G.: Semantic-aware searching over encrypted data for cloud computing. IEEE Trans. Inf. Forensics Secur. **13**(9), 2359–2371 (2018)
62. Wang, D., Jia, X., Wang, C., Yang, K., Fu, S., Xu, M.: Generalized pattern matching string search on encrypted data in cloud systems. In: Proc. of INFOCOM, pp. 2101–2109 (2015)
63. Ding, X., Liu, P., Jin, H.: Privacy-preserving multi-keyword top-k similarity search over encrypted data. IEEE Trans. Dependable Secure Comput. **16**(2), 344–357 (2019)
64. Boldyreva, A., Chenette, N.: Efficient fuzzy search on encrypted data. In: Proc. of FSE, pp. 613–633 (2014)
65. Moataz, T., Ray, I., Ray, I., Shikfa, A., Cuppens, F., Cuppens, N.: Substring search over encrypted data. J. Comput. Secur. **26**(1), 1–30 (2018)
66. Hahn, F., Loza, N., Kerschbaum, F.: Practical and secure substring search. In: Proc. of SIGMOD, pp. 163–176 (2018)
67. Bao, F., Deng, R.H., Ding, X., Yang, Y.: Private query on encrypted data in multi-user settings. In: Information Security Practice and Experience. Springer, Berlin (2008)
68. Li, M., Yu, S., Cao, N., Lou, W.: Authorized private keyword search over encrypted personal health records in cloud computing. In: Proc. of IEEE ICDCS (2011)
69. Zheng, Q., Xu, S., Ateniese, G.: VABKS: verifiable attribute-based keyword search over outsourced encrypted data. In: Proc of IEEE INFOCOM (2014)
70. Hu, B., Liuy, Q., Liu, X., Peng, T., Wu, J.: DABKS: dynamic attribute-based keyword search in cloud computing. In: Proc of IEEE ICC (2017)
71. Peng, T., Liu, Q., Hu, B., Liu, J., Zhu, J.: Dynamic keyword search with hierarchical attributes in cloud computing. IEEE Access **6**, 68948–68960 (2018)
72. Naveed, M.: The fallacy of composition of oblivious ram and searchable encryption. IACR Cryptol. ePrint Arch. **2015**, 668 (2015)
73. Canetti, R., Raghuraman, S., Richelson, S., Vaikuntanathan, V.: Chosen-ciphertext secure fully homomorphic encryption. In: Proc. of IACR International Workshop on Public Key Cryptography, pp. 213–240 (2017)
74. Curtmola, R., Garay, J., Kamara, S., Ostrovsky, R.: Searchable symmetric encryption: improved definitions and efficient constructions. In: Proc. of the 13th ACM Conference on Computer and Communications Security (CCS 2006) (2006)
75. ElGamal, T.: A public key cryptosystem and a signature scheme based on discrete logarithms. In: Proceedings of International Cryptology Conference (CRYPTO), pp. 10–18 (1984)

76. Yu, S., Wang, C., Ren, K., Lou, W.: Attribute based data sharing with attribute revocation. In: Proc. of the 5th ACM Symposium on Information, Computer and Communications Security (ASIACCS 2010), pp. 261–270 (2010)
77. Libert, B., Vergnaud, D.: Unidirectional chosen-ciphertext secure proxy re-encryption. In: Public Key Cryptography–PKC 2008, pp. 360–379. Springer, Berlin (2008)

# Collaborative Intrusion Detection Schemes in Fog-to-Things Computing

**Abebe Diro, Abdun Mahmood, and Naveen Chilamkurti**

# 1 Background

The IoT is an emerging technology that aims to provide a virtual presence for a massive number of intelligent physical objects. In other words, it is a new network of things that extend IT systems to physical environments in order to form a globally connected heterogeneous network of smart objects [1]. This has ushered in the era of smart connectivity, where the power of computing has moved from desktops to pockets and wearable devices [2]. One of the main driving factor for the rapid adoption of the IoT is the miniaturization of hardware. This has enabled the mass production of these smart objects and has also led to its affordability. The potential of IoT technologies and services has prompted wide interest across the globe in the adoption of IoT applications. IoT technologies have been applied to several critical infrastructures, in particular, to devices which are used in smart homes, smart cities and operational technologies [3]. The accumulative effect of this massive connectivity and wide applicability is the surge of data (over 60 ZB data by 2020) at the edge of the network. The explosion of data accounts for about nine trillion USD economic impacts [4, 5]. The big data generated from IoT devices can have a potential influence on people's day-to-day activities such as offering huge profitability and improving quality of life. At the same time, the surge in the traffic from IoT applications increases the potential danger of cyber-attacks as the IoT come with unseen protocols, work flows, interfaces and applications.

Centralized platforms such as cloud suffer from scalability and high delay in data collection, communication and suffer reaction time for IoT applications. Due to this, the ability to leverage embedded intelligence in distributed IoT network is an

A. Diro (✉) · A. Mahmood · N. Chilamkurti
La Trobe University, Bundoora, VIC, Australia
e-mail: a.diro@latrobe.edu.au; a.mahmood@latrobe.edu.au; n.chilamkurti@latrobe.edu.au

© Springer Nature Switzerland AG 2021
W. Chang, J. Wu (eds.), *Fog/Edge Computing For Security, Privacy, and Applications*, Advances in Information Security 83,
https://doi.org/10.1007/978-3-030-57328-7_4

essential architectural component [6, 7]. This necessitates to extend the cloud to a cognitive and distributed network known as fog computing (FC). FC is an emerging architecture of distributed computing [8] that pushes the cloud to the edge of the network for efficient data collection, computation and control, communication and networking, data access and storage [9]. It is a horizontal distributed architecture that distributes the services and resources closer to the origin of the data. FC differs substantially from edge and mobile computing in data processing mechanisms, and the location of intelligence and computational power. It is the push of centralised computing towards the edge of the network, particularly wireless networks, to solve the limitations of the cloud. The emerging usage of FC, however, is not to replace the cloud, but it complements the cloud by bringing intelligence into distributed fog nodes. The FC architecture provides a cloud-to-things continuum which caters for latency, bandwidth and the communication requirements of next generation networks [10]. In addition to controlling and managing the massive amount of IoT data as a mini-cloud at the edge, FC provides a layered architecture as a cloud-to-things continuum for big data storage and analytics.

FC enables the delivery of new types of services closer to the IoT. As it confines data movement, storage and processing to the edge of the network, it offers scalability and low latency for real-time systems [11]. IoT applications benefit from FC services in offloading computation and storage services, and in communicating with nearby nodes for service delivery. In such networks, obtaining networking services from nearby nodes guarantees the IoT urgent response time for supported real-time applications, while delegating closer nodes for computation and storage facilities helps the smart applications to be resource efficient. As it is closer to clients, FC supports IoT applications that leverage artificial intelligence (AI), such as augmented and virtual reality. It is an ideal paradigm for next-generation networks that support the convergence of IoT, 5G and AI applications The shift from centralisation, such as the cloud, to fog networking has also been necessitated by other unique features and advantages in the IoT, such as mobility support, location awareness, scalability, heterogeneity, low latency, and geographic distribution [12]. Because of this, smart city applications such as smart grids, smart transportation and health care systems benefit greatly from fog computing as it provides embedded and distributed intelligence for IoT in data collection and resource utilisation. This reinforces that FC is an ideal architectural element in the application of IoT security, especially cooperative intrusion detection system.

## 2 Intrusion Detection for IoT-Fog Computing

### 2.1 Significance

Cyber-attack detection systems lay a major role in monitoring networks [13] and systems because controls such as cryptographic access controls alone cannot provide the required level of security against the ever-evolving cyber-attacks. This is due to the fact that these systems are limited by aging and design and implementation flaws.

Traditional Internet attack threats and their variants continue to be the major threats for the IoT-Fog computing [14]. The focus of Internet attacks is data manipulation whereas IoT-Fog attacks target controlling actuation. With limited protection, IoT devices are targeted in massive-scale and simplicity than traditional Internet devices. The origin of the attacks could be external adversaries that intend to gain access to the internal network or insiders that have the motive and opportunity to misuse, attack or steal information. The impact and treatment of IoT vulnerabilities is different from the traditional Internet due to resource limitations. The vulnerability of IoT devices can be at the network, device, interface or infrastructure level [11].

With their increased popularity, the adoption and use of IoT applications are highly susceptibility to security attacks [15]. The security of IoT devices is of particular significance as the technology was envisioned without security considerations [11], and the devices are largely engaged in sensitive data collection about humans, critical infrastructures, government secrets and businesses [16]. These assets and infrastructures are an attractive target for state espionage, denial of service, identity theft, fraud and other disruptive attacks. For instance, the cardiovascular measurements of patients can be monitored in real-time by IoT devices integrated in pacemakers, which is extremely sensitive information in terms of privacy, should the readings come under a cyber-attack. IoT attacks can also escalate to massive devices in the IoT network which can be further hired as a botnet for attacks such as DDoS, ransomware, and data ex-filtration. Therefore, these are indications that the IoT environment opens a completely new challenge to security as cyber-attacks propagate from the digital to the physical world.

The importance of securing IoT devices can be witnessed from the recent cyber-attacks on several applications such as the manufacturing industry, connected cars, Internet service providers and eHealth systems. For instance, in an industrial IoT application, a German steel plant experienced a serious cyber-attack that caused significant damage to the production system in 2014 [17]. This attack shows that hackers can succeed in damaging industrial systems [18]. In connected cars, the exploitation of vulnerabilities in the Jeep SUV firmware enabled hackers to hijack its multimedia system over the cellular network in 2015. This incident has serious implications for security controls in smart transportation systems [19]. Furthermore, in 2016, the DNS service provider, Dyn, was attacked by a massive DDoS attack using an IoT botnet as a vector [20–22]. Thus, these security issues motivate the investigation into securing smart devices by employing IoT security controls.

## 2.2 Resource, Scalability and Algorithms Requirements

A secured system guarantees the principles of the widely applicable information security model, namely, the CIA Triad [23] (Confidentiality, Integrity and Availability).

IoT security, however, prioritizes a new Triad named AIC (Authentication, Integrity and Confidentiality). Since security breaches are inevitable despite the mitigation technologies in place, continuous monitoring and detection systems are equally required. These IoT security requirements are complex and competing in nature because of the limited resources, massive connectivity and distributed nature of these devices. Hence, the problems of securing IoT applications against cyberattacks are resource constraints, a lack of scalable architecture, and a lack of existing accurate and robust solutions.

*The Resource Constraints Problem* An exponential increment in the number of heterogeneous IoT devices increases the likelihood of cyber-attacks in various forms and targets. Nevertheless, IoT devices lack support for the evolving and complex security mechanisms such as cryptography and intrusion detection systems due to resource constraints. These small devices cannot compute complex cryptographic elements [24] for device authentication, data confidentiality and deep packet inspection as they are built with limited processors and memory. The same is true for intrusion detection systems, which consume massive storage and processing power for the real-time monitoring and detection requirements of IoT devices. For instance, the meter of a smart grid microcontroller has no capability of computing traditional Internet cryptographic and performing continuous monitoring of intrusion detection operations. Thus, less resource-demanding cryptosystems should be adopted to reduce message size in storage, transit and computing.

*The Architectural Scalability Problem* IoT systems lack well-established computing models unlike traditional networks that make use of the client-server model. However, as a client-server standard model, traditional security solutions such as cloud-based security functions and services cannot be implemented for massive-scale and distributed IoT devices due to high latency, bandwidth inefficiency and a lack of scalability. Fog-IoT communication is predominantly at edge i.e. adhoc communications. These kinds of traffic should be kept to the edge as much as possible. Sending traffic to the cloud suffers from costs such as communication, latency, etc. So, traditional cloud computing need to be extended to the edge to serve the purpose of the IoT. It is impractical to manage the real-time monitoring of billions of IoT devices using the remote cloud. For instance, in smart grids, the deployment of smart meter microcontroller security schemes on the cloud incurs significant overheads. In the IoT environment, most of the produced traffic is not transmitted to the cloud but is rather consumed at the edge of the network. This indicates that there is a strong requirement for a distributed intelligent security architecture to offload security operations as centralized security schemes become inscalable for heterogeneous IoT networks.

*A Lack of Existing Accurate and Robust Intrusion Detection Systems* Intrusion detection systems, such as signatures and rules, have been proposed by several re-searchers [25, 26]. In addition to the human effort invested in updating signatures/rules, these static systems have already failed to detect novel attacks. Furthermore, anomaly detection systems using a machine learning approach have suffered from low accuracy and lack robustness under adversarial attacks. Adversaries attacks can change the state of intrusion detection systems that build on the top of machine learning, specifically traditional neural networks. However, advanced machine learning algorithms have emerged in other fields such as image processing that are robust against perturbations from adversaries. However, traditional ML schemes fail to protect from ever-evolving cyber-attacks.

To sum up, the traditional heavyweight cryptography elements and intrusion detection systems, and the current architecture of securing the Internet is far from delivering the security of the IoT. Thus, there is a need to develop a new lightweight security framework which can efficiently run over the large-scale deployment of small IoT devices.

## 2.3 Challenges and Limitations of the State-of-the-Art IDS for Fog-to-Things Computing

The construction of a resilient IDS is one of the network security challenges faced by organizations and industries despite the advancement of detection techniques [27]. These challenges include decreasing the high false alarm rate (FAR), the detection of novel attacks, the lack of training and test data and dynamic network and system behaviours. These challenges are some of the reasons for the continued widespread use of inefficient and inaccurate signature-based solutions and the reluctance to adopt anomaly-based systems. Thus, as an additional layer of defence to thwart cyber-criminals, effective and robust IDSs should be deployed to monitor malicious activities in real-time for IoT applications. IDSs for the IoT in particular, should be characterized by:

- fast and accurate detection for a timely incident response and recovery with less damage
- deterrence of attacks, acting as prevention systems
- logging of intrusion events for future prevention

However, the detection technologies for IoT devices are challenged by:

1. *big data*- increased connectivity, mainly due to the fact that the IoT needs a high response time for processing and detection of big data volume.
2. *depth of monitoring*- high accuracy of detection requires deep pattern extraction and attribute extraction.

3. *heterogeneity of data and protocols*-heterogeneity introduces complexity of discriminating an attack from normal behaviours.

As a technique of intrusion detection, ML has been applied for cybersecurity applications for over a decade [28]. For instance, spam/phishing detection has extensively used ML techniques where a trained ML classifier distinguishes whether a given email is spam or not by learning from a database of manually labelled emails. As part of the detection process, textual-based content features and meta-features such as message size, IP address and the presence of attachments are extracted from the email to be discriminated as either spam or benign. The classifiers can be classical machine learning classifiers such decision trees (DTs), as support vector machines (SVMs), and artificial neural networks (ANNs). Furthermore, ML has been applied to static malware detection (before execution) using structural information such as the sequence of bytes. Dynamic (during execution) malware detection leverages runtime statistics such as network usage patterns. The classification of malware usually involves feature extraction using n-grams and portable executable (PE). The obtained features can be used as the input of machine learning algorithms for malware detection [29]. Therefore, ML has wide application to intrusion detection systems.

However, the practical usage of ML have been impeded due to the extensive human power required for feature extraction [30]. Though feature extraction constitutes a significant overheads (approx. 80%) of IDS efforts, the engineered features fail to accurately reflect the true patterns of training data [31]. It is likely that these human crafted features miss essential features that help in intrusion detection. This may lead to less scalable, slow and less accurate attack detectors on big data. This means traditional ML struggles to capture the evolving face of cyber-attacks due to incorporated human errors in the process of feature engineering. Furthermore, most zero-day attacks are slight mutants of the previously known threats (approx. 99% the same) [32]. The so-called new attacks (1%) are derived from the [33] already known vectors using similar logic and concepts. Traditional ML algorithms face difficulty in recognizing mutants, mainly for their incapability of extracting deep and abstract features. With the increased power and resources of attackers, traditional ML algorithms are too shallow to detect evolving and complex cyber-attacks. These problems have hindered the adoption of the specific anomaly-based detection of ML to penetrate into security market. This has led to stick with the inaccurate signature-based solutions. Therefore, it is essential to investigate a robust technique for learning such as deep algorithms for the application of intrusion detection.

## 2.4   Leveraging Fog Nodes for IDS

Despite its huge advantages, the centralised model of the cloud cannot satisfy the requirements of data collection, processing, storage and sharing for distributed services and applications such as IoTs [34]. The application of cloud computing for the security of the IoT forms a far apart IoT-Cloud communication architecture that consists of IoT on the users side and the cloud at the remote site. As the distance gap is large between things and the cloud, this two-tier architecture of cloud-to-things computing is not scalable and suffers from a large delay for IoT-based applications. This could be aggravated by the increase in the number of IoT devices, which necessitates an architectural change in the way data processing, storage, control and communication can be handled on the Internet. The architectural arrangement should consider the resource-constraint nature of IoT devices which needs to offload data storage and processing to nodes in proximity [35]. This indicates that the cloud should be extended towards the IoT to form a distributed layer between smart objects and the cloud to monitor resource management and to control IoT devices more closely. The requirements of IoT applications such as latency, location awareness, mobility and resource constraints could be solved using the newly formed distributed architecture formed by pushing the cloud towards the edge. To this end, one of the emerging architectures that supports these requirements is known as fog computing.

The consideration of appropriate network architecture is as important as selecting suitable security algorithms [36]. The individual node security and perimeter security schemes of IT devices cannot be adapted to IoT security because of processing and storage limitations. Simultaneously, the lack of scalability and the remoteness of centralized-cloud hosted security mechanisms cannot satisfy specific requirements such as delay sensitivity and bandwidth efficiency [37]. Such schemes were designed for the traditional Internet where hosts are sparsely connected with sufficient storage, processing and communication resources. It highlights that neither security approaches are suitable for the IoT with limited resources and because they are massively connected to support real-time applications [38, 39]. Apart from offloading computations and storage from both things and the cloud, a fog-based security mechanism seems to be scalable and distributed in the ecosystem that quickly recognizes attacks or suspicious behaviors before escalating to the end devices. This is a crucial property for a fast incident response in IoT without service disruptions, especially for industrial IoT with real-time mission-critical applications. For instance, the service interruption by DDoS on smart grids and connected vehicles can be detected on nearby fog nodes without disrupting the power grids and the car engines, respectively. suspicious events and threats from the Internet are more quickly deterred at fog network than in the cloud, which makes FC-based IDS appealing for critical infrastructures.

## 2.5 Fog Computing IDS Architecture

One of the most important design and implementation aspect of IDS is to choose an architecture that shows the distribution of IDS components. Though evolving continuously, architecturally, there are mainly three approaches of implementing

**Table 1** Resource overheads and scalability of IDS architectures for fog computing

| Design approach | Processing | Storage | Communication latency | Scalability |
|---|---|---|---|---|
| Things-fog-level | Medium | Medium | Low | High |
| Fog-cloud-level | Low | Low | High | Medium |
| Things-fog-cloud-level | Very low | Very low | Very high | Medium |

IDS for IoT in fog computing: *Things-fog-level, Fog-cloud-level and Things-fog-cloud-level*. The resource usage and scalability of each approach is depicted in Table 1.

**Things-Fog-Level IDS Approach** Fog-level is more powerful than things-level in processing, storage and communication resources. Leveraging fog nodes-level together with IoT devices in implementing IDS can bring about a significant efficiency in utilization of resources, response time and threat detection capabilities. This is due to the fact that IoT devices have host-level visibility while the fog nodes have subnetwork-level control and visibility. For instance, IDS monitoring systems in each fog node can collect IoT traffic for malicious event detection which they can share with their neighbourhoods. This brings about intelligence sharing for accurate threat hunting and detection in exchanging AI models and parameters. This architecture is promising in saving resources, fast detection and low false alarm rates, specifically if there is a mechanism of horizontal cooperation among fog nodes [40]. As technological advancement progresses, this architecture will reasonably enable efficient use of computing, processing and communication resources at the edge of the network. Thus, it provides low latency in detecting suspicious behaviours in the IoT network, which is essential for fast incident response (Fig. 1).

Things-fog-level IDS is appropriate for monitoring and logging IoT traffic applications to detect misuses of login and access control policies. The main drawback of this approach is that it is heavy for fog nodes to perform traffic monitoring and intrusion detection at the same time for massive-scale IoT devices. This inculcates that the number of IoT devices monitored under a single fog node should be optimized. As the things-level cannot handle heavy tasks, it also needs a careful design in distributing IDS functions and operations across things-fog nodes. On the other hand, the lack of global network view by the IDS of this architecture can significantly impact the detection rate as it lacks a holistic contextual knowledge. Finally, it is likely that the architecture fails to discover internal attacks if much of the IDS functions reside in fog nodes.

**Fog-Cloud-Level IDS Approach** This approach includes IoT layer to fog environment, and distributes IDS functionalities across fog and cloud levels [41]. The fog level can be leveraged to detect anomalies based on machine learning models. This enables to quickly identify suspicious events before propagating to critical infrastructures. Distributing intrusion detection model across local fog nodes can be taken as a technique of offloading overheads of computing and storage from resource-constrained devices [42]. Simultaneously, fog nodes can share machine learning models and parameters for sharing local experiences. The second level

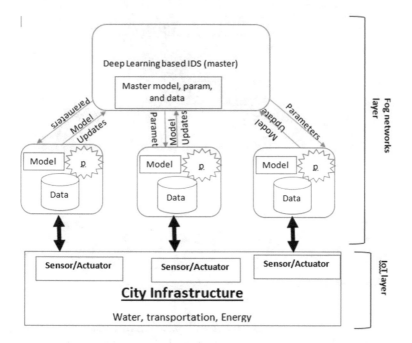

**Fig. 1** Things-fog-level architecture

can leverage the cloud in complementing IDS operation by performing intrusion classification or clustering. The scheme provides higher capacity of processing and storage than the Things-fog-level IDS Approach. In addition, it provides higher privacy protection and global network view. This architecture can be deployed to detect cyber-attacks such as scanning and DDoS as it is equipped with tremendous resources. However, it offers relatively lower response time than the same approach. This indicates that distributed intrusion detection mechanism is more scalable and accurate, and efficient than the cloud approach alone in terms of response time, communication bandwidth.

The architecture is vulnerable to internal attacks as there is no IDS function at IoT level. Nevertheless, there is no way to detect encrypted traffic as the things are without any IDS.

**Things-Fog-Cloud-Level IDS Approach** Intrusion detection can be distributed on three nodes [43]: cloud, fog and edge. The cloud might be responsible for heavyweight functions such as clustering network traffic and training detectors as it is equipped with sufficient resources while the fog level deals with the analysis of alerts. The pre-trained models on the cloud can be deployed on the edge devices to detect malicious events for IoT applications. As things are located at the periphery, the resource efficiency and performance of this approach might lie between the approaches of Things-fog-level and fog-cloud-level IDSs. The global orientation of

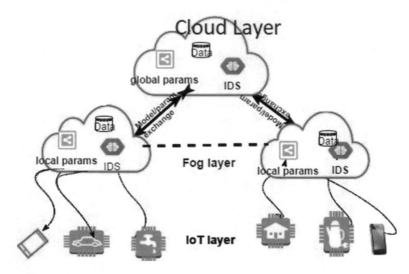

**Fig. 2** Things-fog-cloud-level architecture

the cloud, and the distribution of the fog will make this approach of IDS deployment to have a complete picture of the network. As IDS can be deployed at IoT level, it is extremely useful to detect malicious encrypted traffic. While this approach can provide improved accuracy, detection time, and efficient use of resources it might complicate the design and implementation of IDS. Despite its complexity, three-tier IDS architecture is reliable and secure for fog computing (Fig. 2).

However, the distance of the cloud can expose data communication for security and privacy threats. It is also complicated to coordinate the IDS functionalities distributed across three tiers.

## 3 Emerging Trends and Enablers for Detecting Intrusions in Fog-to-Things Computing

This section discusses the background and theories of emerging technologies/mechanisms such artificial intelligence and fog computing which are used in building the intrusion detection system in fog-to-things.

### 3.1 The Emergence of Deep Learning for Cyber Defense

ML is a field of AI that deals with the mechanism of learning from data. The prevalent machine learning algorithms include linear regression, decision trees, naive-Bayes classifier, logistic regression, SVMs and ANNs [44]. Deep learning (DL) is an advanced machine learning scheme, consisting of deeply layered neural networks(NNs) through which hierarchical and automatical features are learned. It is a ML scheme with several hierarchical layers of complex non-linear processing stages to model complex relationships using multiple levels of representations. DL is a recent breakthrough in ML that mimics the ability of human brain to learn from experience. Similar to the capability of human brain to process raw data using neuron inputs in learning the high-level features, DL enables input data to be fed into deep NNs for hierarchical feature extraction, training and classification [45]. It has a capacity of stabilizing and generalizing model training on big data [46]. The algorithms can discover complex functions mapping input into output without manual human expert intervention. DL algorithms have benefited from the advancement of hardware technologies (e.g. GPU) and algorithms (e.g. deep neural networks). The generation of a massive amount of learning data has also significantly contributed to this evolution as observed in globally dominant companies such as Google and Facebook.

Several deep learning applications have flourished in the field of big data such as image classification, object recognition, natural language processing and computer vision [47]. While large global companies such as Google, Microsoft, and Facebook have already embraced deep learning and have embedded it into their products, the cybersecurity industry is lagging behind in applying this technology to safeguard businesses and products. The success of DL in various areas can be adopted to solve the limitations of traditional ML in combating cyber-threats. For example, mutations due to attacks can be small, for instance, changes to image pixels. This means that DL can bring cybersecurity to a next level by discriminating even small variations of cyber-attacks. Because of the massive amount of data generated by IoT devices and systems, deep learning methods are highly effective in detecting and analyzing intrusions to IoT systems. It has been shown in traffic classification and IDS [40] applications. Hence, the capability of DL in compressed and automatic feature engineering and self-learning empowers IDS in resource efficiency, high cyber-attack detection rate and quick detection.

## 3.2 The Need for Autonomous Cyber Defense

There are several ongoing efforts in academia to bring the success of ML in various fields to cyber security applications. In adopting ML as an underlying technology, cybersecurity can harness the potentials of ML in numerous applications. Malware detection is probably the most prominent application of ML [48, 49]. Network anomaly detection [50] is another widely researched area of ML in cybersecurity. ML is also the backbone technology behind the newly emerged biometric authentication for network and mobile security [51]. The ultimate goal of these efforts is to improve the security posture of industries, businesses and governments by effectively assisting security experts in distinguishing malicious activities from normal network profile at machine speed in big data. This enables to save the time, efforts and cost in making decisions of segregating normal network data from cyber-attacks.

While tremendous achievements have been made in applying ML to cybersecurity, there is a large vacuum to fill the gaps of research in bringing autonomy cybersecurity [52]. The existing ML based security systems have made massive progress in predicting and classifying cyber-attacks using network traffic. However, the slow in speed, and the limited in knowledge of human loop fail to make rapid actionable decisions on the outputs of ML models. To solve this issue, and improve a security posture, ML should be given a permission to take actions and decisions to respond to cyber-attacks at machine speeds. Despite technical and administrative challenges, autonomous and intelligent security systems using AI will revolutionize the security of all sectors.

Classical ML algorithms are constrained by the steps of decisions made to solve a problem [53]. While some problems are a sing-step in nature, others required multi-steps to be solved. The multi-step decision process calls for learning systems which provide feedback, particularly reinforcement learning. Reinforcement learning functions without labelled data (unlike supervised learning), however, it also requires feedback to guide the algorithm (unlike supervised learning). This means it is a semi-supervised learning scheme that needs to understand the consequence of its actions to reinforce the behaviour (positive or negative) to continuously learn to react to its environment. In the efforts to improve a security posture of organisations and governments, embracing autonomy enables to reduce human loops from the security in taking actions and decisions. Reinforcement learning fits perfectly in line of making autonomous decisions in cybersecurity, especially to detect and defend cyber-criminals. Apart from automatic learning like other ML algorithms, it provides algorithms that learn policies of next system/network profile from the observable state of systems/networks. This enables to learn automatically over historical data from past experiences in various actions, conditions and settings using feedback signal to adapt to previously unseen conditions.

With the prevalence of machine learning such as reinforcement learning in playing games, cybersecurity is massively benefited from the autonomy of AI applications. Like playing games, autonomous cybersecurity requires experience gathering, opponent defeating and measurable rewards. In a dynamic network, observing the normal network profiles at various points in time, reinforcement learning can learn general policy about the state of the network through Markov Decision Process (MDP). As a multi-step problem solving process, MDP can be defined using four components: *state-space, action-space, reward function, and discount factor* [54].

Observation Space: Commonly known as the state space, this describes what the reinforcement learning agent can see and observe about the world to use in its decision-making process. In the course of learning to solve a problem, observable relevant features or everything monitored by the algorithm to decide is determined state-space. Actions: This is a set of actions the agent can choose from and execute as part of its policy. In other words, it is the set of actions and conditions under

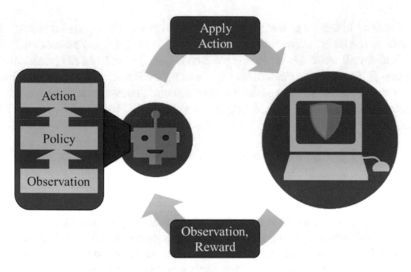

**Fig. 3** Iterative reinforcement learning decision cycle

which an algorithm can be used. Reward Signal: This provides both positive and negative reinforcement to the agent and informs the learning algorithm as to whether actions are leading to favourable or unfavourable outcomes. The reward function is the feedback signal the algorithm uses to learn and optimize policies. After receiving a reward as a consequence of an action, the change in the environment enables to update the policy and maximize the reward for future actions. The Discount Factor specifies the value of long-term vs. short-term rewards.

As shown in Fig. 3, an agent is responsible for updating and optimizing the security program and function of a network. For instance, the agent can perform actions, such as scans, logging, monitoring and classifications. Referring to its policy and observed current state of a network, the agent can decide on the bets next course of action at any given time. The actions taken from the observation, learned from prior experiences, will be a new experience for the next observation, and it is used to update a policy. The feedback is applied as a positive or negative reward in improving the security posture of the network in iterative manner. Hence, reinforcement learning algorithms can adapt to, and autonomously solve complex long-term unseen problems of network security in dynamic environment.

MDP can be formulated by 5-attributes consisting of state at time t $t(s_t)$, an action at time $t(a_t)$, state probability transition at time $t(P(s_t + 1|s_t, a_t)$, discount factor $\gamma$ and reward function at time $t(r_t)$. The objective function is to come up with the appropriate action to be taken at each state (i.e. produce an optimal policy $\pi^*$) to maximize the agent's reward over long period of time).

As a common reinforcement learning algorithm, Q-learning is a temporal difference learning that uses recursive algorithm to estimate a value function. It works by forming a Q table in which states form rows and actions represent columns. The quality function Q(s,a) estimates the maximum overall reward r the agent obtains in the state s and performing the action a to transit to next state $s'$. The solution to the Bellman equation $Q(s, a) = r + \gamma \ max_a Q(s', a')$ provides the value of Q(s,a) for all states and actions. This means Q(s,a) converges to optimal quality function $Q^*(s, a)$ after a random start by iterating the Bellman equation. Practically, however, the derivative form of the Bellman equation, known as temporal difference learning function, is used in updating the Q value to Q', where $\alpha(0 < \alpha < 1)$ is a learning rate and $\gamma(0 < \gamma < 1)$ is a constant for immediate or delayed reward.

$$\hat{Q}(s, a) \leftarrow (1 - \alpha)(Q(s, a) + \alpha(r + \gamma max_a Q(\hat{s}, \hat{a}))$$

RL solves complex problems which requires continuous and real-time learning, particularly for mitigating the evolving cyber-attacks. It is ideal for cybersecurity IoT security domains where construction of security learning models is challenging due to real-time prevention and detection requirements. However, as in other machine learning algorithms, RL suffers from curse of dimensionality to be deployed on a single node as learning agents and spaces increase in real-time. With the advent of deep learning, the curse of dimensionally is significantly reduced for learning from IoT traffic as it provides representation learning with hundreds of state features and actions. This enables, at least partially, to automatically learn policy generalizations from a massive state-space of cyber-domain. Further scalability can be achieved from devising a novel architecture that enables to learn locally, but exchange experiences using coordination. In this regard, the distributed fog nodes significantly reduce the need to train and monitor network traffic on a single node such as cloud.

## 3.3   The Application of Federated Learning-Based Intrusion Detection in Fog Computing

Federated learning is a decentralized and collaborative mechanism of training machine learning models for increased the efficiency and confidentiality of the training [55]. It assumes the existence of N nodes represented by $S_i$, where $i \in [1, N]$, each having training data $D_i$. Each node $S_i$ locally trains on data $D_i$ to produce a model $M_i$ t with a performance of $P_i$. In federated learning, all nodes jointly train a model $M_f$ with a performance of $P_f$ where there exists $i \in [1, N]$ such that $P_f > P_i$.

Intrusion detection system could harness the benefits of FL if blockchain is leveraged as an underlying distributed system to securely and transparently store and exchange ML model updates. The combination of FL with blockchain concept provides auditable, reliable and traceable jointly-learned ML models [56]. In fact, the distributed immutable ledger property of the blockchain can provide a mechanism of tracing ML parameter updates such as weights, gradients and models to mitigate adversarial attacks that can evade or poison ML models. This means blockchain mechanism enables the integrity of incrementally-learned ML models by cryptographically chaining a series of ML models.

FL can reduce the workload of training models on traditionally centralized systems such as cloud by distributing computation, storage and communication of learning. This architecture can be the next generation of anomaly detection mechanism for resource- constrained environments, specifically IoT-fog computing. Deep FL has been investigated for anomaly detection in system logs [57]. It has been indicated that FL minimizes the volume of data storage and movement in both decentralized and centralized systems, which significantly reduces the chance of data security and privacy. However, the it requires extensive studies to prove its data leakage resistance [33], effectiveness, efficiency and performance as the area is new to the scientific community.

## 4 Deep Learning-Based Intrusion Detection Schemes for Fog Computing: A Use Case

### 4.1 Architecture

This section showcases distributed IDS using deep learning scheme.

Fog nodes cam be leveraged to locally train ML models that can monitor IoT applications for suspicious events. The nodes can periodically exchange parameters to jointly establish the best model by model averaging via coordinator node/server. The coordinator node is used as a point of parameter and experience exchange among the fog nodes as it is prohibitively expensive in communication and processing to adopt peer-to-peer architecture in IoT environment. In this case, the diversity of IoT devices under the control of diverse fog nodes plays a pivotal role in providing different local observation. The result is that the coordinator has a better knowledge of the current state of the whole IoT-Fog network though each node is better aware of the local situations. Under this assumption, the coordinator agent learns whether or not to alarm, or take defensive action. With certain degree of accuracy, a fog node can take offensive action such as closing ports of attack or disabling an IoT device. The topology of the proposed IDS is depicted in Fig. 4. It consists of a fog node and IoT devices. Each fog node monitors IoT devices under its control, and receives only partial state of the global IoT network to make decisions. The hierarchical level consists of h tiers (in our case, it is 2 i.e. IoT-fog-Coordinator). Thus, the number of IoT devices under a coordinator node is $\sum_0^h n^{h-1}$.

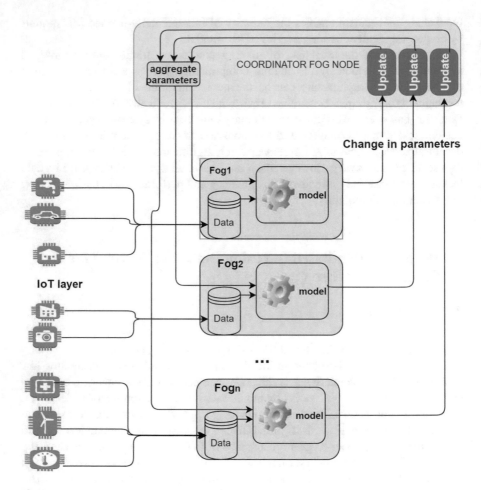

**Fig. 4** Deep model-based attack detection architecture for fog-to-things communication

## 4.2   *Algorithms*

The proposed self-taught system has two learning modules and a classifier which aims at:

1. Reducing dimensionality for lightweight IDS construction
2. Accelerating training and detection
3. Improving the accuracy of detection

Stochastic gradient descent (often abbreviated SGD) is an iterative method for optimizing an objective function with suitable smoothness properties (e.g. differentiable or subdifferentiable). It can be regarded as a stochastic approximation of gradient descent optimization, since it replaces the actual gradient (calculated from the entire data set) by an estimate thereof (calculated from a randomly selected subset of the data) [58]. The training and aggregation algorithms are adopted from sequential SGD for distributed fog network. Having initial training weights $W_n$ , a learning rate $\alpha$ and bias parameters $b_n$, $DATA_{total}$ is the total data space across all distributed m nodes. This means data on a given fog node $DATA_1, DATA_2, \ldots, DATA_m$ are subsets of $DATA_{total}$. The local data $DATA_i$ on the given fog node can be further divided into $DATA_{i_1}, DATA_{i_2}, \ldots, DATA_{i_s}$ samples. The $DATA_{i_s}$ samples on each fog node is further be split into $DATA_{i_{sc}}$ on processor threads $n_c$. Algorithm 1 shows local training and model/parameter updates on each fog node while Algorithm 2 reveals the aggregate function at the coordinator node.

**Algorithm 1** Local-learning ( $\alpha, n_{receive}, n_{send}$ )
step=0
While TRUE
    If(step % $n_{receive}$ ==0 )
      then get-Paramaeters (parameters)
    data $\leftarrow$ get-minibatch (k $\in$ data)
    sgd $\leftarrow$ compute-sgd (parameters, k)
    update weights $W_{ji} := W_{ji} - \alpha \frac{\alpha L(W,b|j)}{\alpha W_{ji}}$
    update bias $b_{ji} := W_{ji} - \alpha \frac{\alpha L(W,b|j)}{\alpha b_{ji}}$
    If(step % $n_{send}$ ==0 )
      send-Paramaeters ($\Delta W_{ji}, \Delta b_{ji}$)


**Algorithm 2** Aggregate ( $p_1, p_2, \ldots, p_n, n_{fog}$)
    receive p from all $n_{fog}$ nodes
    updated parameter=compute $\frac{1}{n_{fog}} \sum_{i=1}^{n_{fog}} p_i$
    send the updated p to all


As shown in the training algorithm, the coordinator server broadcast initial random parameters to other nodes for training. Each node runs multiple threads that are aggregated at master thread. The coordinator, finally, receives update from the master thread of every node asynchronously. The algorithm shows the process of local intrusion detection that is orchestrated by the coordinator server.


## 4.3   Evaluation and Discussions

The primary requirements of an attack detection system in modern networks are automatic feature extraction and high accuracy of detection. We conduct experiments using two forms of semi-supervised cyber-attack detection in our system on the NSL-KDD dataset: *the AE model and DBN model.* In both cases, unsupervised abstract feature extraction is used from unlabeled data using the technique of pre-training. In these self-taught schemes, the pre-training process only considers the unsupervised extraction of features in the half symmetry (encoding) of the encode-decode scheme. This arrangement aims to decrease the computational cost for training and detection while delivering almost the same accuracy as the encode-decode process. The other approach incorporates the supervised learning approach, in particular, the LSTM model. All the unsupervised models use shallow learning systems as the end layer for classification. The rationale behind the choice of these algorithms over other is their effectiveness in other fields, and their ability to produce lightweight features for the resource-constrained IoT environment. The comparison of the deep models is with classical machine learning models such as softmax alone and random forest. It is also essential to discuss the implications of the obtained results.

**The AE Model** In the experimentation phase, the deep learning package Keras on Theano and the distributed processing framework Apache Spark [59] are combined to implement the system Fig. 6. The first learning module in our training procedure gives optimal parameters for the second module as can be seen from the steps involved in our attack detection system in Fig. 5. First, m unlabelled data $x_u^{(1)}, x_u^{(2)}, x_u^{(3)}, \ldots, x_u^{(m)} \in R^n$ are extracted from the training data by removing the labels. These samples are input to the unsupervised module to learn the optimal parameter set such as weight W, and bias b (as shown in Fig 5). Then, m labelled train data $\{(x_l^{(1)}, y^{(1)}), (x_l^{(2)}, y^{(2)}), (x_l^{(3)}, y^{(3)}) \ldots, x_l^{(m)}\} \in R^n$ are fed to the supervised module to produce reduced and latent representation of the same data as $\{(h_l^{(1)}, y^{(1)}), (h_l^{(2)}, y^{(2)}), (h_l^{(3)}, y^{(3)}) \ldots, h_l^{(m)}\} \in R^n$ through the reconstruction of the output values $\{(\overline{x}_l^{(1)}, y^{(1)}), (\overline{x}_l^{(2)}, y^{(2)}), (\overline{x}_l^{(3)}, y^{(3)}) \ldots, \overline{x}_l^{(m)}\} \in R^n$. At this stage, AEs extract layer-wise hierarchical and discriminatory features, which enable the model to learn the complex relationships in the data. The module learns features from d-dimensional input vector $x \in R^d$ to map to latent representation $h_i \in D^{d_i}$ using a deterministic function:

$$h_i = \sigma(W_i.h_{i-1} + b_i); i = \overline{1, n} \tag{1}$$

where $\sigma$ is activation function and n is the number of hidden layers.

The second module enables the original input data to be substituted with vectors of activation $h_l^{(m)}$. The training set becomes $\{(h_l^{(1)}, y^{(1)}), (h_l^{(2)}, y^{(2)}), (h_l^{(3)}, y^{(3)}) \ldots, h_l^{(m)}\}$. Finally, the dimension-reduced input set is applied to Softmax to obtain the prediction functions. The test data passes through the same procedure as the training data in the module 2 for prediction/classification (Figs. 6, 7, and 8).

We compared our AE-Softmax with single-Softmax and single-RF using binary classification as shown in Table 2. Accordingly, the best accuracy of the proposed distributed deep learning model is observed to be 99.20% while the single-Softmax and single-RF have a best accuracy of 97.22% and 84.57%, respectively. In the same experiment, the detection rate of the deep learning mode is superior to its shallow counterparts. It is also observed that our model outperforms both the single-Softmax

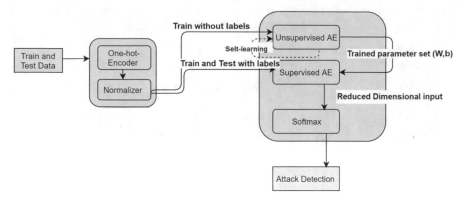

**Fig. 5** Block diagram of the proposed AE based attack detection

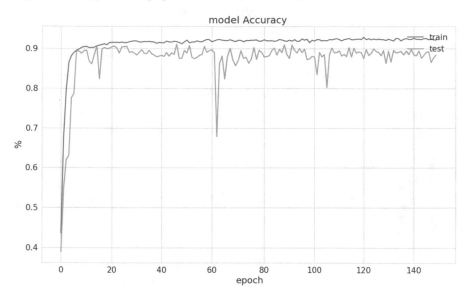

**Fig. 6** AE model training and test curve on 2-class NSL-KDD dataset in distributed settings

and RF in FAR as shown in Table 2. The ROC curve of a shallow model (RF model) lies below that deep model with higher number of FARs.

**The DBN Model** As in the case of AE, the features are learned by layer-wise pre-training using RBM to provide the initial weights. The learning step involves the adjustment of network weights in the hidden successive layers of RBM. After initializing DBN from the weights obtained in the pre-training phase, iterative global parameter adjustment (fine-tuning) is accomplished by a backpropagation algorithm in training the whole network. The weights obtained through this process enhance the discrimination capacity of the model. Cross-entropy is used as the loss function in this unsupervised fine-tuning phase, and regularization techniques are applied to

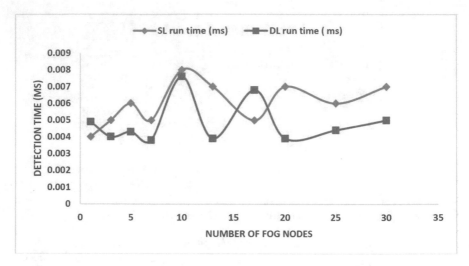

**Fig. 7** Comparison between DL and SL in detection time in distributed settings

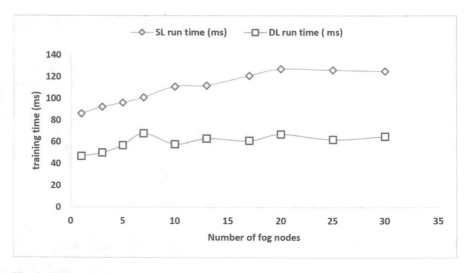

**Fig. 8** Comparison between DL and SL in training time distributed settings

solve the problem of over-fitting. The fine-tuning stage ends with trained model and weights for classification purposes. The classification task inputs the last hidden layer k to discriminate the network traffic into normal/attack or normal/multi-attack classes. In relation to performance measures, unseen test data are chosen to represent zero-day attack detection. The training curves of the accuracy of the algorithm are shown in Fig. 9.

As shown in Table 3, the detection accuracy of DL is better than classic ML model. Though there are performance differences among various deep models, they

**Table 2** The performance of AE-softmax vs Single-Softmax Model on 2-class NSL-KDD dataset in standalone and distributed settings

| Algorithm | Detection rate (DR) | False alarm rate (FAR) |
|---|---|---|
| Distributed AE-softmax | 99.20 | 0.80 |
| Single-softmax | 95.22 | 6.57 |
| Single-RF | 84.57 | 15.43 |
| Standalone AE-softmax | 97.72 | 2.28 |

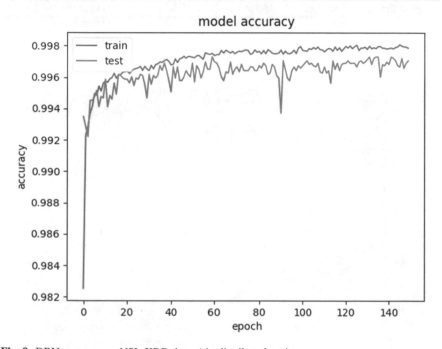

**Fig. 9** DBN accuracy on NSL-KDD dataset in distributed settings

**Table 3** Performance of deep vs shallow models of DBN on 2-class NSL-KDD dataset in standalone and distributed settings

| Algorithm | Detection rate (DR) % | False alarm rate (FAR) % |
|---|---|---|
| Distributed DBN-softmax | 99.78 | 0.22 |
| Standalone DBN-softmax | 97.95 | 2.05 |
| AE-DBN-softmax | 99 | 1 |

achieved better performance than the centralized model. For instance, in binary classification, Table 3 shows the FAR of the DL (0.22%) is much less than that of the ML model (6.57%). In multi-classes, as shown in Table 4, the performance of DL is also better than the normal ML model for most attack categories. For instance, the R2L.U2R recall of the DL is 91%, while the traditional model has a recall of

**Table 4** Performance of deep vs shallow models of DBN on 4-class NSL-KDD dataset in distributed settings

| Model type | Class | Precision (%) | Recall (%) | F1 measure (%) |
|---|---|---|---|---|
| DBN-softmax | Normal | 99.52 | 97.43 | 98.47 |
| | DoS | 97 | 99.5 | 98.22 |
| | Probe | 98.56 | 99 | 98.78 |
| | R2L.U2R | 71 | 91 | 80 |
| Single-softmax | Normal | 99.35 | 95 | 97 |
| | DoS | 96.55 | 99 | 97.77 |
| | Probe | 87.44 | 99.48 | 93 |
| | R2L.U2R | 42 | 82.49 | 55.55 |

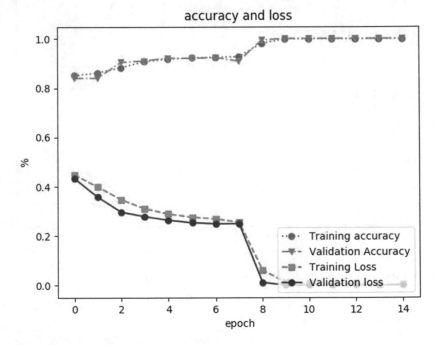

**Fig. 10** Performance of binary classification

82.49%. Similarly, the average recall of DL is 96.5% whereas SL scores an average recall of 93.66% in multi-classification.

**The LSTM Model** As shown in Table 6 and Fig. 10, the overall detection rate (99.91%) of the DL is over 9% higher than the ML (90%). It is evident that the high values of precision and recall are related to low false positives and false negatives, respectively.

The training and detection times of the LSTM model are also compared with the LR model, as shown in Table 5. The average number of instances trained on the LR

**Table 5** Performance results of models on ISCX dataset

| Algorithm | Average instances trained per sec | Average test instances per sec | Memory space (MB) |
|---|---|---|---|
| LSTM | 374.8 | 80.2 | 5.6 |
| RF | 54.3 | 1177.8 | 286 |

**Table 6** The performance of LSTM vs RF on ISCX and AWID datasets in binary classification

| | ISCX dataset | AWID dataset | | | | |
|---|---|---|---|---|---|---|
| Algorithm | ACC | Recall | Precision | ACC | Recall | Precision |
| LSTM | 99.91 | 99.96 | 99.85 | 98.22 | 98.9 | 98.5 |
| RF | 90.92 | 99 | 89.11 | 84.87 | 90 | 85 |

model per second is 54.3 while LSTM required 374.6 training instances per second. It can be seen that the training time od LSTM network is considerably greater than training the LR model. Moreover, it has been observed LR model has detected 1177.8 instances per second while LSTM model has detected 80.2 instances per second. However, the memory storage of the LR model is almost 50 times greater than that of the LSTM model. This indicates the compactness features and models of deep learning.

LSTM model has been compared with softmax in multi-class classification. It has been shown that the overall detection rate of LSTM model (98.85%) is greater than the rate of softmax (86%) similar to binary classification scheme (Table 6).

## 5 Conclusion and Further Research

IDSs have a significant importance for monitoring IoT systems against internal and external cyber-attacks. The emergence of fog computing has brought about new alternative distributed architectures for implementing IDS in the context of IoT applications. The choice of an architecture in which IDS can be deployed depends on the parameters such as resource efficiency, scalability and response time.

Pre-training has been effective in accelerating data training and attack detection. For instance, the training times of SL are higher than DL, and increasing sharply as the number of nodes increase while the training times of DL are almost constant and below that of SL. Similarly, the detection speed of DL is less than that of SL as the node increases. The experiment results show clearly that the reduction of dimensions gained through pre-training techniques enabled to build lightweight IDS as shown in Figs. 8 and 7, respectively. The results indicate that the deep learning model is superior in detection accuracy, detection rate and false alarm rate over the shallow model. In the proposed distributed and parallel learning approach, every fog node trains its data locally using AE and shares the best parameters with the other nodes via the coordinator node. The details of the architecture are depicted in Fig. 4. The architecture enabled to contain training data to the edge which would have been trained on a single location, the cloud. It distributes storage, processing and controls to local nodes so that each fog node can detect cyber-threats for the nearby IoT devices while best model is exchanged via the coordinator node. As shown in Table 2, the boost in accuracy of deep model could be a result of model and parameter sharing among fog network computing nodes. This is compelling evidence to show that the parameter-sharing approach results in the scalability and increased accuracy of the detector compared to the standalone detector.

From the detailed analysis, it seems that things-fog architecture provides a significant reduction in the latency of communication between IoT devices and fog nodes for detecting suspicious events and cyber-attacks. Though it also enables to offload computational and processing activities, it doesn't offer a massive save of the resources in contrast to the things-fog-cloud scheme. However, distributing detection schemes across things-fog-cloud enables to save processing and storage resources as the cloud offloads most of the operations. Nevertheless, this approach complicates IDS design and incurs relatively higher latency of communication than the first scheme.

With regard to improving IDS algorithms, a limited number of works in the literature indicate that there is a strong need for lightweight IDS that can reduce false alarm rate. In this regard, the things-fog architecture ideal in hosting learning algorithms in such a way that fog nodes can share best parameters or models in peer-to-peer or via coordinator.

It is wise to explore the application of blockchain-based fog architecture for implementing IDS as it has a potential benefit of overcoming the poisoning and evasion attacks of machine learning models. Technologies such as software defined networking (SDN) can also provide a global view of fog networks by utilizing its controller in cooperating fog nodes.

# References

1. Diro, A.A., Reda, H.T., Chilamkurti, N.: Differential flow space allocation scheme in SDN based fog computing for IoT applications. J. Ambient Intell. Humaniz. Comput. 1–11 (2018). https://doi.org/10.1007/s12652-017-0677-z
2. David, M., Murman, C.: Designing Apps for Success: Developing Consistent App Design Practices. Focal Press, Burlington (2014)
3. CISCO: Cisco visual networking index predicts near-tripling of IP traffic by 2020. https://newsroom.cisco.com/press-release-content?type=press-release&articleId=1771211 (2016). Accessed on June 2018
4. Caron, X., Bosua, R., Maynard, S.B., Ahmad, A.: The internet of things (IoT) and its impact on individual privacy: an Australian perspective. Comput. Law Secur. Rev. 32(1), 4–15 (2016)
5. Louis, C.: 2017 roundup of internet of things forecasts. https://www.forbes.com/sites/louiscolumbus/2017/12/10/2017-roundup-of-internet-of-things-forecasts/#3a6fc1051480 (2017). Accessed on July 2018

6. Frahim, J., Pignataro, C., Apcar, J., Morrow, M.: Securing the internet of things: a proposed framework. https://www.cisco.com/c/en/us/about/security-center/secure-iot-proposed-framework.html (2015). Accessed on June 2018

7. Diro, A.A., Chilamkurti, N., Veeraraghavan, P.: Elliptic curve based cybersecurity schemes for publish-subscribe internet of things. In: International Conference on Heterogeneous Networking for Quality, Reliability, Security and Robustness, pp. 258–268. Springer, Berlin (2016)

8. Vaquero, L.M., Rodero-Merino, L.: Finding your way in the fog: towards a comprehensive definition of fog computing. ACM SIGCOMM Comput. Commun. Rev. 44(5), 27–32 (2014)

9. Bonomi, F., Milito, R., Zhu, J., Addepalli, S.: Fog computing and its role in the internet of things. In: Proceedings of the First Edition of the MCC Workshop on Mobile Cloud Computing, pp. 13–16. ACM, New York (2012)

10. Consortium, O.: Definition of fog computing. https://www.openfogconsortium.org/resources/definition-of-fog-computing (2015) Accessed on Dec 2017

11. Stojmenovic, I., Wen, S.: The fog computing paradigm: Scenarios and security issues. In: 2014 Federated Conference on Computer Science and Information Systems (FedCSIS), pp. 1–8. IEEE, Piscataway (2014)

12. Klas, G.I.: Fog computing and mobile edge cloud gain momentum open fog consortium, etsimec and cloudlets (2015)

13. Almseidin, M., Alzubi, M., Kovacs, S., Alkasassbeh, M.: Evaluation of machine learning algorithms for intrusion detection system. In: IntelligentSystemsandInformatics (SISY) (2017)

14. Andrea, I., Chrysostomou, C., Hadjichristofi, G.: Internet of things: security vulnerabilities and challenges. In: 2015 IEEE Symposium on Computers and Communication (ISCC), pp. 180–187. IEEE, Piscataway (2015)

15. Solutions, C.F.C.: Cisco Fog Computing Solutions: Unleash the Power of the Internet of Things. Cisco Systems Inc, San Jose (2015)

16. Diro, A., Chilamkurti, N., Kumar, N.: Lightweight cybersecurity schemes using elliptic curve cryptography in publish-subscribe fog computing. Mob. Netw. Appl. 22, 848–858 (2017)

17. Ab Rahman, N.H., Glisson, W.B., Yang, Y., Choo, K.-K.R.: Forensic-by-design frame-work for cyber-physical cloud systems. IEEE Cloud Comput. 3(1), 50–59 (2016)

18. Pajic, M., Weimer, J., Bezzo, N., Sokolsky, O., Pappas, G.J., Lee, I.: Design and implementation of attack-resilient cyberphysical systems: with a focus on attack-resilient state estimators. IEEE Control Syst. 37(2), 66–81 (2017)

19. Ring, T.: Connected cars–the next target for hackers. Netw. Secur. 2015(11), 11–16 (2015)

20. McDermott, C.D., Petrovski, A., Majdani, F.: Towards situational awareness of botnet activity in the internet of things (2018)

21. Kolias, C., Kambourakis, G., Stavrou, A., Voas, J.: Ddos in the IoT: Mirai and other botnets. Computer 50(7), 80–84 (2017)

22. Packard, H.: Internet of Things Research Study 2015, vol. 2. http://www8.hp.com/h20195 (2014)

23. Fenrich, K.: Securing your control system: the "cia triad" is a widely used benchmark for evaluating information system security effectiveness. Power Eng. 112(2), 44–49 (2008)

24. Banks, A., Gupta, R.: Mqtt version 3.1. 1. OASIS standard, vol. 29 (2014)

25. Wu, H., Schwab, S., Peckham, R.L.: Signature based network intrusion detection system and method, 9 Sept 2008. US Patent 7,424,744

26. Garcia-Teodoro, P., Diaz-Verdejo, J., Maciá-Fernández, G., Vázquez, E.: Anomaly-based network intrusion detection: techniques, systems and challenges. Comput. Secur. 28(1–2), 18–28 (2009)

27. Lee, B., Amaresh, S., Green, C., Engels, D.: Comparative study of deep learning models for network intrusion detection. SMU Data Sci. Rev. 1(1), 8 (2018)

28. Deepak, B., Pavithra, H.: Applications of machine learning in cyber security. Digit. Image Process. 10(5), 93–98 (2018)

29. Ford, V., Siraj, A.: Applications of machine learning in cyber security. In: Proceedings of the 27th International Conference on Computer Applications in Industry and Engineering (2014)

30. Guy, C.: Introducing deep learning: boosting cybersecurity with an artificial brain. http://www.darkreading.com/analytics/introducing-deep-learning-boosting-cybersecurity-with-an-artificial-brain/a/d-id/1326824 (2016). Accessed on July 2018
31. Diro, A., Chilamkurti, N.: Distributed attack detection scheme using deep learning approach for internet of things. Future Gener. Comput. Syst. **82**, 761–768 (2018)
32. Diro, A., Chilamkurti, N.: Deep learning: the frontier for distributed attack detection in fog-to-things computing. IEEE Commun. Mag. **56**(2), 169–175 (2018)
33. Hitaj, B., Ateniese, G., Perez-Cruz, F.: Deep models under the GAN: information leakage from collaborative deep learning. In: Proceedings of the 2017 ACM SIGSAC Conference on Computer and Communications Security, Dallas, TX, 30 Oct–3 Nov 2017, pp. 603–618. ACM, New York (2017)
34. Botta, A., De Donato, W., Persico, V., Pescapé, A.: Integration of cloud computing and internet of things: a survey. Future Gener. Comput. Syst. **56**, 684–700 (2016)
35. El Jaouhari, S.: A secure design of WoT services for smart cities. PhD thesis, Ecole nationale supérieure Mines-Télécom Atlantique (2018)
36. Al Shayokh, M., Abeshu, A., Satrya, G., Nugroho, M.: Efficient and secure data delivery in software defined wban for virtual hospital. In: 2016 International Conference on Control, Electronics, Renewable Energy and Communications (ICCEREC), pp. 12–16. IEEE, Piscataway (2016)
37. Ni, J., Zhang, A., Lin, X., Shen, X.S.: Security, privacy, and fairness in fog-based vehicular crowdsensing. IEEE Commun. Mag. **55**(6), 146–152 (2017)
38. Yi, S., Li, C., Li, Q.: A survey of fog computing: concepts, applications and issues. In: Proceedings of the 2015 Workshop on Mobile Big Data, pp. 37–42. ACM, New York (2015)
39. Yi, S., Qin, Z., Li, Q.: Security and privacy issues of fog computing: a survey. In: International Conference on Wireless Algorithms, Systems, and Applications, pp. 685–695. Springer, Berlin (2015)
40. Diro, A., Chilamkurti, N.: Leveraging LSTM networks for attack detection in fog-to-things communications. IEEE Commun. Mag. **56**(9), 124–130 (2018)
41. Illy, P., Kaddoum, G., Moreira, C.M., Kaur, K., Garg, S.: Securing fog-to-things environment using intrusion detection system based on ensemble learning (2019). Preprint, arXiv:1901.10933
42. Prabavathy, S., Sundarakantham, K., Shalinie, S.M.: Design of cognitive fog computing for intrusion detection in Internet of Things. J. Commun. Netw. **20**(3), 291–298 (2018)
43. Hosseinpour, F., Vahdani Amoli, P., Plosila, J., Hämäläinen, T., Tenhunen, H.: An intrusion detection system for fog computing and IoT based logistic systems using a smart data approach. Int. J. Digit. Content Technol. Appl. **10**, 34–46 (2016)
44. Berman, D.S., Buczak, A.L., Chavis, J.S., Corbett, C.L.: A survey of deep learning methods for cyber security. Information **10**(4), 122 (2019)
45. Deng, L.: A tutorial survey of architectures, algorithms, and applications for deep learning. APSIPA Trans. Signal Inf. Process. **3**, e2 (2014)
46. Vincent, P., Larochelle, H., Lajoie, I., Bengio, Y., Manzagol, P.-A.: Stacked denoising autoencoders: learning useful representations in a deep network with a local denoising criterion. J. Mach. Learn. Res. **11**, 3371–3408 (2010)
47. Najafabadi, M.M., Villanustre, F., Khoshgoftaar, T.M., Seliya, N., Wald, R., Muharemagic, E.: Deep learning applications and challenges in big data analytics. J. Big Data **2**(1), 1 (2015)
48. Li, J., Sun, L., Yan, Q., Li, Z., Srisaan, W., Ye, H.: Significant permission identification for machine learning based android malware detection. IEEE Trans. Industr. Inf. **14**, 3216–3225 (2018)
49. Bartos, K., Sofka, M., Franc, V.: Optimized invariant representation of network traffic for detecting unseen malware variants. In: USENIX Security Symposium (2016)
50. Tuor, A., Kaplan, S., Hutchinson, B., Nichols, N., Robinson, S.: Deep learning for unsupervised insider threat detection in structured cybersecurity data streams (2017). arXiv:1710.00811
51. Staff, M., Flieshman, G.: Face ID on the iPhone X: everything you need to know about Apple's facial recognition. Macworld, 25 Dec 2017. https://www.macworld.com/article/3225406/iphone-ipad/face-id-iphone-x-faq.html. Accessed 12 June 2018

52. Buczak, A.L., Guven, E.: A survey of data mining and machine learning methods for cyber security intrusion detection. IEEE Commun. Surv. Tutor. **18**(2), 1153–1176 (2016)
53. Mnih, V., Kavukcuoglu, K., Silver, D., Rusu, A., Veness, J., Bellemare, M., Graves, A., et al.: Human-level control through deep reinforcement learning. Nature **518**(7540), 529 (2015)
54. Wright, R., Dora, R.: Learning to win: making the case for autonomous cyber security solutions, Aug 2018. https://www.csiac.org/csiac-report/learning-to-win-making-the-case-for-autonomous-cyber-security-solutions/. Accessed 16 Jan 2020
55. Qinbin, L., Wen, Z., Bingsheng, H.: Federated learning systems: vision, hype and reality for data privacy and protection (2019). Preprint, arXiv:1907.09693
56. Preuveneers, D., Vera, R., Ilias, T., Jan, S., Wouter, J., Elisabeth, I.: Chained anomaly detection models for federated learning: an intrusion detection case study. Appl. Sci. **8**(12), 2663 (2018)
57. Du, M., Li, F., Zheng, G., Srikumar, V.: DeepLog: anomaly detection and diagnosis from system logs through deep learning. In: Proceedings of the 2017 ACM SIGSAC Conference on Computer and Communications Security, Dallas, TX, 30 Oct–3 Nov 2017, pp. 1285–1298. ACM, New York (2017)
58. Bottou, L., Bousquet, O.: The tradeoffs of large scale learning. In: Suvrit, S., Nowozin, S., Wright, S.J. (eds.) Optimization for Machine Learning, pp. 351–368. MIT Press, Cambridge (2012). ISBN 978-0-262-01646-9
59. Zaharia, M., et al.: Apache spark: a unified engine for big data processing. Commun. ACM **59**(11), 56–65 (2016)

# On the Feasibility of Byzantine Agreement to Secure Fog/Edge Data Management

Ali Shoker and Houssam Yactine

## 1 Introduction

Fog and edge computing extend the cloud computing model by offloading some of the storage and computation closer to the data source or user [5, 50, 57, 61]. This is achieved through introducing extra storage and computing layers between the cloud data center and fog/edge applications, often latency-sensitive. The benefits are many: reduced response time, less bandwidth utilization, higher security and privacy, among others [5, 56, 75].

In particular, security (and privacy) can be improved since data is kept close to its source rather than being exposed to vulnerable environments all the way to the cloud center. However, this form of isolation makes the fog/edge nodes prone to data integrity problems due to potential malicious attacks or arbitrary faults usually tolerated via replication, e.g., Byzantine fault tolerance [8, 42]. On the other hand, applications that benefit from replicated or decentralized data across fog/edge nodes have to face the challenge of data management in untrusted environments.

The need for trusted systems against malicious and arbitrary, a.k.a., Byzantine [19, 41, 69], behaviors is gaining a lot of attention given the rise of the blockchain technology. Beyond its prime application, i.e., cryptocurrency [48, 73], blockchains proposed a generic solution to decentralized systems whose nodes are not (all) trusted. This has recently caused disruption in several fields of computer technology and economy [51], and revived a new wave of Byzantine fault tolerance (BFT) research [3, 7, 9, 21, 35, 37, 43, 44], as traditionally known in academia for

A. Shoker (✉)
VORTEX Colab, Porto, Portugal

H. Yactine
HASLab, INESC TEC and Universidade do Minho, Braga, Portugal

© Springer Nature Switzerland AG 2021
W. Chang, J. Wu (eds.), *Fog/Edge Computing For Security, Privacy, and Applications*, Advances in Information Security 83,
https://doi.org/10.1007/978-3-030-57328-7_5

121

decades [8, 42]. This raises the question whether BFT and blockchain protocols can be leveraged to empower fog/edge applications in untrusted environments.

A notable observation is that fog/edge computing and Blockchain technologies introduce complementary security solutions and challenges. While fog/edge computing improves the edge node security and privacy by reducing the attack surface, i.e., through limiting the exposure to external edge domains, blockchain is geared towards remote data access in untrusted environments.

In particular, blockchain ensures the integrity of the system against Byzantine behaviors through employing some variants of agreement protocols to maintain a total order on operations. This brings the classical CAP and PACELC theorems [1, 22] tradeoffs between availability and consistency, which requires trading one for the other in geo-replicated systems, where network partition-tolerance is often impossible.

In this work, we shed the light on the feasibility of BFT/blockchain protocols to fog/edge computing from a data management perspective. Our aim is to explore (in Sect. 4) the available BFT approaches and analyze their application to fog/edge computing considering the different security, consistency, and availability tradeoffs. While we do not intend to do an exhaustive survey of existing techniques, we try to ease the understanding of the existing synergies among different areas like fog/edge computing, blockchain, security, and data management. Our study addresses three different tradeoff approaches:

1. The first is using the Strong Consistency (SC) model as those Byzantine fault tolerant State-Machine Replication (SMR) protocols [8, 28, 76]. In general, these solutions are not effective in the geo-replicated fog/edge setting due to their high latency on both *read* and *write* operations [71].
2. The second is an Eventual Consistency (EC) approach that makes use of blockchain protocols in a peer-to-peer fashion while using a variant of proof-of-something protocols (e.g., Proof-of-Work, Proof-of-Stake, etc. [15, 20, 34, 45]. Such protocols provide low response time on *read* (although stale) operations, but they are blocking on *write* operations.
3. The third approach is based on Strong Eventual Consistency (SEC) [58, 62, 79] that allows (stale) *reads* as well as immediate writes. This works as long as operations are designed to be commutative, provided with a conflict-resolution technique as in Conflict-free Replicated DataTypes or Cloud types [6, 59]. This approach has recently gained great adoption in industry, in the fault-recovery model, which encouraged the development of solutions in the Byzantine fault model as well [62, 79].

Our findings, presented in Sect. 5, demonstrate that addressing Byzantine behaviors in fog/edge computing is in its infancy. Despite the recent development of BFT/blockchain solutions, they are yet to meet the needs of latency-sensitive fog/edge applications. Interestingly, there is significant overlap in different aspects of these technologies and their application requirements which encourages further research in this direction.

## 2    Background on Fog/Edge Computing, BFT, and Blockchains

Fog/edge computing and blockchain emerged in the last decade as two distinct technologies with different purposes [48, 57]. However, they currently share common fundamentals and applications. We first overview these technologies demonstrating their definitions.

### 2.1    Fog/Edge Computing

Edge computing was proposed in 2009 [57] as an intermediate layer, originally the Cloudlet, between the cloud and the application to reduce the response time especially in latency-sensitive and computation-hungry applications, e.g., speech recognition, natural language processing, computer vision and graphics, machine learning, augmented reality, planning, and decision making. This later evolved to the foundation of different hierarchical layers between the application and the cloud, giving rise to what is known as fog computing. Fog computing targeted the Internet of things applications [46, 53, 64, 78] where data generated by smart things is aggregated and preprocessed before off-loading it to the cloud or edge layers.

More recently, as used in the OpenFog Standard [31, 50], the benefits of edge computing in terms of reduced latency, bandwidth use, and security evolved to lateral (distributed) layer in the fog close to the data source and fog/edge applications [50, 61].

In this work, we adopt the OpenFog Standard [50] definition of for and edge computing. As depicted in Fig. 1, the fog is a hierarchy of layers located between the application and the cloud data center. In this architecture, the network edge represents the horizontal layer close to the applications. However, we generalize the edge to support more definitions where the network edge depends on the context and applications [32, 55, 78]. For instance, a MEC server is the edge in mobile computing, the router/gateway layer is the edge in smart networking [32, 55], whereas the smart boards are the edge in smart IoT applications [13, 46, 50, 61, 64, 78] (e.g., agriculture, industrial manufacturing, etc.). Therefore, we assume the existence of different edge layers that constitute a horizontal layer in the fog architecture, as in Fig. 1.

Last, but not least, security is fundamental in fog/edge computing. Huge amounts of sensitive (personal, public, or governmental) data is collected via smart things [30, 65, 66], i.e., sensors, cars, networks, cameras, etc. More seriously, this data is the knowledge base for decision making in smart and safety-critical systems where attacked actuators can be catastrophic [54, 67]. Fog/edge computing mitigates these threats by not exposing the data to remote sites. However, this is a double-edged sword: it reduces the attack surface to external threats, but makes the edge node less tolerant to Byzantine attacks that can besiege the node making it a single

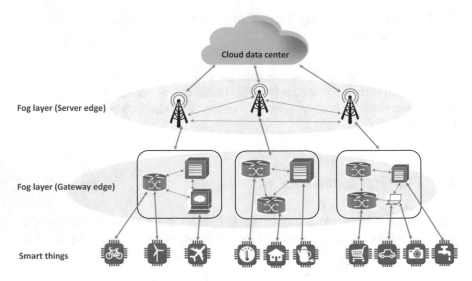

**Fig. 1** Fog/edge computing architecture

point of failure. In addition, access to remote data beyond the local edge domain becomes more challenging since edge nodes and peers are often less protected or equipped compared to cloud data centers.

## 2.2  Blockchain and BFT

The *Blockchain* was first proposed as an underlying layer for the Bitcoin cryptocurrency [48]. It proposed a distributed peer-to-peer infrastructure that implemented distributed consensus in untrusted environments, based on cryptography and heavy computation. Any two untrusted peers can transact over the network that guarantees total order on transactions and immutability via a Proof-of-Work (PoW) leader election protocol: the peer that solves a cryptographic puzzle [16] is chosen as leader and has the privilege to commit a set of transactions in a block (in the blockchain). This has later progressed in three interesting directions.

The first is abstracting the blockchain as an infrastructure in untrusted environments for applications where a trusted centralized authority cannot be assumed or guaranteed [7, 37, 73]. This gave rise to a plethora of applications in Fintech, supply chain, social networks, industry, IoT, economy, etc.

The second direction is the observation that the blockchain security model is equivalent to the Byzantine model that targets malicious or arbitrary behaviors, and has been studied for three decades in academia [8, 36, 42]. This suggested new sustainable alternatives to the PoW protocols that are criticized for being extremely

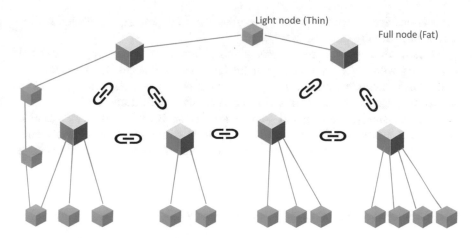

**Fig. 2** A common Blockchain architecture

energy-hungry[1] (a process called *mining*). In particular, the Practical Byzantine Fault Tolerance (PBFT) [8] protocol represented a gold standard that inspired most modern BFT and blockchain protocols (see Sect. 4).

The third direction is moving towards hybrid P2P architectures. Indeed, it was observed that pure P2P blockchain protocols hit the practical limits [28, 52, 71] towards finality latency (e.g., time to confirm transactions), scalability (storage and computation limits), and energy (due to the PoW puzzle solving). Therefore, most recent blockchains [7, 37, 48, 73] adopt the architectures depicted in Fig. 2. Thin nodes have constrained resources and thus offload the storage and heavy computation to fat nodes, with abundant resources.

## 3 Fog/Edge Computing Meets Blockchain

Despite growing in two distinct areas, the fog/edge computing and blockchain technologies intersect in several aspects. For instance, they both (1) embrace distribution or decentralization as a means to boost the availability of the deployed service; and (2) they try to reduce the latency on applications by locating computation and storage closer to the data source or user.

More interestingly, blockchain can be complementary to fog/edge computing in security [33, 80]: Although keeping the data close to its source in edge computing avoids vulnerabilities outside the local edge domain, it makes it a single point of failure and prone to malicious attacks. This is common in edge applications where nodes are *fat edge* nodes with abundant resources that can empower

---

[1] https://www.bbc.com/news/technology-48853230.

latency-sensitive computation-hungry applications like gaming, augmented reality, autonomous driving safety assistance, etc.

Modern edge systems are becoming more distributed across edge nodes not necessarily within the same edge domain [50, 61]. This necessitates secure data sharing in vulnerable environments. This is common in *thin edge* networks as in smart cities, agriculture, manufacturing, collaborative applications, etc.

Not surprisingly, the standard way to tolerate such attacks is via Byzantine Agreement, e.g., using BFT protocols which is a main foundation of blockchain. This makes the latter a potential candidate to bridge the security gap of fog/edge systems in untrusted settings. It is therefore interesting to discuss the existing potential and challenges towards this direction.

In the next sections, we tackle this topic through focusing on the Byzantine resilient data management perspective and its availability and consistency tradeoffs.

## 4 The Scope of Byzantine-Resilient Data Management Approaches

The benefits of fog/edge computing come at the price of new challenges at the data management level. Indeed, bringing storage and computation close to the data source, e.g., smart things, is deemed efficient as long as the data queried or updated is recent and local. Once access to remotely geolocated data sources or more historical data is sought, requests (Writes and Reads) are either offloaded close to the data storage or the data is pulled to close-by edge nodes. This is controversial with the low latency and autonomy requirements of fog/edge applications. The problem is further aggravated in the presence of faults and Byzantine behaviors or malicious attacks where an agreement is usually required to maintain a replicated state and thus exclude such non-benign behaviors.

These challenges suggest some tradeoffs specified by the CAP theorem [22]: one has to trade availability for consistency or vice versa, given that partition-tolerance cannot be guaranteed in geo-replicated systems. Consequently, there are three main data consistency approaches that are usually followed in such systems:

1. **Strong Consistency:** this trades availability for consistency through ensuring atomic writes and reads on data before delivering to the application or end user. This is often implemented through a State-Machine-Replication (SMR) consensus protocol, like Paxos or Raft [40, 49] under crash-recovery faults, and PBFT [8] under Byzantine faults.
2. **Eventual Consistency:** this trades consistency for availability to ensure low response time (mainly on reads) and high autonomy of fog/edge applications [63, 70]. However, these approaches suffer from finality delays or rollbacks on writes, which cannot be afforded in most fog/edge applications. Blockchain's proof-of-something (PoX), e.g., proof-of-stake and proof-of-work [15, 20, 34] or a combination between them, is the prominent approach here.

3. **Strong Eventual Consistency:** this is a tradeoff model of the preceding ones that allows low response time on reads and writes and eventually resolve potential conflicts maintaining the expected fog/edge application semantics [58, 79]. Well-known approaches here using Last-Writer-Wins (LWW) [74] or Conflict-free Replicated DataTypes (CRDTs) [59] to resolve conflicts in decentralized data management e.g., Cassandra [18, 38], OBFT [79], and ByzEc [62].

In the following section, we overview the current approaches in the three categories with more emphasis on the secure fog/edge data management solutions in face of Byzantine/malicious behaviors.

## 4.1   Approach 1: Strong Consistency (BFT-SMR)

### 4.1.1   System and Faults Models

Fog/edge nodes are asynchronously connected by a network or the Internet. Each node retains a full replica of the data in the system. Replicas communicate through messaging over the network. The network may fail to deliver messages, delay them, duplicate them, or deliver them out of order. The system follows the Byzantine failure model where Byzantine nodes may behave arbitrarily or maliciously (e.g., controlled by a strong adversary).

However, replicas are assumed to fail independently. This assumes a high degree of diversity in different nodes, e.g., hardware, operating system, libraries, implementations, etc. All nodes use cryptographic techniques to prevent spoofing and replays and to detect corrupted messages. In particular, messages are signed via public-key signatures or message authentication codes, and message integrity is verified via digests produced by collision-resistant hash functions. Finally, an adversary is assumed to be computationally bound, i.e., it is unable to break the aforementioned cryptographic techniques.

### 4.1.2   Problem Definition

In fog/edge computing, an application sends requests to the closest possible edge node in the system. If this node is malicious or Byzantine, the data integrity could be violated and thus the fog application semantics is broken. A BFT-SMR system solves this by having the replicas act as single state machines where all requests, reads and writes, are confirmed by the entire machine before delivery [8, 42]. This is done by solving the standard problem of Byzantine Agreement (or consensus) to ensure the integrity of the system despite the presence of a minority of Byzantine replicas (usually up to one third). The Byzantine consensus problem [24, 42] has the following general requirements:

- **Agreement:** Correct nodes $i$ output the same value $o$.
- **Validity:** If all or majority of nodes have the same input $b$, then $o = b$.
- **Termination:** All correct nodes decide on an output and terminate.

### 4.1.3  Desired Safety and Liveness Properties

Safety in such systems requires the system to satisfy linearizability (or serializability), i.e., execute operations atomically one at a time [42]. In particular, this ensures that data invariants in the service cannot be broken by Byzantine clients or replicas. Safety is usually maintained despite network partitions, and thus the protocols run under partial synchrony: there is an unknown bound where the system eventually synchronizes [8]. Liveness on the other hand means that clients eventually receive replies to their requests, provided that at most one third of the replicas is Byzantine. Due to the impossibility of implementing consensus in an asynchronous system [17], such systems rely on a weak form of synchrony, i.e., partial synchrony [8], to maintain liveness.

### 4.1.4  Byzantine Agreement

To demonstrate the Byzantine agreement technique, we focus on the Practical Byzantine Fault Tolerance (PBFT) protocol [8] that is considered a gold standard for BFT- SMR. Since our purpose is to demonstrate the concept, we opt to exclude the pedantic details that can be consulted in the references.

In this paradigm, replicas form a deterministic state machine. On each request, either read or write, a Byzantine consensus protocol is run to ensure the atomicity across all (usually $3f + 1$) replicas despite the presence of $f$ Byzantine replicas. The consensus protocol is usually quorum-based: a request is confirmed if it has a majority quorum of matching replies. The purpose is to ensure that the intersection of read-write and write-write quorums does not correspond to a Byzantine replica as shown in Fig. 3.

Figure 4 summarizes the phases of PBFT which are often used in Byzantine agreement algorithms. In this figure, replica 0 is the primary and replica 4 is Byzantine. In a nutshell, the protocol works as follows:

- The replicas are arranged in a sequence of configurations, called *views*.
- In each view, a primary node plays the role of the *leader* to other *backup* nodes.

**Fig. 3** The intersection of read-write and write-write quorums

RQ ∩ WQ > 1

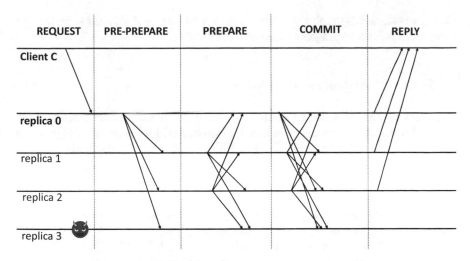

**Fig. 4** PBFT Byzantine agreement phases in normal case operation

- When a client sends a request to a primary node, it starts a three-phase protocol to atomically multicast the request to the backup nodes.
- in the *pre-prepare* phase, the primary multicasts a pre-prepare message to all other backup replicas.
- When receiving the pre-prepare message, the backup node enter the *prepare* phase by multicasting a prepare message to all other replicas including the primary and add both messages pre-prepare and prepare to its log.
- When prepared correctly, the node multicasts a commit message to other replicas. This starts the *commit* phase. When replicas accept commit message, execute it and send a reply to the primary or the client directly.
- The recipient (the primary or the client) waits for a majority quorum of matching replies from different replicas.

Since BFT-SMR replicas are assumed to be deterministic and start in the same initial state, the system satisfies the safety property by guaranteeing that all non Byzantine replicas agree on a total order for the execution of requests despite Byzantine failures. Although the system can achieve safety under asynchrony, partial synchrony is required in some cases to detect a misbehaving primary to ensure progress across views.

In the case where a Byzantine primary is detected, a primary election phase [8] is started by the backups (upon agreement) and a new view with a new primary is started (we omit these details for simplicity). This is necessary to maintain the liveness of the system. Another liveness requirement is that the intersection of quorums is $f + 1$ nodes instead of $f$ as is usually considered in non-Byzantine consensus protocols, e.g., Paxos and Raft [40, 49]. The reason is that it is impossible to differentiate a Byzantine node from a slow one, e.g., under network partitioning.

This guarantees that despite network partitions and up to $f$ Byzantine nodes, the system can still make progress and commit requests.

### 4.1.5 Current State of the Art Solutions

The practical Byzantine fault tolerant protocol PBFT is the seminal practical BFT protocol from which its successors inspire. Before the blockchain era, this protocol has been considered very costly due to its high latency, low throughput and fault-scalability (up to $f = 5$). Consequently, several optimizations have been introduced in other protocols as in Zyzzyva, Q/U, HQ, UpRight [2, 10, 12, 36], etc.

After the introduction of blockchain's Proof-of-Work protocol variants, known to be heavily computation-hungry, PBFT is now referred to as a sustainable protocol. This urged another round of optimizations that lead to modern protocols like Tendermint, HyperLedger, HotStuff, SBFT, FBFT [3, 7, 25, 37, 44], etc. Although these protocols managed to reduce the latency, throughput, and scalability of PBFT, they could not completely avoid the synchronization overhead of consensus [28, 71].

In particular, the best-case latency where no Byzantine faults exists can range from one, e.g., SBFT, to three messaging round-trips, HotStuff. On the other hand, Tendermint and HotStuff make use of primary node rotation (which makes fair use of primary node features), and thus optimize for the view-change protocol. Finally, HyperLedger and FBFT try to divide the load over federated groups or channels to improve the scalability of the system, but they are backed by protocols similar to the previous ones.

### 4.1.6 Feasibility Discussion and Limitations

Having a consensus protocol at its core, a BFT-SMR approach implements the strong consistency (serializability) model: all writes and reads are totally ordered and atomically confirmed after a Byzantine majority agreement is reached. Similar to the classical non-Byzantine consensus protocols, like Paxos and Raft, BFT-SMR protocols exhibit high synchronization overhead at scale. The overhead is much higher in the BFT-SMR case due to the high replication factor ($3f + 1$ replicas to tolerate $f$ Byzantine replicas), extensive cryptography use, and the complex consensus messaging pattern. This incurs high latency and low throughput and scalability. Given this, BFT protocols could not cope with the demands of fog/edge computing for two reasons: latency and scalability.

In particular, edge servers are usually located close to the end clients to avoid the network delays to remote nodes. Once a BFT-SMR protocol is used, this overhead is even higher since the system is as slow as the slowest (likely far-most) replica. This is controversial to the purpose of edge computing that optimizes for low-latency cases. The second problem is that the limited scalability of such protocols constrains the number of edge replicas that can be deployed. This means that upon using a BFT protocol, it is not possible to take full advantage of edge computing as edge nodes cannot always be located close to many users (assumed to be geographically dispersed).

Finally, it is worthwhile to mention that several cryptocurrency platforms are using Blockchain protocols that are inspired from BFT-SMR. However, most of these protocols are optimized to be more scalable in public settings [3, 7, 25, 37, 44].

## 4.2 Approach 2: Eventual Consistency (BFT-P2P)

### 4.2.1 System and Faults Models

The system model in this category assumes that edge nodes are peers connected through a peer-to-peer (P2P) network. Nodes communicate through messaging that are propagated via a gossip protocol in an asynchronous multi-hop manner. Membership can be private or public. Whereas in the former case nodes' identities are known a priori, nodes have to establish Sybil-resistant identities (i.e., public/private keys) in the latter [48, 73].

A standard method is to solve a computationally-hard puzzle on their locally-generated identities (i.e., public keys) verified by all other non-Byzantine nodes. All messages sent in the network are authenticated with the sender's private key to prevent spoofing. (To avoid repetition, all assumptions on the network and cryptographic techniques in the BFT-SMR case hold here as well.) To improve the scalability, the system can be sharded in such a way several shards can commit transactions in parallel. Another possibility is to arrange nodes in committees to assist in committing atomic transactions. Nodes can be fat or thin nodes. Thin nodes that cannot afford retaining the entire state can hold a cache, and request the missing data from a fat node.

### 4.2.2 Problem Definition

The ultimate goal of BFT-P2P systems is to *eventually* achieve agreement on a total order. Since full zero-hop membership is not possible, nodes are allowed to submit update requests that are eventually confirmed by the system. This allows a huge number of concurrent updates that is hard to maintain conflict resolution. Therefore, it is common to arrange several requests in more coarse-grained transactions confirmed in *blocks*. Total order is then achieved as long as a trusted leader exists to commit valid transactions. Therefore, the problem in such systems is often a leader election problem [26], whose specifications are:

- **Uniqueness**: There is exactly one non-faulty process that considers itself the leader.
- **Agreement**: All non-faulty group members know this leader.

### 4.2.3 Desired Safety and Liveness Properties

The safety properties of BFT-P2P are often more relaxed than in BFT-SMR. In particular, serializability is only enforced eventually. This means there is an instability period after the execution of a transaction and before it is confirmed. The guarantee here is that all confirmed transactions are totally-ordered. This however only holds for update requests. Read requests are delivered locally or requested from a remote fat node in the case of a cache miss. Importantly, concurrent read requests are not observed until they get confirmed (which reduces freshness).

On the other hand, the liveness property requires transactions to be eventually committed. This necessitates the successful election of a leader which is the core of such systems as explained next. In general, despite this guarantee, such systems suffer from *finality* delays due to the starvation caused by concurrent transactions at scale.

### 4.2.4 Blocks, Transactions, and the (Block) Chain

The BFT-P2P approach optimizes for scalability and decentralization. It allows for thousands to millions of nodes across the globe to participate in the system through asynchronous messaging via a P2P gossip protocol.

In particular, any node can issue write requests in transactions that are eventually confirmed by the system. Confirmation is crucial to maintain a consistent state through ensuring total order of write operations. Since consensus is not practical at this (up to millions) scale, as discussed in BFT-SMR, this approach relies on a single leader at a time to confirm requests. For security and availability reasons, this requires a fair (e.g., random or contribution-based) leader election technique. The uniqueness property of leader election ensures that no concurrent confirmations are happening at a time.

Another scalability technique in such systems is segregating transactions (of requests) into coarse-grained batches, called blocks. In its epoch, a leader can confirm one block instead of a single request or transaction. The purpose is to reduce the overhead of the (costly) leader election process in favor of a larger number of requests confirmed on each epoch. The ultimate goal is to reduce the finality of the system that is sometimes orders of magnitude higher than BFT-SMR.

A security requirement in blockchain applications is to ensure the immutability of confirmed blocks against a strong adversary (e.g., that controls half of the network). To prevent this, confirmed blocks are linked in a chain of blocks, i.e., a blockchain, by including the reference (i.e., a hash digest) of the last globally committed block in the blockchain in a newly committed block. This makes it extremely hard for an adversary to revert back confirmed blocks (unless it controls more than 50% of the network).

### 4.2.5 Leader Election

Leader election in a BFT-P2P system is a daunting task due the huge number of peers in the decentralized, likely globally-scaled, system. Consequently, choosing a leader though atomic broadcast is impractical. There are two main approaches to do leader election, i.e., via Proof-of-Something (PoX) or Committees. Proof-of-Something is a generalization of Proof-of-Work (PoW), originally used in the Bitcoin [48] underlying blockchain leader election (a process called *mining*).

In PoW, leader election is a competition through which peers try to solve an extremely hard-to-solve cryptographic puzzle [16]: to find a salt that when hashed with a digest of the last block in the chain satisfies a publicly known *difficulty* property (e.g., ends with a leading number of zeros). The salt stands as a PoW for the node to confirm a new block (i.e., implicitly chosen as leader for that epoch). This technique boosts the security since it is arguably very hard to control 51% of the computing power of the network [72]. Solving the puzzle is also key to reduce Sybil attacks [14] in public settings. For these reasons, PoW is currently advocated for scenarios that require high decentralization and security (e.g., Cryptocurrencies).

Unfortunately, the PoW approach is known to be extremely energy-consuming and not green-friendly [60, 68]. This encouraged the research for several less-costly alternatives like Proof-of-Stake [34] that is biased to peers that contribute more to the network. Although PoS is less secure than PoW, it has many benefits over PoW such as faster processing of transactions lows energy consumption. In addition to PoS, there are Proof-of-Elapsed-Time (PoET) that uses random waiting time protocol using trusted execution environment like Intel's Software Guard Extension (SGX), and many others [4, 11]. However, it argued that none of the new alternatives can achieve the security level of PoW. Regardless of the PoX method use, these protocols are often criticized for having finality delays.

In particular, the time to provide a PoX incurs some waiting time by design, that ranges from seconds to minutes [48, 73]. This imposes direct impact on write requests that cannot be committed in a low latency as required in most fog/edge applications.

An alternative leader election method is to use BFT-SMR protocols within peers organized in *Committees* [37]. A committee is a group of elected peers that have some extra privileges. Since the committee size is usually orders of magnitude smaller than the entire network size, it is possible to run a consensus protocol. The committee that runs the consensus protocols can thus choose a leader and commit transactions. This solution reduces the fairness in the system given the committee selection technique, and diminishes decentralization that is sought in several applications. In addition, this technique reduces the finality of transactions, but it is as best as the BFT-SMR protocols used in a committee.

### 4.2.6 Current State of the Art Solutions

Common BFT-P2P are mainly those developed for blockchain applications. All protocols inspired from the Bitcoin blockchain that proposed the Proof-of-Work approach [48]. However, Bitcoin's protocol only considered cryptocurrency applications. The blockchain protocol used in Ethereum [73] is similar to that of Bitcoin, however it generalizes the blockchain for any turing-machine style application, among them supply chain, decentralized organizations, etc. These two protocols imposed a tuned latency on issuing new blocks (on average 10 min in Bitcoin and several seconds in Ethereum) and required extensive computation power (e.g., equivalent to those of complete countries [27]). The underlying Blockchain protocols of both Bitcoin and Ethereum are BFT-P2P. These protcols however do not use the classical BFT-SMR consensus to reach agreement. They follow an asynchronous majority agreement with Game Theory [47] approach in PoW to decide on a leader that confirms transactions (in a block).

These protocols can maintain security if up to half of the network nodes are Byzantine [72].

Several protocols have been found later to address these caveats. Among them the Proof-of-Stake variants suggested freezing a stake (i.e., an amount of cryptocoins) for an amount of time to have the chance of block confirmation [34]. Other Proof-of-Something protocols, like Proof-of-Storage, Proof-of-Stake-Velocity, Proof-of-Credit, Proof-of-Exercise [60], followed the same concept without requiring solving a crypto-puzzle. Proof-of-Elapsed-Time (PoET) proposed replacing the puzzle with a time-delay proposed by a protocol in trusted execution environment like Intel's Software Guard Extension (SGX), and many others [9, 11].

Alternative protocols that are based on BFT protocols are considered hybrid protocols of PoX and scalable versions of PBFT (or its variants discussed in Sect. 4.2). For instance, Tendermint extends the PBFT protocol with Proof-of-Stake approach, i.e., by giving weights to the peer's vote. Algorand [21] uses Pure Proof-of-Stake (PPoS) consensus protocol built on Byzantine Agreement (BA) similar to PBFT. Each block is approved by a unique committee of nodes that are randomly selected by a lottery-based verifiable random function based on public information from the blockchain.

A similar randomness approach used by Dfinity [29] is to have a decentralized randomness beacon to produce random outputs as the seed. Similarly, Omniledger [35] used a committee selection process in a similar fashion as the lottery algorithm of Algorand in addition to sharding: splitting the overheads of processing trans-actions among groups of nodes, called shards. In a similar fashion, Elastico [43] uses shards processed by smaller BFT-based committees whose number grows near linearly in the total computational power of the network.

All of these protocols are considered less secure than PoS methods as they can only tolerate up to one third of Byzantine nodes in addition to sometimes being vulnerable to DoS attacks. The advantage is that finality latency, e.g., few seconds, is often less than that of PoX-based protocols.

### 4.2.7  Feasibility Discussion and Limitations

BFT-P2P protocols are geared for large numbers of peers communicating through gossip protocols. This limits their finality speed and thus imposes high latency on write requests. In this paradigm, applications are required to expect some time (seconds or minutes) to make sure a request has been eventually confirmed. Although this can be acceptable in some fog/edge applications that aim at reducing the overhead on cloud datacenters, it is not feasible for fog/edge applications where timely responses are crucial.

In addition, some PoW variants allow for forks that are only resolved later, by rolling back confirmed blocks. This can have serious implications on some fog applications where actuators are controlled based on this data. These very solutions are also very expensive to implement due to the extensive energy consumption that is scarce—especially—in thin edge networks.

Nevertheless, these protocols are typical for cryptocurrencies due to their high scalability in public permissionless settings.

## 4.3   Approach 3: Strong Eventual Consistency (BFT-SEC)

### 4.3.1  System and Faults Models

This approach assumes a system of edge nodes that are loosely coupled in an asynchronous network. Nodes communicate through messaging via a reliable broadcast or gossip P2P protocol. The system assumes that at most $f$ fog/edge servers out of $3f + 1$ can be Byzantine. In addition, a Strong Eventual Consistency [58] (SEC) is assumed on data: to reduce response time, edge nodes can serve (concurrent) read and write requests of clients immediately, and they eventually synchronize their states in the background. Therefore, a client's read request retrieves the value of the locally observed state where write requests may yield conflicts with remote nodes. This has two implications at the data level: client applications are assumed to tolerate stale reads, and employ a conflict-resolution technique to ensure convergence. Examples are the use of Last Writer Wins in Cassandra [38], Cloud Types [6], or Conflict-free Replicated DataTypes where operations are assumed/made commutative [59].

Some hybrid variants adopt a two-tier structure where a separate BFT cluster is used in the backend. In this case, the same assumptions used in Sect. 4.1 holds. In addition, the cryptographic assumptions made in Sect. 4.1 are also assumed here.

### 4.3.2  Problem Definition

Although in the [58] (SEC) model concurrent writes can lead to divergence, a conflict resolution technique can ensure system convergence under benign faults.

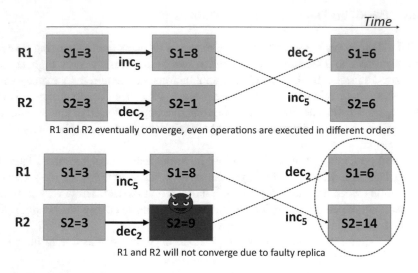

**Fig. 5** The problem of divergence in the presence of Byzantine node(s)

However, as exemplified in Fig. 5, convergence is violated if fog/edge servers are Byzantine. In fact, an attacked fog/edge server can apply local operations incorrectly. Thus, even after the propagation of remote operations to all other fog/edge servers, the Byzantine replicas will diverge forever as there is no chance to "pause" or halt the running system. On the other hand, the possibility to use a classical BFT protocol is also unacceptable. Indeed, since clients cannot tolerate the delays of integrity checking with other fog/edge servers prior to the reply, as in BFT protocols [8, 36], a Byzantine fog/edge server will "free-ride" and diverge from other correct fog/edge servers.

### 4.3.3 Desired Safety and Liveness Properties

Regarding the safety property, the system requires that all resulting states generated at all correct edge/fog servers are eventually equivalent. This assumes a commutativity property of used datatypes which leads to equivalent state upon executing the same log of operations [79]. Furthermore, the fog/edge nodes' agreement on that stable state is required to be final.

As for liveness, all correct fog/edge nodes should always agree on a stable state once operations are delivered thereof. This is often guaranteed by having the distinct write quorums intersect in more than $2f + 1$ nodes despite network partitions.

For the BFT cluster side, when used, liveness and safety properties follow the same rules described in the BFT-SMR case in Sect. 4.1.

### 4.3.4 Current Solutions

This approach is based on the Optimistic Byzantine Fault Tolerance model (OBFT) [79]. Such approach does not require total order on operations as long as the datatypes used are commutative. The idea is that applying the same set of commutative operations in any order will lead to the same final state if Byzantine nodes are tolerated. This allows for concurrent writes without synchronization with other nodes which reduces the response time significantly. As explained above, the disadvantage is that read requests only observe the locally delivered operations (i.e., the remote concurrent writes are temporally not locally reflected in the state).

In OBFT, Byzantine agreement is used only to establish a common state synchronization point. The set of individual states is needed to resolve conflicts passing through a state synchronization round, which typically involves the following steps:

1. Byzantine agreement on the set of individual states to be used for construction of the final state.
2. Construct the final state given the set of individual states.
3. Resolve any inconsistency at each individual replica based on the final state.

In this solution, the algorithms used to compute the final state must meet the following two requirements: (1) The algorithms must be deterministic to ensure that all non-faulty replicas adopt the same final state; and (2) the final state must be a valid state in that it should not include changes introduced by Byzantine faulty entities (such as clients or replicas). Unfortunately, this solution is not suitable for edge computing due to the blocking synchronization phase and the implementation complexity, as the authors declare [77].

An alternative solution [62], called ByzEc, assumes a three-tier system model (sketched in Fig. 6) that is composed of a frontend and backend. The backend deploys a BFT cluster (that runs a PBFT-like protocol) which serve as validation service for frontend requests. The frontend follows the geo-replicated system model in which fog/edge servers are geographically located and fully replicated. A client application, such as an IoT device, issues its requests to a single fog/edge node, preferably the closest one. As shown in Fig. 6, the frontend of ByzEc is composed of the clients layer and fog/edge servers layer. The clients layer can have thousands of simultaneous clients with different capacities. This is common in available systems [23] where application semantics are exploited to improve performance.

The fog/edge servers layer runs the service and can include tens of fog/edge servers geo-graphically distributed as points of presence close to clients in various regions. Similar to OBFT case, since applications tolerate reading stale data, fog/edge servers follow a relaxed data consistency model (e.g., eventual consistency or causal consistency [39]) in which they serve client's replies without prompt synchronization with other fog/edge servers.

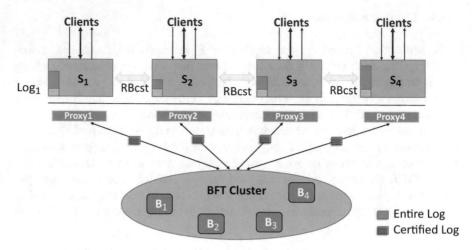

**Fig. 6** ByzEc architecture

### 4.3.5 Feasibility Discussion and Limitations

This approach is promising as it does not violate the latency requirements in fog/edge systems. However, it is limited to applications that make use of commutative operations and can tolerate some window of unverified state via the BFT cluster. This practically means the existence of a backlog and recovery scheme at the client to rollback malicious updates. For instance, edge applications can implement Ads, membership counters, recommender-based sets using this approach. Since this data is not critical, it may be desired to take advantage of the low-latency. To the contrary, this is not feasible to cases where actuators exists since physical actions cannot be rolled back.

In addition, the OBFT [79] solution is impractical since it requires blocking state synchronization to ensure consistent state for all non-Byzantine replicas. To the contrary, ByzEc [62] avoid blocking by running Byzantine agreement in the background, off the critical path of clients requests. Although promising, these solutions are not mature enough to be used in fog/edge computing.

## 5   Findings and Concluding Remarks

Fog/edge computing is a promising model that is optimized for data- and latency-sensitive applications as it keeps computation and data close within the secure edge domain, i.e., close to data source and user-end [50, 57]. This mitigates the potential threats that otherwise arise along the way to the cloud center. It however raises new challenges on the availability of the system against malicious attacks that can exploit this form of isolation.

More recently, fog/edge computing systems are becoming more distributed [50, 61] which requires secure data management in untrusted environments: different edge nodes belong to untrusted owners or domains. We therefore believe that making fog/edge systems more resilient to Byzantine behaviors is of high relevance that is worth more research.

In this work, we analyzed the relevance of Blockchain or Byzantine resilient solutions to fog/edge computing focusing on the availability and consistency tradeoffs—raised by the CAP theorem in scalable and available systems [22]. Our goal is to make it easier for researchers in the different areas to make synergies between these topics. A summary of the main characteristics and differences between these approaches is presented in Table 1.

This study lead to these following findings and conclusions:

- Blockchain and BFT protocols have the potential to bridge the distributed fog/edge systems gap in untrusted environments. This is referred to the commonalities in fog/edge and blockchain architectures and applications (as discussed in Sects. 2 and 3).
- Blockchain solutions are still immature to be used in fog/edge computing due to their limitations on finality latency. However, there are promising advancements that encourage further research.
- BFT-SMR are strongly consistent solutions that have limited scalability and high latency on reads and writes. This is not suitable for many latency-sensitive fog/edge applications. However, applications that can afford a latency up to one second, e.g., no use of actuators, may benefit from these solutions to improve their tolerance to Byzantine attacks.
- BFT-P2P PoX are eventually consistent solutions that have high finality latency on writes (seconds or minutes) which is unacceptable in fog/edge applications.
- BFT-P2P are hybrid (PoX-BFT-based) eventually consistent solutions that have the potential in the future to support scalable fog/edge applications where a finality latency of less than a second is acceptable. We encourage more research in this direction.

**Table 1** Comparison of the three different Byzantine-resilient data management approaches for fog/edge

| Approach | Considered solutions | Consistency | Latency | Security | Fog/edge compatibility |
|---|---|---|---|---|---|
| BFT-SMR | PBFT, Zyzzyva, Q/U, HQ, UpRight, BFTSMaRt, | Strong | Medium | Medium | Low |
| BFT-P2P | HotStuff, SBFT, FBFT... PoW, PoS, DPoS, PoE, PoET, PoSV, SGX, PPoS... | Eventual | High | High | Medium |
| BFT-SEC | OBFT, ByzEc | Strong eventual | Low | Medium | Medium |

- BFT-SEC is a tradeoff between Byzantine resilience and availability. It is tailored to distributed fog/edge applications where read and write latency is critical (e.g., in milliseconds) but freshness is not and recovery is possible. Current solutions are limited to applications and data types supported by conflict resolution techniques like CRDTs and Cloud Types [6, 59].

**Acknowledgments** The research leading to these results has received funding support from the project "NORTE-06-3559-FSE-000046—Emprego altamente qualificado nas empresas—Contratação de Recursos Humanos Altamente Qualificados (PME ou CoLAB)", financed by the Norte's Regional Operational Programme (NORTE 2020) through the European Social Fund (ESF).

# References

1. Abadi, D.: Consistency tradeoffs in modern distributed database system design: cap is only part of the story. Computer **45**(2), 37–42 (2012)
2. Abd-El-Malek, M., Ganger, G.R., Goodson, G.R., Reiter, M.K., Wylie, J.J.: Fault-scalable byzantine fault-tolerant services. ACM SIGOPS Oper. Syst. Rev. **39**(5), 59–74 (2005)
3. Abraham, I., Malkhi, D., Nayak, K., Ren, L., Yin, M.: Sync HotStuff: simple and practical synchronous state machine replication
4. Aissi, S.: Method and apparatus for secure application execution. US Patent 9,317,689, 19 Apr 2016
5. Bonomi, F., Milito, R., Zhu, J., Addepalli, S.: Fog computing and its role in the internet of things. In: Proceedings of the First Edition of the MCC Workshop on Mobile Cloud Computing, pp. 13–16 (2012)
6. Burckhardt, S., Gotsman, A., Yang, H., Zawirski, M.: Replicated data types: specification, verification, optimality. In: ACM Sigplan Not., vol. 49, pp. 271–284. ACM, New York (2014)
7. Cachin, C.: Architecture of the hyperledger blockchain fabric. In: Workshop on Distributed Cryptocurrencies and Consensus Ledgers, vol. 310 (2016)
8. Castro, M., Liskov, B.: Practical Byzantine fault tolerance and proactive recovery. ACM Trans. Comput. Syst. **20**(4), 398–461 (2002)
9. Chen, L., Xu, L., Shah, N., Gao, Z., Lu, Y., Shi, W.: On security analysis of proof-of-elapsed-time (PoET). In: International Symposium on Stabilization, Safety, and Security of Distributed Systems, pp. 282–297. Springer, Berlin (2017)
10. Clement, A., Kapritsos, M., Lee, S., Wang, Y., Alvisi, L., Dahlin, M., Riche, T.: Upright cluster services. In: Proceedings of the ACM SIGOPS 22nd Symposium on Operating Systems Principles, pp. 277–290. ACM, New York (2009)
11. Costan, V., Devadas, S.: Intel SGX explained. IACR Cryptol. ePrint Arch. **2016**(086), 1–118 (2016)
12. Cowling, J., Myers, D., Liskov, B., Rodrigues, R., Shrira, L.: HQ replication: a hybrid quorum protocol for Byzantine fault tolerance. In: Proceedings of the 7th symposium on Operating Systems Design and Implementation, pp. 177–190 (2006)
13. Da Xu, L., He, W., Li, S.: Internet of things in industries: a survey. IEEE Trans. Ind. Inform. **10**(4), 2233–2243 (2014)
14. Douceur, J.R.: The Sybil attack. In: International Workshop on Peer-to-Peer Systems, pp. 251–260. Springer, Berlin (2002)
15. Duong, T., Fan, L., Zhou, H.-S.: 2-hop blockchain: combining proof-of-work and proof-of-stake securely. Cryptol. ePrint Arch. Report 2016/716 (2016)

16. Dwork, C., Naor, M.: Pricing via processing or combatting junk mail. In: Annual International Cryptology Conference, pp. 139–147. Springer, Berlin (1992)
17. Fischer, M.J., Lynch, N.A., Paterson, M.S.: Impossibility of distributed consensus with one faulty process. J. ACM 32(2), 374–382 (1985)
18. Friedman, R, Licher, R.: Hardening cassandra against Byzantine failures. arXiv preprint arXiv:1610.02885 (2016)
19. Gashi, I., Popov, P., Strigini, L.: Fault tolerance via diversity for off-the-shelf products: a study with SQL database servers. IEEE Trans. Dependable Secure Comput. 4(4), 280–294 (2007)
20. Gervais, A., Karame, G.O., Wüst, K., Glykantzis, V., Ritzdorf, H., Capkun, S.: On the security and performance of proof of work blockchains. In: Proceedings of the 2016 ACM SIGSAC Conference on Computer and Communications Security, pp. 3–16 (2016)
21. Gilad, Y., Hemo, R., Micali, S., Vlachos, G., Zeldovich, N.: Algorand: scaling Byzantine agreements for cryptocurrencies. In: Proceedings of the 26th Symposium on Operating Systems Principles, pp. 51–68 (2017)
22. Gilbert, S, Lynch, N.: Brewer's conjecture and the feasibility of consistent, available, partition-tolerant web services. SIGACT News 33(2), 51–59 (2002)
23. Gilbert, S, Lynch, N.: Brewer's conjecture and the feasibility of consistent, available, partition-tolerant web services. ACM SIGACT News 33(2), 51–59 (2002)
24. V. Gramoli, From blockchain consensus back to Byzantine consensus. Future Gener. Comput. Syst. 107, 760–769 (2020)
25. Gueta, G.G., Abraham, I., Grossman, S., Malkhi, D., Pinkas, B., Reiter, M.K., Seredinschi, D.-A., Tamir, O., Tomescu, A. Sbft: a scalable decentralized trust infrastructure for blockchains. arXiv preprint arXiv:1804.01626 (2018)
26. Gupta, I., Van Renesse, R., Birman, K.P.: A probabilistically correct leader election protocol for large groups. In: International Symposium on Distributed Computing, pp. 89–103. Springer, Berlin (2000)
27. https://digiconomist.net/bitcoin-energy-consumption
28. Han, R., Gramoli, V., Xu, X.: Evaluating blockchains for IoT. In: 2018 Ninth IFIP International Conference on New Technologies, Mobility and Security (NTMS), pp. 1–5. IEEE, Piscataway (2018)
29. Hanke, T., Movahedi, M., Williams, D.: Dfinity technology overview series, consensus system. arXiv preprint arXiv:1805.04548 (2018)
30. Hashem, I.A.T., Chang, V., Anuar, N.B., Adewole, K., Yaqoob, I., Gani, A., Ahmed, E., Chiroma, H.: The role of big data in smart city. Int. J. Inf. Manag. 36(5), 748–758 (2016)
31. IEEE Standards Association: IEEE standard for adoption of openfog reference architecture for fog computing (2019). https://standards.ieee.org/standard/1934-2018.html
32. Hu, Y.C., Patel, M., Sabella, D., Sprecher, N., Young, V.: Mobile edge computing-a key technology towards 5G. ETSI White Paper 11(11), 1–16 (2015)
33. Kang, J., Yu, R., Huang, X., Wu, M., Maharjan, S., Xie, S., Zhang, Y.: Blockchain for secure and efficient data sharing in vehicular edge computing and networks. IEEE Internet of Things J. 6(3), 4660–4670 (2018)
34. Kiayias, A., Russell, A., David, B., Oliynykov, R.: Ouroboros: a provably secure proof-of-stake blockchain protocol. In: Annual International Cryptology Conference, pp. 357–388. Springer, Berlin (2017)
35. Kokoris-Kogias, E., Jovanovic, P., Gasser, L., Gailly, N., Syta, E., Ford, B.: Omniledger: a secure, scale-out, decentralized ledger via sharding. In: 2018 IEEE Symposium on Security and Privacy (SP), pp. 583–598. IEEE, Piscataway (2018)
36. Kotla, R., Alvisi, L., Dahlin, M., Clement, A., Wong, E.: Zyzzyva: speculative Byzantine fault tolerance. ACM Trans. Comput. Syst. 27(4), 7:1–7:39 (2010)
37. Kwon, J.: Tendermint: consensus without mining. Draft v. 0.6, fall 1(11) (2014)
38. Lakshman, A., Malik, P.: Cassandra: a decentralized structured storage system. ACM SIGOPS Oper. Syst. Rev. 44(2), 35–40 (2010)
39. Lamport, L.: Time, clocks, and the ordering of events in a distributed system. Commun. ACM 21(7), 558–565 (1978)

40. Lamport, L., et al. Paxos made simple. ACM Sigact News **32**(4), 18–25 (2001)
41. Lamport, L., Shostak, R., Pease, M.: The Byzantine generals problem. ACM Trans. Program. Lang. Syst. **4**(3), 382–401 (1982)
42. Lamport, L., Shostak, R., Pease, M.: The Byzantine generals problem. In: Concurrency: The Works of Leslie Lamport, pp. 203–226 (2019)
43. Luu, L., Narayanan, V., Zheng, C., Baweja, K., Gilbert, S., Saxena, P.: A secure sharding protocol for open blockchains. In: Proceedings of the 2016 ACM SIGSAC Conference on Computer and Communications Security, pp. 17–30 (2016)
44. Mazieres, D.: The Stellar consensus protocol: a federated model for internet-level consensus. Stellar Dev. Found. **32** (2015).
45. Milutinovic, M., He, W., Wu, H., Kanwal, M.: Proof of luck: an efficient blockchain consensus protocol. In: Proceedings of the First Workshop on System Software for Trusted Execution, pp. 1–6 (2016)
46. Mohanraj, I., Ashokumar, K., Naren, J.: Field monitoring and automation using IoT in agriculture domain. Proc. Comput. Sci. **93**, 931–939 (2016)
47. Myerson, R.B.: Game Theory. Harvard University Press, Cambridge (2013)
48. Nakamoto, S.: Bitcoin: a peer-to-peer electronic cash system (2008)
49. Ongaro, D., Ousterhout, J.: In search of an understandable consensus algorithm. In: USENIX Annual Technical Conference, pp. 305–319 (2014)
50. OpenFog Consortium Architecture Working Group et al. Openfog reference architecture for fog computing. OPFRA001 **20817**, 162 (2017)
51. Pilkington, M.: Blockchain technology: principles and applications. In: Research Handbook on Digital Transformations. Edward Elgar Publishing, Cheltenham (2016)
52. Poon, J., Dryja, T.: The bitcoin lightning network: scalable off-chain instant payments (2016)
53. Ren, J., Guo, H., Xu, C., Zhang, Y.: Serving at the edge: a scalable IoT architecture based on transparent computing. IEEE Netw. **31**(5), 96–105 (2017)
54. Ronen, E., Shamir, A.: Extended functionality attacks on IoT devices: the case of smart lights. In: 2016 IEEE European Symposium on Security and Privacy (EuroS&P), pp. 3–12. IEEE, Piscataway (2016)
55. Sabella, D., Vaillant, A., Kuure, P., Rauschenbach, U., Giust, F.: Mobile-edge computing architecture: the role of MEC in the Internet of Things. IEEE Consum. Electron. Mag. **5**(4), 84–91 (2016)
56. Satyanarayanan, M.: The emergence of edge computing. Computer **50**(1), 30–39 (2017)
57. Satyanarayanan, M., Bahl, P., Caceres, R., Davies, N.: The case for VM-based cloudlets in mobile computing. IEEE Pervasive Comput. **8**(4), 14–23 (2009)
58. Shapiro, M., Preguiça, N., Baquero, C., Zawirski, M.: A comprehensive study of convergent and commutative replicated data types (2011)
59. Shapiro, M., Preguiça, N., Baquero, C., Zawirski, M.: Conflict-free replicated data types. In: Proceedings of the 13th International Conference on Stabilization, Safety, and Security of Distributed Systems, SSS'11, pp. 386–400. Springer, Berlin (2011)
60. Shoker, A.: Sustainable blockchain through proof of exercise. In: 2017 IEEE 16th International Symposium on Network Computing and Applications (NCA), pp. 1–9. IEEE, Piscataway (2017)
61. Shoker, A., Leitao, J., Van Roy, P., Meiklejohn, C.: Lightkone: towards general purpose computations on the edge. White Paper published on http://www.lightkone.eu 40 (2016)
62. Shoker, A., Yactine, H., Baquero, C.: As secure as possible eventual consistency: work in progress. In: Proceedings of the Third International Workshop on Principles and Practice of Consistency for Distributed Data, PaPoC '17, pp. 5:1–5:5. ACM, New York (2017)
63. Singh, A., Fonseca, P., Kuznetsov, P., Rodrigues, R., Maniatis, P.Singh, A., Fonseca, P., Kuznetsov, P., Rodrigues, R. and Maniatis, P.: Zeno: eventually consistent Byzantine-fault tolerance. In: Proceedings of the Sixth USENIX Symposium on Networked Systems Design and Implementation, NSDI'09, pp. 169–184. USENIX Association, Berkeley (2009)

64. Soliman, M., Abiodun, T., Hamouda, T., Zhou, J., Lung, C.H.: Smart home: integrating internet of things with web services and cloud computing. In: 2013 IEEE Fifth International Conference on Cloud Computing Technology and Science, vol. 2, pp. 317–320. IEEE, Piscataway (2013)
65. Sun, X., Ansari, N.: EdgeIoT: mobile edge computing for the Internet of Things. IEEE Commun. Mag. **54**(12), 22–29 (2016)
66. Sun, Y., Song, H., Jara, A.J., Bie, R.: Internet of Things and big data analytics for smart and connected communities. IEEE Access **4**, 766–773 (2016)
67. Sundmaeker, H., Guillemin, P., Friess, P., Woelfflé, S.: Vision and challenges for realising the Internet of Things. Cluster Eur. Res. Projects Internet of Things Eur. Commission **3**(3), 34–36 (2010)
68. Truby, J.: Decarbonizing bitcoin: Law and policy choices for reducing the energy consumption of blockchain technologies and digital currencies. Energy Res. Soc. Sci. **44**, 399–410 (2018)
69. Vandiver, B., Balakrishnan, H., Liskov, B., Madden, S.: Tolerating Byzantine faults in transaction processing systems using commit barrier scheduling. SIGOPS Oper. Syst. Rev. **41**(6), 59–72 (2007)
70. Vogels, W.: Eventually consistent. Commun. ACM **52**(1), 40–44 (2009)
71. Vukolić, M.: The quest for scalable blockchain fabric: Proof-of-work vs. BFT replication. In: International Workshop on Open Problems in Network Security, pp. 112–125. Springer, Berlin (2015)
72. Watanabe, H., Fujimura, S., Nakadaira, A., Miyazaki, Y., Akutsu, A., Kishigami, J.: Blockchain contract: securing a blockchain applied to smart contracts. In: 2016 IEEE International Conference on Consumer Electronics (ICCE), pp. 467–468. IEEE, Piscataway (2016)
73. Wood, G.: Ethereum: a secure decentralised generalised transaction ledger. Ethereum Project Yellow Paper **151**, 1–32 (2014)
74. Yarabarla, S.: LWW (2017). https://www.oreilly.com/library/view/learning-apache-cassandra/
75. Yi, S., Hao, Z., Qin, Z., Li, Q.: Fog computing: platform and applications. In: 2015 Third IEEE Workshop on Hot Topics in Web Systems and Technologies (HotWeb), pp. 73–78. IEEE, Piscataway (2015)
76. Zhang, Y., He, D., Choo, K.K.R.: BaDs: blockchain-based architecture for data sharing with ABS and CP-ABE in IoT. Wirel. Commun. Mobile Comput. **2018**, 2783658 (2018)
77. Zhao, W.: Application-aware Byzantine fault tolerance. In: 2014 IEEE 12th International Conference on Dependable, Autonomic and Secure Computing, pp. 45–50. IEEE, Piscataway (2014)
78. Zhao, Z., Lin, P., Shen, L., Zhang, M., Huang, G.Q.: IoT edge computing-enabled collaborative tracking system for manufacturing resources in industrial park. Adv. Eng. Inf. **43**, 101044 (2020)
79. Zhao, W., Chai, H.: Byzantine fault tolerance for services with commutative operations. In SCC. IEEE, Piscataway (2014)
80. Zhaofeng, M., Xiaochang, W., Jain, D.K., Khan, H., Hongmin, G., Zhen, W.: A blockchain-based trusted data management scheme in edge computing. IEEE Trans. Ind. Inf. **16**(3), 2013–2021

# Part III
# Privacy in Fog/Edge Computing

# Privacy Issues in Edge Computing

Qi Xia, Zeyi Tao, and Qun Li

## 1 Introduction

With the quick development of Internet of Things (IoT), massive data is produced every day. For example, photographers can easily take 100 Mb photos per day; a surveillance camera can easily take 20 Gb video record per day. Considering the numerous amount of IoT devices, the total amount of data is beyond imagination. The computational limitation of an IoT device makes it almost impossible to process and analyse surveillance camera videos and photos in real time. With the help of cloud computing, a centralized server with sufficient computational power is able to process this data. However, limited by the low bandwidth and high latency, cloud computing is not efficient enough to deal with this large amount of data in real time. Therefore, edge computing has emerged as an effective technology to reach high bandwidth and low latency [1–4]. By offloading some of the computational power and storage to the edge of the network, edge computing is capable to deliver new services and applications to billions of IoT devices, such as augmented reality, video analytics, smart home, smart hospital, Internet of vehicles, etc.

Figure 1 shows a simple structure of cloud edge infrastructure. Cloud server, which has sufficient computational resources and storage space, is usually in a data center and far away from most of end users. At the edge of the network, edge servers are geographically close to end devices to ensure high bandwidth and low latency. Edge servers usually have considerable computational resources and storage space than end devices, but not as many as cloud server. The end device usually communicates with the edge server to get a quick response.

Q. Xia (✉) · Z. Tao · Q. Li
The College of William and Mary, Williamsburg, VA, USA
e-mail: qxia@cs.wm.edu; ztao@cs.wm.edu; liqun@cs.wm.edu

© Springer Nature Switzerland AG 2021
W. Chang, J. Wu (eds.), *Fog/Edge Computing For Security, Privacy, and Applications*, Advances in Information Security 83,
https://doi.org/10.1007/978-3-030-57328-7_6

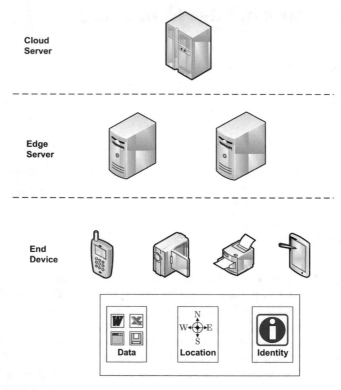

**Fig. 1** Cloud edge infrastructure

However, privacy is an important issue of edge computing [5, 6]. In summary, there are three kinds of privacy concerns in edge computing.

- **Data Privacy** Because of the high bandwidth of edge computing, more data are transmitted between end devices and edge servers. This allows more private information to be transmitted. On the other hand, unlike cloud computing, which has a central data center and is usually strictly supervised, edge servers are hard to control. Therefore there may be some edge servers that are curious about the sensitive data of end users. While the end device is usually connected to the nearest edge server and may be migrated from one edge server to another edge server for a better quality of experience, it is easy to leak private data during this process. In order to protect sensitive data, in edge computing, a privacy-preserving algorithm may be run between the cloud server and the edge server or the end device and the edge server.
- **Location Privacy** In edge computing, the location privacy mainly refers to the location privacy of the edge device users. Since edge devices are usually connected to the geographically closest edge server to offload tasks for a better experience, a curious edge server, can easily infer that the location of the end device user is not far from the edge server. In addition to this concern, once the

user with the end device moves and the end device shifts service from one edge server to another edge server, by communicating with each other, the active path information of this user may even be disclosed to the curious edge server. This will pose a big challenge to protect our location information.

- **Identity Privacy** Identity privacy is also a very serious privacy issue. It occurs in some practical applications in real life. For instance, when connect with the edge server, it is very common to fill some online forms or provide sensitive personal information. This information may be stored in the edge server and later be requested for authorization. Since this information is related to end user's identity and may connect with user's payment or other sensitive information, for most people, this identity privacy is more sensitive.

Apart from these three concerns, recent research has focuses more on the combination of edge computing and emerging technologies, such as big data and machine learning. Big data and machine learning are the hottest areas recently. Edge computing is an ideal platform for big data and machine learning because of its high bandwidth and low latency. It can also arrange resource allocation to maximize the use of the computational resources from end devices, edge servers and cloud servers. However, some wireless big data and machine learning applications such as smart city, online business, smart hospital, etc., contain a lot of sensitive information, which brings new challenge to user privacy. In fact, most of the privacy issues with big data and machine learning are still about data privacy, location privacy and identity privacy. However, since the scale of the data set in these areas is much larger than the conventional privacy problem, we will introduce this problem in a separate section as well as differential privacy, which is a popular technique for big data privacy issues on the edge.

In this chapter, we will first introduce three conventional privacy issues and some existing algorithms for these issues in edge computing: data privacy, identity privacy, location privacy. Then we will discuss the privacy issues in edge computing and emerging technologies big data and machine learning. The definition, implementations and properties of a widely used technique differential privacy are introduced. Then we talk about several proposed algorithms using differential privacy in big data and machine learning privacy. In the end of this chapter, we discuss about some future work and summarize this chapter.

## 2 Conventional Privacy Issues in Edge Computing

### 2.1 Overview

The high-frequency interactions between edge users and service providers (such as data transmission, information query and online transaction) continue to arouse people's extensive attention to various privacy requirements in data, identity and location. Users want to store their private data on a data server or cloud with cheap maintenance fees and access it anywhere at anytime with any device. Users can also collect useful information such as working hours around facilities, gasoline prices, attractions etc. around them. In addition, users can benefit from fast online payment through their portable devices. However, there exists many honest but curious adversaries such as edge data centers, infrastructure providers, service providers etc. They greatly challenge the privacy of edge computing paradigms. Compared with traditional privacy issues in cloud schemes, privacy in edge computing is more difficult to protect because for users, they do not even know whether a service provider is trustworthy. Therefore, preserving the privacy of users is a huge challenge that must be carefully considered. In this section, similar to the privacy issues in cloud computing, we consider three conventional privacy issues on data privacy, identity privacy and location privacy in edge computing area.

## 2.2 Data Privacy

Storing data into cloud/edge storage is attractive for most end users due to the facts that:

- users can access their data remotely and share data easily;
- users can avoid capital expenditure on physical hardware costs;
- users do not have to worry about file and storage management issues and leave this burden to the cloud/edge service.

Although cloud/edge services provide users with convenience and value, the data privacy issue remains a big challenge. Sensitive information such as photos, personal health records, and even government data may be leaked to unauthorized users and third party companies or even be hacked. While cloud service providers are usually under strict supervision, edge service providers are not trustworthy at all time, which brings more serious privacy challenges. The edge service providers therefore use firewall or virtualization to prevent data leakage. However, these mechanisms can not protect users' privacy because of untrusted edge storage services.

In cloud computing, in order to preserve the privacy of the data stored in the cloud, the conventional approach is to encrypt users' sensitive data before loading it into the cloud. Then users can retrieve the data back via keyword search or ranked keyword search. Several keyword search based encryption schemes have been proposed to ensure the privacy of data, including [7–9]. However, the main drawback of the encryption methods is their high computational cost and computational overhead. When using resource-constrained mobile devices, it may not be feasible to encrypt large size data. Another type of scheme uses symmetric-key cryptography and public-key cryptography respectively such as [10–12]. Similar to keyword based search scheme, high CPU usage and memory requirement during the encryption and decryption process become a bottleneck.

In order to mitigate the problem that the traditional encryption schemes in cloud computing cannot work well in the mobile edge cloud environment, there are several solutions.

### 2.2.1 Hybrid Architecture for Privacy-Preserving Data Utilization

Li et al. [13] proposed a practical hybrid architecture, in which a private cloud is introduced as an access interface between the data owner and the public cloud. This private cloud can be used as an extension for the resource-constrained mobile devices. Under this architecture, a data utilization system is provided to achieve both exact keyword search and fine-grained access control over encrypted data.

Although this architecture is introduced in cloud computing, it is suitable for resource-constrained mobile devices. This hybrid architecture has four entities in the system: data owners/users, attribute authority, private cloud and public cloud. The attribute authority is a key authority to generate public and private parameters and public cloud is used to store the data. Private cloud is a new entity to solve the problems that the computing resource of end devices is restricted. This system supports key authorization and revocation such that the access control of users is easy to manage. In addition, keyword-based query is supported so that an authorized user is able to use individual private key to generate a query for certain keywords. With the symbol-based trie, it can improve keyword search efficiency.

### 2.2.2 Pseudo-Random Permutation Based Method for Mobile Edge Computing

Later Bahrami and Singhal [14] proposed a new light-weight method for mobile clients to store data on one or more clouds by using pseudo-random permutation (PRP) based on chaos systems. The biggest advantage of using PRP is that this method does not require too much computational power and therefore can be run in mobile devices with low overhead.

There are two phases in this method: disassembly phase and assembly phase, which are similar to the encryption and decryption. In the disassembly phase, we split the files into several parts, which includes one file that contains the header of the original file and multiple files that contain the content of the original file. This process is based on a pseudo-random based pattern and the chunks in each file are also pseudo-randomly scramble with a chaos system. Then in the assembly phase, the end user can use the stored chaos system to reorder the chunks, and then use the stored pattern to decrypt the original file.

### 2.2.3 ESPPA

In 2016, Pasupuleti et al. [15] proposed a method named efficient and secure privacy-preserving approach (ESPPA). This approach utilizes the probabilistic public key encryption technique and ranked keyword search, which reduces the processing overhead of data owners while encrypting files. This also provides an efficient solution for resource-constrained mobile devices.

Unlike the solution we introduced in Sect. 2.2.1, this algorithm does not require an additional private cloud. However, it can still achieve the ranked keyword search for effective data utilization reasons. The ranked keyword search means that when the authorized user queries for a keyword, the edge cloud server finds matching files, ranks the matching files by the relevance scores and send back the top-$K$ relevant files. In order to preserve the privacy on resource-constrained mobile devices, instead of using homomorphic encryption [7], they choose to use probabilistic public key encryption technique [16]. Therefore, the proposed method can achieve the integrity of encrypted files and index stored in the edge cloud.

## 2.3 Location Privacy

In recent years, more and more applications have been adopted for location-based services (LBSs) and have achieved success in many aspects such as improving traffic, road planning, finding the nearest points-of-interest (POIs) etc. Edge computing is a natural and perfect system for LBS, since the end devices usually connect to the geographically nearest edge server for a better quality of experience. To enjoy such conveniences provided by LBS, users have to send queries to the LBS server. However, these queries contain massive information such as users' locations, interests, hobbies etc. Untrusted LBS servers can easily access these sensitive personal data and release these data to third parties such as advertisers. There are two types of location based privacy issue.

- **Restricted Space Identification** For example, the disclosure of a user's location may reveal the user's real-world identity and it has the potential problems to allow an adversary to locate the subject and cause physical harm.
- **Observation Identification** For example, if a LBS provider frequently observes the user's queries for bar and liquor, the adversary may infer the user is alcoholic.

Although distinct, the above two types of privacy issues are closely related.

In order to address privacy issues in LBS and avoid personal information abuse, many approaches have been proposed over recent years. In general, they all share a simple principle of *k-anonymous*. $K$-anonymity was firstly introduced by Gruteser and Grunwald [17]. The location information is represented by a tuple with three intervals $([x_1, x_2], [y_1, y_2], [t_1, t_2])$. The first two intervals describe an area $\mathcal{A}$ where the specific user is located. And $[t_1, t_2]$ indicates a time period of the user being present in such an area. When the user submits the query with a location tuple to the

LBS server, the $k$-anonymous protocol requires an area $\mathcal{A}$ containing at least $k - 1$ neighbors. Therefore $k$-anonymity prevents disclosure of user location by ensuring that user location information can only be accessed by LBS servers if there are at least $k - 1$ distinct associated locations which are indistinguishable. Generally speaking, the larger the anonymity set $k$ is, the higher is the degree of anonymity. To achieve $k$-anonymity, there are three types of solution: trusted anonymization server-based schemes, mobile device-based schemes and caching schemes.

### 2.3.1 Trusted Anonymization Server-Based Schemes

This method is based on a trusted central server called centralized location anonymizer [18, 19]. The goal of this anonymizer is to randomly arrange queries from end devices to one of several edge servers for protecting location privacy.

In detail, to achieve $k$-anonymity, a query is submitted to the LBS server via a centralized location anonymizer. The centralized location anonymizer enlarges the queried location into a Cloaking Region (CR) where the other $k - 1$ neighbors are also covered. As a result, it is difficult for the untrusted LBS server to distinguish the user's real location from other users. These are simple, straightforward and effective methods. However, they suffer from a single point of failure. Since these methods heavily rely on the location anonymizers, once adversary gains control of them, the privacy of all users will be compromised. At the same time, they have a performance bottleneck because all the submitted queries have to go through a single location anonymizer.

### 2.3.2 Mobile Device-Based Schemes

In order to avoid the problems in Sect. 2.3.1, instead of using a centralized anonymizer, [20, 21] suggest using dummy locations, which are randomly selected from the user's mobile device to achieve $k$-anonymity. This can apparently solve the privacy leakage risk in centralized location anonymizer.

However, the side information (e.g., query probability) can be utilized by adversaries, and hence reducing the anonymity degree of $k$-anonymity. By carefully selecting dummy locations, one can potentially eliminate side information leakage [22, 23]. The drawback of these kind of solutions is quite obvious, that is, the communication and storage cost is pretty high. Some compromising solutions focus on decreasing the computational and storage overhead by using VHC mapping [24], encountered-based solution [25], and $k$-anonymous cloaking box [26]. In addition, it is also a heavy computational cost for resource-constrained mobile devices.

### 2.3.3 Caching Schemes

There are more and more research on caching schemes recently. All these schemes are based on pre-fetching the useful location-based information in cache of end devices. We list some recent works below.

Shahriyar et al. proposed Caché to improve user location privacy [27]. The core idea of this method is to periodically pre-fetch potentially useful location-enhanced content well in advance. Thus the end devices can retrieve the location-enhanced content when they need it. This protects the precise location of end users.

Xiaoyan et al. proposed an alternative method called MobiCache [28]. Their method combines $k$-anonymity and Dummy Selection Algorithm (DSA). In order to increase the cache hit ratio, they generate some dummy locations that have not been queried before and choose them to query. In addition, they proposed an enhanced-DSA to choose dummy locations from cells which can make more contributions to both the cache hit ratio and user's location privacy even if they are not cached before.

In 2015, Ben et al. [53] proposed another caching-based solution to protect location privacy in LBSs. In their method, they propose an entropy-based privacy metric to measure the relation between cache hit ratio and the achieved privacy. Then based on this metric, they propose Caching-aware Dummy Selection Algorithm to achieve location privacy.

Although these methods can somehow protect the location privacy of end users, end users still need to store a huge amount of service data for a large area. Besides, to cache data, the end devices need more communications and computations.

## 2.4  Identity Privacy

Personal Identifiable Information (PII) or user identity is information about a person which has been collected, assessed or used by edge cloud services on demand. For example, when users establish their new edge service, they usually fill out an online form and provide sensitive personal information (e.g., name, gender, address, phone number, credit card number, etc.). This information may be stored in a central Identity Provider (IdP) and may be disseminated to service providers (SPs) later for the use of authorizing requests, completing payment, customizing services and so on. In early 2018, the Facebook data scandal caused 50 million users' PII to be disclosed to third party company, Cambridge Analytica, for "analysis" purposes via SP. This practical example tells us to stay alert to protect our personal information properly. Identity privacy issues are highly related to the problem of Identity Management (IDM) in the past decade. There are several solutions recently about identity privacy in edge computing.

In 2013, Khan et al. [29] proposed a light-weight identity protection scheme for Cloud-based mobile users for dynamic credential generation instead of the digital credential method. It uses a trusted entity to offload frequently occurring dynamic credential generation operations to reduce the computational cost on resource-limited mobile devices.

Then Park et al. [30] proposed an Improved Identity Management Protocol (I2DM) by using Pretty Good Privacy (PGP) that is based on Public Key Infrastructure (PKI) for secure mobile cloud computing. I2DM aims to find the weakest point in the network, maximize the load balance at this point to reduce the communication cost. This helps end device users manage their identity information easier.

Consolidated Identity Management (CIDM) system [31] aimed at mitigating three possible vulnerabilities: IDM server compromise, mobile device compromise, and network traffic interception. In practice, privacy is a challenge in IDM. According to [32, 33], identity management must meet the following challenges: *undetectability* aims to hide users' transactions and any other actions in a system; *unlinkability* aims to disconnect user identities and their history of transactions; *confidentiality* aims to enable users' controls. CIDM also uses a trusted third-party authorization server of IDM to manage the sensitive identity information of users. This method can distribute authorization credentials in the token into two related but different parts to countermeasure illegal access vulnerabilities. They also add a human interaction layer before each access is granted which can help defeat mobile devices that are compromised by adversaries.

# 3 Privacy in Edge Computing with Emerging Technologies

## 3.1 Overview

The emerging technology big data [34, 35] and machine learning [36, 37] have brought convenience to people's daily work and life. For example, people can talk to smart home assistant such as Siri or Alexa to get daily temperature, traffic information, control of lights and TV in their home, etc., instead of doing so by themselves; smart city can collect people's public safety, health, utility, and transportation data to help organize the city and make decisions; video analytics use machine learning and especially deep learning to analyze, classify, and process the video in real time; hospital can collect patients' symptoms and disease data and use machine learning to help them understand the disease and diagnosis more effectively.

In order for all these applications to be implemented in practice, a powerful and efficient infrastructure must be provided. Edge computing is naturally a good solution. With the benefit of high bandwidth, edge computing is capable to transmit large amounts of data to edge server for processing to help those end devices with limited computational resources. On the other hand, in some areas that require real time response such as deep neural network training and smart hospital diagnosis, low latency of edge computing is a big advantage.

However, there are several privacy concerns when we talk about big data and machine learning with edge computing. Since the high bandwidth brings more data exchange between the end device and the edge server, the data privacy concern is more serious than ever. As we cannot guarantee if the edge server is trustworthy because of the various edge server providers, the edge server may be curious about end user's private data, causing the leakage of the sensitive information. In summary, the privacy issues in machine learning usually occurs in the training phase and inference phase and the privacy issues in big data usually occurs in the data collection and data mining process.

To protect the users from these kinds of privacy concerns. A lot of privacy-preserving algorithms have been proposed recently. In summary, a good privacy-preserving algorithm must have the following requests.

- **Private** Privacy is obviously the core requirement of the algorithm. We must have privacy guarantee to prevent malicious or curious edge servers.
- **Effective** Here the effectiveness does not mean how effective this algorithm protects the privacy. It means that after we make privacy-preserving modifications to the algorithm, it should not lose the effectiveness performance a lot than state-of-the-art algorithm without preserving the privacy. For example, assuming that a face recognition model with normal training can have 95% accuracy, when we use privacy-preserving training, the accuracy should not have a big drop. If the performance drops a lot, it loses the meaning of training.
- **Scalability** The scale of the dataset in big data and machine learning is usually considerably large. Regardless of the smart city, smart hospital, or any other applications in big data, a big amount of data is the necessary foundation of big data for future analysis and implementation. Thus, the privacy-preserving algorithm must be able to deal with a large scale of dataset while the time complexity should not increase significantly with the increasing scale of the dataset.
- **Lightweight** Since one advantage of edge computing is the low latency, preserving the privacy should not bring more computational or communication overhead to the edge computing system. When we need to make considerable computations in edge server and end devices or make several data communications between edge server and end device to ensure privacy, the quality of experience will reduce significantly. So lightweight is a reasonable condition for privacy-preserving algorithms.

There are several existing techniques to preserve the privacy of big data and machine learning in edge computing. We will first introduce preliminary knowledge of differential privacy, which is a common technique in practice, then talk about some recent algorithms to protect the privacy.

## 3.2 Differential Privacy Preliminary

In this subsection, we will introduce some preliminary knowledges of differential privacy to help understand the recent privacy-preserving algorithms in big data and machine learning along with edge computing.

### 3.2.1 Definition

Differential privacy is one of the most important techniques to protect privacy recently. It is a practical technique to protect the privacy leakage problem in small perturbation of dataset. In another word, it provides a constraint to conventional algorithms that are used to analyze the dataset such that in statistics, it can limit the private information disclosure for whose information is in the dataset.

For example, there is a database that records the salary of each employer. We have a query to get the total compensation of a group of employers. Then we can query the database twice: one is for all data and the other is for all data but a specific employer. Then the difference of the two queries is the exact salary of this specific employer. We can find out that even if the query does not leak any information of a single user, the difference of subdataset can leak the private information. Theoretically we have following definitions.

**Definition 1 (Query)** A Query $f_i$ is a mapping function defined on a database $D$. Denote $F = \{f_1, f_2, \cdots, f_n\}$ as a group of queries.

**Definition 2 (Adjacent Databases)** Assuming database $D$ and $D'$ have the same attribute structure and they differ at most one element, in another word, one database is the proper subset of the other database and the larger database contains one more additional data, we call $D$ and $D'$ adjacent databases to each other.

For example, if $D = \{1, 2, 3, 4, 5, 6, 7\}$ and $D' = \{1, 2, 3, 5, 6, 7\}$, $D$ and $D'$ differ only one element 4, then $D$ and $D'$ are adjacent databases.

The first theoretical definition $\epsilon$-differential privacy is from [38] as following.

**Definition 3 ($\epsilon$-Differential Privacy)** A randomized function query $f$ gives $\epsilon$-differential privacy if for all datasets $D_1$ and $D_2$ who are adjacent databases, and all $S \subseteq Range(f)$,

$$Pr(f(D_1) \in S) \le e^{\epsilon} \cdot Pr(f(D_2) \in S) \tag{1}$$

The probability is taken over the coin tosses of $f$.

Definition 3 is a little bit abstract. Let's talk about it in details. In fact, this definition uses $\epsilon$ to control the difference of the query output distribution in those two adjacent databases. In some papers we also call this query a mechanism or algorithm. They output of the query has randomness. This definition guarantees that even if some data are added or removed from the database, the output of the query should not be significantly changed. For example, when we want to query the total compensation of a group of people, just returning the exact total compensation is at privacy leakage risk. An $\epsilon$-differential private query should return a range or a random number around the total compensation, so even if we run this query a lot of times to its adjacent databases, the exact salary for an individual is still protected.

$\epsilon$-differential privacy is a powerful way to protect privacy, but sometimes Definition 3 is too strong to achieve, there is a general $(\epsilon, \delta)$-differential privacy defined in [39] as following.

**Definition 4 ($(\epsilon, \delta)$-Differential Privacy)** A randomized function query $f$ gives $(\epsilon, \delta)$-differential privacy if for all datasets $D_1$ and $D_2$ who are adjacent databases, and all $S \subseteq Range(f)$,

$$Pr(f(D_1) \in S) \leq e^\epsilon \cdot Pr(f(D_2) \in S) + \delta \tag{2}$$

The probability is taken over the coin tosses of $f$.

Definition 4 is more general than Definition 3 since there is an added $\delta$ on the right of (2). In practice, we always choose $\delta$ as a very small constant. $\epsilon$-differential is a special case of $(\epsilon, \delta)$-differential privacy when we choose $\delta = 0$.

### 3.2.2 Implementation

After we know definitions of the classic $\epsilon$-differential privacy and the generalized $(\epsilon, \delta)$-differential privacy, we know that the differential privacy is to make modifications to query results such that query results are accurate in general but ambiguous enough to protect the privacy of the individual data. However, how to implement the differential privacy is a problem.

A natural idea to implement differential privacy is to add the noise to the query result so that the result can be more ambiguity. For example, in the previous example about the total compensation, we can add some noise to the total compensation result such that the query result is not exact number of total compensation but an approximate result around the exact number. Then we can use this result as the query output and it can protect the privacy of a single data.

Practical ways to implement differential privacy are based on this idea of adding noise. Basically there are three ways to add the noise: Laplace mechanism, Gaussian mechanism and exponential mechanism. We will talk about them one by one. Before we talk about these three mechanisms, we first define the global and local sensitivity of a query, or function.

**Definition 5 (Global Sensitivity)** For a query function $f : D \to \mathbb{R}^d$ where $D$ is a database and $\mathbb{R}^d$ is a $d$-dimensional real number vector, its global sensitivity in any adjacent databases $D$ and $D'$ is

$$GS_f = \max_{D, D'} ||f(D) - F(D')|| \tag{3}$$

**Table 1** Laplace, Gaussian and exponential mechanisms

| Mechanism | Modified query result | Noise PDF |
|---|---|---|
| Laplace | $f(D) + N(0, GS_f^2 \delta^2)$ | $\frac{1}{2b} e^{-\frac{|x-\mu|}{b}}$ |
| Gaussian | $f(D) + N(0, GS_f^2 \delta^2)$ | $\frac{1}{b\sqrt{2\pi}} e^{-\frac{(x-\mu)^2}{2b^2}}$ |
| Exponential | $Pr(M_{ex}(D) = r) = \dfrac{e^{\frac{\epsilon q(D,r)}{2 \cdot GS_{q(\cdot,r)}}}}{\sum_r e^{\frac{\epsilon q(D,r)}{2 \cdot GS_{q(\cdot,r)}}}}$ | N/A |

**Definition 6 (Local Sensitivity)** For a query function $f : D \to \mathbb{R}^d$ where $D$ is a database and $\mathbb{R}^d$ is a $d$-dimensional real number vector, its local sensitivity in a given database $D$ and all its adjacent database $D'$ is

$$LS_{f(D)} = \max_{D'} ||f(D) - F(D')|| \tag{4}$$

The Laplace, Gaussian and exponential mechanisms are listed in Table 1.

According to [39–41], we have the following Theorem 1 to introduce the privacy guarantee of these three mechanisms.

**Theorem 1** *If we use the Laplace mechanism and exponential mechanism as the query result, then it satisfies $\epsilon$-differential privacy. If we use Gaussian mechanism as the query result, then it satisfies $(\epsilon, \delta)$-differential privacy.*

### 3.2.3 Properties

When we have several different differential private algorithms, one question is how the privacy will change after we combine them. Here we introduce two major combination methods: sequential and parallel. Theoretically, we have following theorems.

**Theorem 2 (Sequential Composition)** *If $M_1, M_2, \cdots, M_n$ are algorithms or queries that access a private databse $D$ such that $M_i$ satisfies $\epsilon_i$-differential privacy, then the combination of their outputs satisfies $\epsilon$-differential privacy where $\epsilon = \epsilon_1 + \epsilon_2 + \cdots + \epsilon_n$.*

**Theorem 3 (Parallel Composition)** *If $M_1, M_2, \cdots, M_n$ are algorithms or queries that respectively access disjoint database $D_1, D_2, \cdots, D_n$ such that $M_i$ satisfies $\epsilon_i$-differential privacy, then the combination of their outputs satisfies $\epsilon$-differential privacy where $\epsilon = \max(\epsilon_1, \epsilon_2, \cdots, \epsilon_n)$.*

These properties provides the privacy guarantee for multiple combinations of algorithms. Therefore, once we have more than one differential private algorithms, we can also achieve differential privacy with sequential or parallel combinations of them.

## 3.3    Privacy-Preserving Algorithms

In this section, we will introduce some privacy preserving algorithms in big data and machine learning along with edge computing in details. It should be noted that the privacy problems in this area are at a very new stage and still have a lot of work to do, some pioneers have already found some important topics along with a series of techniques to solve these problems. The two major problems in this area are:

- Machine learning model training and prediction privacy.
- Big data privacy.

There are also some other privacy topics in this area. We will talk about those problems in the following sections.

### 3.3.1    Preserving Privacy in Machine Learning

The model training and prediction are the most important part of machine learning. They are also two of the major procedures in big data analysis. A good machine learning model can provide very accurate fitting and prediction results, which brings a lot of useful information to help us study the dataset and make a convenience to human beings. However, the model training of deep learning or big data analysis always requires a large demand of computational resources. To distribute the computational resources, edge computing provides a natural platform to offload the heavy computation from the end devices to edge servers or cloud servers.

The data privacy is always a big concern in this process. From the trained model or the training process, a lot of information can be leaked. Recent research can even rebuild the dataset from the training model by generative adversarial networks [42]. Therefore it is very important to protect the privacy while training the model. In summary, it has two phases of privacy leakage concerns:

- **Training Phase** Training a machine learning model encourages the machine to learn the explicit knowledge of the training dataset. With edge computing, there are two ways of training. The first one is the distributed machine learning. Multiple end devices or edge servers work together to train a model. In this scenario, privacy may leak when the centralized parameter server is malicious or some peer workers are curious. The second case is leveraging the training phase to the edge server by end device. Because some machine learning models are complicated such as deep neural networks, it is a common way to leverage part of the training to the edge server. There are lots of interactions between end devices and edge server such as parameter transmission and weight update, which can also leak private information.
- **Inference Phase** After the machine learning model is trained, the application is to use the trained model to do inference. While the most of the models are deployed in the edge server, end user can communicate with the edge server to implement the inference phase. The input data are uploaded to the edge server

and the inference results will be returned to the end device. This can cause data or identity leakage.

Differential privacy is a very common technique to use in this area. One of the problems in differential privacy is that it will lose the accuracy of the query if we want to protect more privacy since we add noise to query results. However in machine learning training, this is not a very big concern because in machine learning and big data, the model training is usually based on numerous data and to learning the statistics regulation among them. It also usually needs a lot of epochs of training to get a good performance, so an individual noise is not a big deal and those noise is assumed in the statistical model itself. Since in differential privacy, the noise we added usually satisfies a distribution of mean value 0, after several epochs of training, the interference of the noise will be offset. Therefore, differential privacy is a good way to protect privacy in machine learning model training. There are several existing works in this area.

## Model Partition

In 2018, Yunlong et al. [54] proposed a model partition based privacy preserving model training algorithm in edge computing. Their algorithm mainly focuses on deep neural network model training.

In deep neural network, there are multiple layers in the neural network. Training a whole network on the end device is a cumbersome work and usually consumes a lot of time. A natural idea is to offload this work to edge server. To protect the privacy, they split the whole neural network into two parts. The first part contains only the first layer and the activation function. The other part contains all rest layers. To reasonably distribute computational resources and protect the privacy, they let the first part run on end devices and the second part run on the edge server. To protect the privacy, in the forward propagation process, Gaussian noise is added to activation results of the first part so that the adversary cannot refer the original input data by activation results. In the backward propagation, the backward loss change information is transmitted from the edge server to end devices to finish one iteration.

It is apparent that activation results that is transmitted from end devices to the edge server is $(\epsilon, \delta)$-differential private because of the Gaussian noise added to results. However, to ensure the privacy, it is also important to ensure that when all activation results pass through the second part of neural networks on the edge server, the output is still private. We can take the output as a combination of input activation results. From properties of differential privacy, since the total loss of network prediction can be seen as a composed mechanism of multiple differential private mechanisms, the final output also satisfies differential privacy. This means that this model partition way can preserve the privacy in neural network model training on edge computing.

## Output Perturbation and Objective Perturbation

Output perturbation and objective perturbation are two algorithms that was proposed by Miao et al. in 2018 [43]. Unlike the model partition, output perturbation and objective perturbation can not only be used in deep neural networks, but also extend to other machine learning techniques.

The difference between those two algorithms is that output perturbation is adding Laplace noise to the output prediction to preserve the privacy of the model prediction and objective perturbation is adding Laplace noise to the objective function or loss function to preserve the privacy of the model training.

Assuming the modified loss function is $K(u, D)$, we have:

$$K(u, D) = \frac{1}{n} \sum_i Loss(u(x_i), y_i) + \lambda Z(u) \tag{5}$$

where $u(\cdot)$ is the function of the prediction model, $x_i$, $y_i$ are the data and target in database $D$, $Loss(\cdot)$ is a loss function and $Z(\cdot)$ represents the smoothness of the function. Then the prediction model $U$ is to minimize this modified loss function to get a good model.

$$U = \arg\min_u K(u, D) \tag{6}$$

Output perturbation algorithm is adding a random Laplace noise $q$ to the result of $U$:

$$U'(x) = U(x) + q \tag{7}$$

This can protect the privacy of prediction results.

On the other hand, objective perturbation algorithm adds Laplace noise to the modified loss function $K(u, D)$:

$$K'(u, D) = K(u, D) + \frac{1}{n} q^T u \tag{8}$$

By minimizing the $L_2$ normalization of $K'(u, D)$, we can get the prediction model $U$ by:

$$U = \arg\min_u K'(u, D) + \frac{1}{2} ||u||^2 \tag{9}$$

From Du et al. [43], the privacy analysis is given to ensure the $\epsilon$-differential privacy in those two algorithms. They can respectively protect the data privacy during the training and prediction processes.

## Separate Training

Separate training is another algorithm proposed by Mengmeng et al. to protect the data privacy [44]. It uses two edge servers and assumes that there are no collusion between both of them.

Within the separate training aggregation framework, two edge servers collect data from sensors in each area. Once a sensor collect some data, it will randomly split the collected data into two parts and randomly send each part with Laplace noise to an edge server. The process will continue until edge servers collect enough data. Both edge servers will aggregate the received data and train a machine learning model based on those data. With the assumption that there is no collusion between edge servers, each edge server cannot get or infer the completed data by itself. When executing queries, both of edge servers compute queries based on their models. To protect the privacy, Laplace noises are added to query results. They by aggregating prediction results from both edge servers, a privacy-preserving prediction result is obtained.

The authors of this paper proved that if each record is independent in this dataset, the aggregated result can also provide $\epsilon$-differential privacy. Meanwhile, from properties of differential privacy, we know that even we have a series of private mechanisms, the composition of them can still provide a $\epsilon$-differential privacy.

### 3.3.2 Preserving Privacy in Big Data

The emerging development of internet of things, massive data are produced and shared everyday. Big data is a science to study and analyze those large scales of data. Edge computing has accelerated the speed of collection data. In edge computing, end devices including various kinds of sensors are collecting numerous amount of data and transmit to edge servers everyday.

Among big data one important privacy issue is that how to provide privacy guarantee when publish or release those data. For example, when the data publisher releases various statistics of the data, the adversary can query those statistics to recover the original data, which will significantly disclose some important message or private information. So how to preserve the data privacy while publishing the big data statistics is a big problem recently. On the other hand, since the data are usually stored scatteredly on edge servers. Then how to prevent privacy breaches from honest but curious edge servers is a challenging issue. In the implementation of big data techniques, there are two phases that may leak the privacy:

- **Data Collection** Data collection and aggregation are the most intuitive way to leak the sensitive information. The first step of big data is to collect a large scale of data and aggregate them, among which a lot of data are sensitive. Meanwhile, the uncertainty of the credibility in edge servers increases the risk of privacy leakage. Once the sensitive data is taken advantage by malicious edge servers, the privacy issue can be really severe.

- **Data Mining** Data mining and data analysis are the core step of big data. Data mining uses some techniques such as statistics learning, machine learning to study the implicit information of the data. These information usually include the identity, location, preference, habit information of a single person or a group of people. While users usually leverage those data mining procedures to the edge server, privacy-preserving algorithms must be used here to protect against information leakage by untrusted edge servers.

Differential Privacy is still a common technique in the big data privacy area. Since for the large scale of data, one row of data is not that important. People are more willing to care about the approximate result. Therefore, adding noise to the data will not harm the total performance. In this section, we will summarize some recent privacy preserving algorithms in big data.

### Partitioned Histogram Data Publishing Algorithm

Partitioned histogram data publishing algorithm is proposed by Yi et al. in [45]. It is an algorithm to protect the data privacy when the edge server release the data. On one hand, they add noise to the data to keep data private between end devices and edge server against privacy leakage. On the other hand, during the data transmission, only the partition histogram of the data statistics will be transmitted to cancel the impact of the noise.

The algorithm contains four steps. First of all, we need to divide the original dataset into several histogram bins and add Laplace noise to each bin. Secondly, cluster partition operation is used to obtain new partition histogram. Thirdly, we are going to use this new partition histogram to build a wavelet tree using wavelet transform and add Laplace noise to the wavelet tree. Lastly, we can restore the histogram partition and publish the private histogram.

The privacy analysis is still based on the differential privacy. Since this algorithm has added the Laplace noise twice, authors proved that the partition histogram publishing algorithm based on wavelet transform satisfies $\epsilon$-differential privacy.

### Content-Based Publish–Subscribe Scheme

The content-based publish-subscribe scheme is proposed by Qixu et al. in [46]. It provides privacy protection to brokers in the publish-subscribe system. The publish-subscribe system is widely used in modern applications. It is a scheme to categorize the published messages and send categorized messages to subscribers by brokers. Edge computing provides a effective platform for this system. However, privacy issues are in the implementation of publish-subscribe scheme in edge computing because there may be unethical brokers or brokers are facing risk of hacking, sniffing and corrupting, which may cause private information leakage.

This algorithm includes three major steps. The publish-subscribe system firstly generated notification messages by using top-$K$ U-FIM algorithm based on user's dataset to min top-$K$ most frequent itemsets and adding exponential mechanism to results. Then based on the results, Laplace noise is added to the operated dataset. In the end, the broker uses attributes of top-$K$ most frequent itemsets to match corresponding events to publish.

Exponential and Laplace mechanisms are used here to ensure the $\epsilon$-differential privacy. So it is both suitable for numeric data and non-numeric data.

### 3.3.3 Other Topics

There are a lot of other topics in this area. In fact, since this is a very new research area, the topics are various.

- Smart home hub privacy [47]. They propose a smart home system called HomePad, which uses elements in a directed graphs to represent applications in this system and use module functions to isolate the usage of the data. It can achieve user defined privacy policy by modeling elements and the flow graph using Prolog rules.
- Differential privacy–based location privacy [48]. A differential privacy based framework is proposed by Qiucheng et al. to protect location privacy. They build a noise quadtree to map two-dimensional spatial data into an interval tree, and nodes in the tree correspond to a certain sub-area of the two-dimensional spatial data. Then they used Hilbert curve to reduce the retrieval computation cost.
- Online social multimedia big data retrieval privacy [49]. To support big data analytics while preserving privacy in edge computing, they can build multimedia content cluster tree from top to the bottom to handle the dynamically varying cached MC datasets and add noise to the cluster tree. In addition, they propose an evaluation method to measure the credibility of edge nodes.
- Smart city privacy [50]. The authors talk about modeling the privacy content among the big data including data graph, information graph and knowledge graph of smart city in DIKW (Data, Information and Knowledge) architecture. They categorize context graphs into target resources and add privacy guarantee to the conversion for IoT devices to process.
- Internet of connected vehicles privacy [51]. They design a V2V (Vehicle-to-Vehicle) communication-based route-obtaining algorithm to offload the computation to edge nodes. Then they propose NSGA-II (non-dominated sorting genetic algorithm II) to reduce the computational cost and preserve privacy of vehicles tracking, identity tampering and virtual vehicle hijacking.
- Federated learning privacy [52]. They adopt a blockchain to replace the centralized server in the classic federated learning system to reduce the privacy leakage risk. In addition, they also add differential privacy noise to extracted features and intermediate computational results in order to protect end user's privacy and enhance test accuracy.

All these topics are new and interesting. In reality, many work of privacy problems in edge computing are still at an early stage and have a lot of work to explore. We will talk about them in the next section.

# 4 Future Work

In the previous section, we mentioned that the privacy problems in edge computing are still in a very early stage. In fact, with the rapid development of edge computing and the gradual demand of privacy, more and more people have started to pay attention to the privacy issues. In addition to the topics we discussed in previous sections, we believe that there will be more interesting privacy problems in edge computing area. Here are some examples.

- Real time analysis. In edge computing, low latency brings a lot of opportunities on real time analysis. While sensors and end devices collect a lot of dynamic data and need to analyze them in real time, privacy guarantee should also be provided in a timely manner. Therefore, there must be efficient and effective privacy-preserving algorithms in real time.
- Privacy overhead. Nowadays, privacy-preserving algorithms are more focused on how to preserve privacy effectively. However, in order to obtain privacy, there are more computational overheads, which may introduce significant latency to the edge computing environment. Therefore, it is also important to preserve privacy while saving computational costs.
- Privacy accuracy balance. At present, many privacy-preserving algorithms, especially those that use differential privacy, will introduce noise into query results, which may lead to a decrease in accuracy. How to balance the privacy guarantee and the accuracy reduction will be an interesting problem.
- Smart vehicle. Self-driving vehicle is a trend of the future car. A lot of companies and research institutions are paying attention to it. Edge computing provides convenience on sensing data and vehicle communication. However, driving information is important and sensitive. How to provide the privacy for self-driving car will be a very popular privacy problem.

# 5 Summary

In this chapter, we have talked about some existing privacy-preserving algorithms and useful techniques in conventional data, location and identity privacy and new technologies machine learning and big data, but there are a lot of more interesting open privacy problems in edge computing. It is a growing demand for people to protect their private information in this age of information explosion. The privacy problem in edge computing must be paid more attention in the future.

# References

1. Satyanarayanan, M.: The emergence of edge computing. Computer **50**(1), 30–39 (2017)
2. Shi, W., et al.: Edge computing: vision and challenges. IEEE Internet of Things J. **3**(5), 637–646 (2016)
3. Yi, S., Li, C., Li, Q.: A survey of fog computing: concepts, applications and issues. In: Proceedings of the 2015 Workshop on Mobile Big Data (2015)
4. Yi, S., et al.: Fog computing: platform and applications. In: 2015 Third IEEE Workshop on Hot Topics in Web Systems and Technologies (HotWeb). IEEE, Piscataway (2015)
5. Zhang, J., et al.: Data security and privacy-preserving in edge computing paradigm: survey and open issues. IEEE Access **6**, 18209–18237 (2018)
6. Yi, S., Qin, Z., Li, Q.: Security and privacy issues of fog computing: a survey. In: International Conference on Wireless Algorithms, Systems, and Applications. Springer, Cham (2015)
7. Yu, J., et al.: Toward secure multikeyword top-k retrieval over encrypted cloud data. IEEE Trans. Dependable Secur. Comput. **10**(4), 239–250 (2013)
8. Liu, Q., Wang, G., Wu, J.: Secure and privacy preserving keyword searching for cloud storage services. J. Netw. Comput. Appl. **35**(3), 927–933 (2012)
9. Kuzu, M., Islam, M.S., Kantarcioglu, M.: Efficient similarity search over encrypted data. In: 2012 IEEE 28th International Conference on Data Engineering. IEEE, Piscataway (2012)
10. Cao, N., et al.: Privacy-preserving multi-keyword ranked search over encrypted cloud data. IEEE Trans. Parallel Distrib. Syst. **25**(1), 222–233 (2013)
11. Örencik, C., Savaş, E.: An efficient privacy-preserving multi-keyword search over encrypted cloud data with ranking. Distrib. Parallel Databases **32**(1), 119–160 (2014)
12. Wang, C., et al.: Enabling secure and efficient ranked keyword search over outsourced cloud data. IEEE Trans. Parallel Distrib. Syst. **23**(8), 1467–1479 (2011)
13. Li, J., et al.: Privacy-preserving data utilization in hybrid clouds. Future Gener. Comput. Syst. **30**, 98–106 (2014)
14. M. Bahrami, M. Singhal, A light-weight permutation based method for data privacy in mobile cloud computing. In: 2015 Third IEEE International Conference on Mobile Cloud Computing, Services, and Engineering, San Francisco, pp. 189–198 (2015)
15. Pasupuleti, S.K., Ramalingam, S., Buyya, R.: An efficient and secure privacy-preserving approach for outsourced data of resource constrained mobile devices in cloud computing. J. Netw. Comput. Appl. **64**, 12–22 (2016)
16. Menezes, A.J., et al.: Handbook of Applied Cryptography. CRC Press, Boca Raton (1996)
17. Gruteser, M., Grunwald, D.: Anonymous usage of location-based services through spatial and temporal cloaking. In: Proceedings of the First International Conference on Mobile Systems, Applications and Services (2003)
18. Mokbel, M.F., Chow, C.-Y., Aref, W.G.: The new casper: query processing for location services without compromising privacy. In: Proceedings of the 32nd International Conference on Very Large Data Bases (2006)
19. Chow, C.-Y., Mokbel, M.F., Aref, W.G.: Casper* Query processing for location services without compromising privacy. ACM Trans. Database Syst. **34**(4), 1–48 (2009)
20. Kido, H., Yanagisawa, Y., Satoh, T.: An anonymous communication technique using dummies for location-based services. In: ICPS'05. Proceedings. International Conference on Pervasive Services, 2005. IEEE, Piscataway (2005)
21. Lu, H., Jensen, C.S., Yiu, M.L.: PAD: privacy-area aware, dummy-based location privacy in mobile services. In: Proceedings of the Seventh ACM International Workshop on Data Engineering for Wireless and Mobile Access (2008)
22. Niu, B., et al.: Achieving k-anonymity in privacy-aware location-based services. In: IEEE INFOCOM 2014-IEEE Conference on Computer Communications. IEEE, Piscataway (2014)
23. Niu, B., Li, Q., Zhu, X., Li, H.: A fine-grained spatial cloaking scheme for privacy-aware users in location-based services. In: 2014 23rd International Conference on Computer Communication and Networks (ICCCN), Shanghai, pp. 1–8 (2014)

24. Pingley, A.: CAP: a context-aware privacy protection system for location-based services. In: 2009 29th IEEE International Conference on Distributed Computing Systems. IEEE, Piscataway (2008)

25. Manweiler, J., Scudellari, R., Cox, L.P.: SMILE: encounter-based trust for mobile social services. In: Proceedings of the 16th ACM Conference on Computer and Communications Security (2009)

26. Hu, H., Xu, J.: Non-exposure location anonymity. In: 2009 IEEE 25th International Conference on Data Engineering. IEEE, Piscataway (2009)

27. Amini, S., et al. Caché: caching location-enhanced content to improve user privacy. In: Proceedings of the Ninth International Conference on Mobile Systems, Applications, and Services (2011)

28. Zhu, X., Chi, H., Niu, B., Zhang, W., Li, Z., Li, H.: MobiCache: when k-anonymity meets cache. In: 2013 IEEE Global Communications Conference (GLOBECOM), Atlanta, pp. 820–825 (2013)

29. Khan, A.N., Mat Kiah, M.L., Madani, S.A., et al.: Enhanced dynamic credential generation scheme for protection of user identity in mobile-cloud computing. J. Supercomput. **66**, 1687–1706 (2013)

30. Park, I., Lee, Y., Jeong, J.: Improved identity management protocol for secure mobile cloud computing. In: 2013 46th Hawaii International Conference on System Sciences, Wailea, pp. 4958–4965 (2013)

31. Khalil, I., Khreishah, A., Azeem, M.: Consolidated Identity Management System for secure mobile cloud computing. Comput. Netw. **65**, 99–110 (2014)

32. Birrell, E., Schneider, F.B.: Federated identity management systems: a privacy-based characterization. IEEE Secur. Privacy **11**(5), 36–48 (2013)

33. Werner, J., Westphall, C.M., Westphall, C.B.: Cloud identity management: a survey on privacy strategies. Comput. Netw. **122**, 29–42 (2017). ISSN 1389-1286

34. Chen, M., Mao, S., Liu, Y.: Big data: a survey. Mobile Netw. Appl. **19**(2), 171–209 (2014)

35. McAfee, A., et al.: Big data: the management revolution. Harv. Bus. Rev. **90**(10), 60–68 (2012)

36. Alpaydin, E.: Introduction to Machine Learning. MIT Press, Cambridge (2020)

37. Murphy, K.P.: Machine Learning: A Probabilistic Perspective. MIT Press, Cambridge (2012)

38. Dwork, C.: Differential privacy: a survey of results. In: International Conference on Theory and Applications of Models of Computation. Springer, Berlin (2008)

39. Dwork, C., Roth, A.: The algorithmic foundations of differential privacy. Found. Trends Theor. Comput. Sci. **9**(3–4), 211–407 (2014)

40. Dwork, C., et al.: Calibrating Noise to Sensitivity in Private Data Analysis. In: Theory of Cryptography Conference. Springer, Berlin (2006)

41. McSherry, F., Talwar, K.: Mechanism design via differential privacy. In: 48th Annual IEEE Symposium on Foundations of Computer Science (FOCS'07). IEEE, Piscataway (2007)

42. Goodfellow, I., et al.: Generative adversarial nets. In: Advances in Neural Information Processing Systems (2014)

43. Du, M., et al.: Differential privacy preserving of training model in wireless big data with edge computing. IEEE Trans. Big Data **6**(2), 283–295 (2018)

44. Yang, M., et al.: Machine learning differential privacy with multifunctional aggregation in a fog computing architecture. IEEE Access **6**, 17119–17129 (2018)

45. Qiao, Y., et al.: An effective data privacy protection algorithm based on differential privacy in edge computing. IEEE Access **7**, 136203–136213 (2019)

46. Wang, Q., et al.: PCP: a privacy-preserving content-based publish–subscribe scheme with differential privacy in fog computing. IEEE Access **5**, 17962–17974 (2017)

47. Zavalyshyn, I., Duarte, N.O., Santos, N.: HomePad: a privacy-aware smart hub for home environments. In: 2018 IEEE/ACM Symposium on Edge Computing (SEC). IEEE, Piscataway (2018)

48. Miao, Q., Jing, W., Song, H.: Differential privacy–based location privacy enhancing in edge computing. Concurrency Comput. Pract. Experience **31**(8), e4735 (2019)

49. Zhou, P., et al.: Differentially-private and trustworthy online social multimedia big data retrieval in edge computing. IEEE Trans. Multimedia **21**(3), 539–554 (2018)
50. Duan, Y., et al.: Data privacy protection for edge computing of smart city in a DIKW architecture. Eng. Appl. Artif. Intell. **81**, 323–335 (2019)
51. Xu, X., et al.: An edge computing-enabled computation offloading method with privacy preservation for internet of connected vehicles. Future Gener. Comput. Syst. **96**, 89–100 (2019)
52. Zhao, Y., et al.: Mobile edge computing, blockchain and reputation-based crowdsourcing IoT federated learning: a secure, decentralized and privacy-preserving system. arXiv preprint arXiv:1906.10893 (2019)
53. Niu, B., et al.: Enhancing privacy through caching in location-based services. In: 2015 IEEE Conference on Computer Communications (INFOCOM). IEEE, Piscataway (2015)
54. Mao, Y., et al.: Learning from differentially private neural activations with edge computing. In: 2018 IEEE/ACM Symposium on Edge Computing (SEC). IEEE, Piscataway (2018)

# Privacy-Preserving Edge Video Analytics

Miao Hu, Yao Fu, and Di Wu

## 1 Introduction

Edge video analytics (EVA) shows great potential to be applied in artificial intelligence-driven system design, e.g., autonomous driving and smart city, and has been an area of intense research in the past few years. Given a video clip taken from cameras, edge video analytics is the process of extracting features from images/videos and applying these features to emerging applications, e.g., finding the criminal from the crowd. Apart from surveillance, it has applications in robotics, multimedia and forensics. The design and implementation of an efficient EVA model has always been an essential issue, which has received significant attention from both the computer vision research community and the computer system research community due to its wide applicability and utility.

In recent years, the privacy issues in edge video analytics aroused tremendous attentions. In summary, the privacy issues can be classified into the following categories: (1) *Privacy in video collection*. In most video analytics applications, the video collector (e.g., the camera) does not locally conduct the tasks due to resource limitation. Generally, the videos will be offloaded to other computation-powerful servers for model training and task execution. In the video collection and offloading process, video contents are inevitably at the risk of private information leakage [12, 28, 30, 33]. (2) *Privacy in video storage*. It is also a big concern to determine where should the collectors store their video contents with private information. Reliable data storage is important for video analysis tasks, especially

M. Hu · Y. Fu · D. Wu (✉)
The School of Data and Computer Science, Sun Yat-sen University, Guangzhou, China

Guangdong Key Laboratory of Big Data Analysis and Processing, Guangzhou, China
e-mail: humiao5@mail.sysu.edu.cn; fuyao7@mail2.sysu.edu.cn; wudi27@mail.sysu.edu.cn

© Springer Nature Switzerland AG 2021
W. Chang, J. Wu (eds.), *Fog/Edge Computing For Security, Privacy, and Applications*, Advances in Information Security 83,
https://doi.org/10.1007/978-3-030-57328-7_7

for a video retrieval application. Some studies (e.g., [7] and [28]) applied a trusted edge server performing video denaturing and providing data storage. (3) *Privacy in video analytics.* The privacy issues in video analytics-related applications attract more and more attentions, such as [14, 15, 18]. The video analytics applications are generally built on neural network models. Intuitively, some works persist that the privacy issues are solved as only network parameters are shared. However, network parameters can be leveraged to reveal private information of video contents. The privacy issues still cannot be ignored.

Despite the privacy issues mentioned above, the design of a privacy-preserving edge video analytics system is especially challenging: *First*, the goals of improving EVA model accuracy and protecting client privacy are somewhat conflicted with each other. Commonly, the model performance relies on the accuracy of model parameters trained on each dataset. A higher EVA model accuracy is always at the cost of more disclosure of privacy on clients' video contents. *Second*, existing crypto-based privacy-preserving approaches (e.g., homomorphic encryption) are too compute-intensive, which are not practical to be implemented in the real systems [14, 15, 18]. A light-weight privacy-preserving EVA model with low complexity is more desirable. *Third*, most popular privacy-preserving schemes (e.g., differential privacy) usually face with serious performance degradation [9, 31]. Besides, it is also unclear how the EVA model accuracy varies under different privacy conditions. *Last but not the least*, the EVA model training condition might change with time. It is expected to design a flexible EVA model training framework that can be dynamically tuned to fit all scenes. To solve the privacy issues in EVA systems, researchers have made significant progress in the past years. However, the above four key challenges have not been well addressed.

In this chapter, our objective is to address the above mentioned challenges and achieve both high accuracy and privacy preservation simultaneously. To this purpose, we propose a light-weight federated learning framework for EVA model training, which is called *FedEVA*. The main idea of *FedEVA* is to perturb clients' private model parameters with publicly available model parameters using a local perturbation operation, while the model training algorithms can still be run over perturbed parameters. There is no need for the coordinator server to recover the private model parameters. Note that, compared with compute-intensive crypto operations, both perturbation and model aggregation in our *FedEVA* framework are linear operations with low time and space complexity. Thus, *FedEVA* is light-weight and highly efficient. In addition, our framework can ensure the same recommendation accuracy on the perturbed parameters as that on the original parameters. It is possible for adversaries to capture perturbed model parameters, but they cannot easily recover the private model parameters of clients due to the lack of perturbation key. Thus, *FedEVA* can effectively prevent the leakage of user privacy.

Overall, our main contributions in this chapter can be summarized as below:

- We propose a light-weight privacy-preserving EVA framework called *FedEVA*. By conducting perturbation over private EVA model parameters, our proposed framework can prevent the leakage of client personal information. We also verify that the model accuracy will not be affected by introducing *FedEVA* into the EVA model training systems.

- Different from previous crypto-based methods, *FedEVA* is light-weight and highly efficient. In the meanwhile, the EVA model training can still work well on the perturbed parameters. We can also dynamically change the weight parameters of the *FedEVA* framework to achieve different degree of privacy preservation, which improves the scalability and flexibility of a privacy-preserving EVA system.
- We evaluate our *FedEVA* framework through large-scale real-world datasets. The experiment results show that our *FedEVA* framework achieves a significant improvement in protecting client privacy compared to baselines, meanwhile the accuracy and efficiency of the EVA model can still be guaranteed.

The remainder of this chapter is organized as follows. Section 2 introduces the historical backgrounds for edge video analytics and standard privacy-preserving solutions. Section 3 presents the preliminary information. Section 4 proposed the system model, where the detailed algorithms are shown in Sect. 5. Section 6 evaluates the performance of the proposed algorithms. Section 7 concludes this chapter and outlooks future research directions.

## 2 Related Works

### 2.1 Edge Video Analytics

Utilization of edge computing in video analytics can help save cost, bandwidth and energy. To design an efficient edge computing system for real-time video stream analytics, characteristics of video contents should also be taken into consideration. Bilal and Erbad [3] highlighted potentials and prospects of edge computing for interactive media, and presented some preliminary works on how edge computing can be used to tackle video analytics challenges. Ran et al. [22] designed a framework tied together front-end devices with more powerful backend "helpers" to allow deep learning to be executed locally or remotely in the cloud/edge. They considered the complex interaction among model accuracy, video quality, battery constraints, network data usage, and network conditions to determine an optimal offloading strategy. Wang et al. [29] proposed an adaptive wireless video transcoding framework based on the edge computing paradigm by deploying edge transcoding servers close to base stations. It is essential to efficiently extract video characteristics, and then apply them into edge computing architecture design.

Some special characteristics from video content or edge servers can be further exploited to enhance the EVA system efficiency. Zhang et al. [34] presented Vigil, which contributed on frame selection to suppress redundancy. When multiple cameras capture different views of an object or person of interest to the user query,

Vigil uploads only the frames that best capture the scene. Chowdhery and Chiang [6] proposed a model predictive compression algorithm that uses predicted drone trajectory to select and transmit the most important image frames to the ground station to maximize the application utility while minimizing the network bandwidth consumption.

The model training is one of the most important issues for EVA system design, which has also attracted tremendous attentions. Liu et al. [20] proposed a general deep neural network (DNN) architecture, and used it in their edge-based video analytics system. They also implemented a buffer management scheme, which uses Nvidia CUDA mapped memory feature to simplify the memory movement between CPU and GPU. Tarasov and Savchenko [25] proposed to apply the multi-task cascaded convolutional networks to obtain facial regions on each video frame. Then image features are extracted from each located face using preliminary trained convolutional neural networks (CNNs). Yaseen et al. [32] focused on tuning hyper-parameters associated with the deep learning algorithm to construct the video analytics model. Uddin et al. [27] proposed SIAT, which provided basic distributed video processing APIs and distributed dynamic feature extraction APIs which extract prominent information from the video data. In this chapter, we focus on the privacy-preserving EVA model training framework design.

## 2.2 Privacy-Preserving Video Analytics

For video analytics applications, a large number of surveillance cameras installed at different places capture video data for further analytics. The privacy issues have been focused in the video analytics framework design, which allows users to have specific control over their sensitive information, preventing from being abused by third parties.

With the advance of computer version (CV) technologies, it is possible to achieve a more granular privacy protection in the initial stage of data collection. Wang et al. [28, 30] developed OpenFace, a mechanism for privacy-preserving data collection, which denatures captured video data based on user-defined privacy policy. By applying video denaturing technology, OpenFace can selectively blur faces that occur in video frames, greatly alleviating the privacy concern. Zarepour et al. [33] blurred or eliminated the sensitive subjects with image processing technologies. Jana et al. [12] proposed DARKLY, which is integrated with OpenCV and replace the raw input feature with opaque references, which cannot be directly dereferenced by an untrusted application. However, the characteristics of the pictures or videos may be changed and cannot be applied into the video analytics.

During the data analysis phase, sensitive information may need to be processed on untrusted platforms due to the limited computing power of edge nodes. During the analysing stage, crypto technologies can reduce unauthorized data access. Fully homomorphic encryption allows data analytics being performed on encrypted data directly, instead of applying an additional decryption operation on the video. Jiang et al. [14] used level homomorphic encryption and performed privacy preserving scale-invariant feature transform on encrypted images. To enable encryption operations to run on resource constrained nodes, [15] proposed TargetFinder, which applies homomorphic encryption to search for images that include the target on encrypted data. Besides, TargetFinder used optimization technologies to reduce computation overhead of cryptographic primitives on energy and computation-constrained edge devices. Besides, [18] proposed a novel privacy preserving computing framework, where the terminal devices perform the light-weight permutation-substitution encryption and edge nodes adopt the homomorphic encryption. The edge-assisted framework can greatly reduce the computational, communication and storage burden while ensuring data security.

Federated learning facilitates the collaborative training of models without the sharing of raw data. By averaging local gradient updates, federated averaging (FedAvg) proposed by McMahan et al. [21] performs well on federated learning with non-iid data. However, [2, 17] demonstrated that simply maintaining data locality during training processes does not provide sufficient privacy guarantees. Recently, some privacy-preserving frameworks were proposed to hide clients' contributions during training, balancing the trade-off between privacy loss and model performance. Bonawitz et al. [4] designed a communication-efficient, failure-robust protocol for secure aggregation of high-dimensional data.

To obscure an individual's identity, differential privacy (DP) adds mathematical noise to a small sample of the individual's usage pattern. For example, [10] proposed an algorithm for client sided differential privacy preserving federated optimization. Wu et al. [31] used noisy differentially-private gradients to minimize the fitness cost of the federated learning model using stochastic gradient descent. However, differential privacy might lead to slow convergence on model training, and possibly low accuracy given a large number of parties with relatively small amounts of data.

It is non-trivial to design a federated learning system capable of preventing interference over training on the distributed datasets while ensuring the resulting model also has acceptable predictive accuracy. Some federated learning approaches use secure multiparty computation (SMC) to accelerate the model convergence rate, however this might introduce a high burden on the communications between clients. Moreover, SMC is vulnerable to interference. Truex et al. [26] proposed a scalable approach to protect against interference threats, which combines differential privacy with secure multiparty computation. However, it is unclear how to realize the SMC operations without introducing high computation overhead. To the authors' best knowledge, it is still a great challenge to balance the tradeoff between the training performance and the privacy-preserving level. This chapter aims to propose a privacy-preserving federated learning system without affecting the performance of video analytics model.

# 3 Technical Preliminary

Before presenting the system model and the proposed privacy-preserving edge video analytics scheme, we first introduce some preliminary knowledge.

## 3.1 Federated Optimization

Federated learning, as a federated optimization paradigm, makes multiple devices to jointly learn a global objective model without sharing their local data. The training data is stored on the local clients or edge devices, which can prevent the leakage of user privacy. Similar to distributed optimization, federated learning allows clients or edge devices to compute local updates and a coordinator server makes use of the gradients sent by user devices to aggregate global model parameters.

Following the pioneer work proposed by McMahan et al. [21], we also assume a synchronous update scheme that proceeds in rounds of communication. There is a fixed set of $K$ clients with fixed datasets $\mathscr{D}_1, \mathscr{D}_2, \cdots, \mathscr{D}_K$ where $n_k = |\mathscr{D}_k|$. At the beginning of each round, a random fraction $C$ of clients is selected, and the server sends the current global algorithm state to each of these clients (e.g., the current model parameters). We only select a fraction of clients for efficiency. Each selected client then performs local computation based on the global state and its local dataset, and sends an update to the server. The server then applies these updates to its global state, and the process repeats (Fig. 1).

While we focus on non-convex neural network objectives, the algorithm we consider is applicable to any finite-sum objective of the form

$$\min_{\omega} \quad f(\omega), \tag{1}$$

where

$$f(\omega) = \frac{1}{n} \sum_{i=1}^{n} f_i(\omega). \tag{2}$$

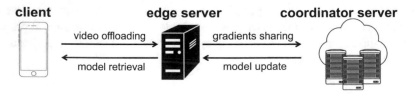

**Fig. 1** Federated learning driven edge video analytics model training framework

For a machine learning problem, we have $f_i(\omega) = l(x_i, y_i; \omega)$, where the loss of the prediction on example $(x_i, y_i)$ made with model parameters $\omega$. Thus, we can re-write the objective in Eq. (1) as

$$f(\omega) = \sum_{k=1}^{K} \frac{n_k}{n} F_k(\omega), \tag{3}$$

where

$$F_k(\omega) = \frac{1}{n_k} \sum_{i \in \mathscr{D}_k} f_i(\omega). \tag{4}$$

The recent video analytics applications of deep learning have almost exclusively relied on variants of stochastic gradient descent (SGD) for optimization. SGD can be applied naively to the federated optimization problem, where a single batch gradient calculation (say on a randomly selected client) is done per round of communication. This approach is computationally efficient, but requires very large numbers of rounds of training to produce good models.

## 3.2 Federated Averaging

Federated Averaging (FedAvg) is the most common algorithm for federated learning, by optimizing the local objective $F_k$ on client $k$ in each round $t$. $T$ is defined to indicate the number of total rounds of client-server communications and parameter updates to produce global model. In FedAvg, when each client locally executes the procedure of gradient descent on the current model using its local data, we have

$$\omega_{t+1}^k = \omega_t - \eta_t \nabla F_k(\omega_t, \xi_t^k), \tag{5}$$

where $\omega_t$ are the current model parameters, $\eta_t$ is the learning rate, $\xi_t^k$ and $\omega_{t+1}^k$ are local data used and new parameters, respectively. The global iteration step in the server is to take an average of the locally updated results and obtain a new global model, namely,

$$\omega_{t+1} = \sum_{k \in S_t} \frac{n_k}{n} \omega_{t+1}^k. \tag{6}$$

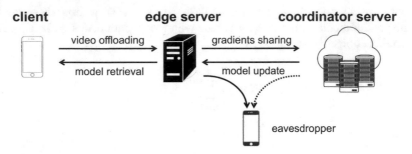

**Fig. 2** The general eavesdropper attack model

## 3.3 Attack Model

The attacker takes advantage of unsecured network communications to access data as it is being sent or received by its user. This chapter mainly focuses on the eavesdropping attack, also known as the sniffing or snooping attack. As shown in Fig. 2, an eavesdropper might monitor the information transmitted over a network by a computer, smartphone, or another connected device. Thus some key features of the local captured videos are at risk of information disclosure.

## 4 *FedEVA* Framework Overview

The focus of the *FedEVA* framework is to protect the private information (e.g., video analytics results) in clients' video contents from being leaked. We utilize a helper with publicly available training datasets to guarantee the privacy-preserving property while not introducing much communication or computation overhead. To the authors' best knowledge, this concept is first introduced in this work. We will present more design details in the following.

The *FedEVA* framework contains four key components, i.e., the client, the edge server, the helper and the coordinator server.

- **Client**. The client is responsible for collecting the video contents and offloading them to the edge server. As the video source, the client can be the camera, the cellphone, the laptop, and so on. Generally, the client side does not have much computation resource for model training.
- **Edge server**. The edge servers train the EVA model based on the video contents collected from its connected clients. Generally, the edge server is in the proximity of its connected clients. We assume that the video offloading process is executed on a dedicated communication channel, which cannot cause the leakage of information. This work focuses on the possible leakage of the learned EVA model parameters from the edge server to the coordinator server.

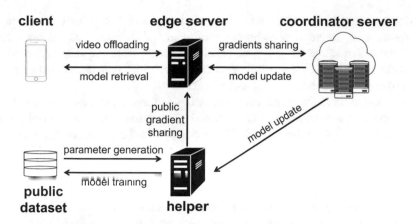

**Fig. 3** *FedEVA*: a federated learning driven edge video analytics framework

- **Helper**. The helper is associated with pre-downloaded video datasets. The implementation of the *helper* is one of the key contributions of the proposed *FedEVA* framework. Since the training results from the helper will be shared to other edge servers, the helper only runs on the public datasets to prevent the disclosure of private information.
- **Coordinator server**. The coordinator assists all the edge servers to collaboratively train models using *FedAvg* or *FedSGD* algorithm.

As shown in Fig. 3, the *FedEVA* framework contains data or information workflow between the modules. We introduce the key procedures as follows.

- *Video offloading*. Due to the limited computation resources, the client does not locally execute the model training function. Instead, each client periodly offloads its generated video clips regarded as the model inputs to the nearby edge server in connection.
- *Distributed model training*. The model training is conducted on the edge server with image/video inputs collected from clients. In this work, we do not propose a new model training algorithm, and any model training schemes can be applied to our *FedEVA* framework. In our *FedEVA* framework, the model training process can be divided into two categories.

  – *Model training on the private datasets*. The private clients are recruited to help enhance the performance of the trained EVA model. The incentive mechanisms are out of the scope of this work, which can be found in some other literatures, e.g., [24].
  – *Model training on the public datasets*. The helper proposed in our *FedEVA* framework is to help guarantee the privacy-preserving level of the analytics results from the private clients.

- *Public gradients sharing.* In the *FedEVA* framework, the privacy issues on the client sides are guaranteed by the setting of "public gradients sharing". Each client will receive the identical public gradients from the helper. By adding different weights to the public gradients, the actual gradient update based on the private video dataset can thus be guaranteed.
- *Model update.* In the coordinator server, the perturbed gradients will be summed to obtained a synchronized EVA model. The "model update" procedure will continue to update until the EVA model approaches a rather good level of performance.

The *FedEVA* framework can be integrated into any edge-based video analytics system. The model training framework can work well on the perturbed parameters without recovering the practical parameters from each local user. As modifications are mostly made at the user side, it is easy to rapidly deploy the *FedEVA* framework over the existing edge-based video analytics systems. One limitation is that the perturbation operation requires the use of model parameters trained on the public dataset, which may not be always available in certain scenarios. One feasible solution is to design incentive mechanism to promote some users denoting their datasets as the public ones.

## 5 *FedEVA* Algorithm Design and Analysis

This section will introduce the designed algorithm in our *FedEVA* framework for each module. As the client will not need to conduct a lot of operations, we will present the algorithms designed for the helper, the edge server, and the coordinator server, respectively.

### 5.1 Algorithm Design

#### 5.1.1 Algorithm 1: Helper Algorithm

The operation on the helper is similar to that in standard federated learning, as illustrated in Algorithm 1.

---

**Algorithm 1** Helper algorithm

---
1: **HelperUpdate**($\omega$): // run on the helper
2: $\mathscr{B} \leftarrow$ (split $\mathscr{D}'$ into batches of size $B$)
3: **for** batch $b \in \mathscr{B}$ **do**
4:    $\omega' \leftarrow \omega - \eta \nabla l(\omega; b)$
5: **end for**
6: return $\omega'$ to each edge server

---

### 5.1.2   Algorithm 2: Edge Server Algorithm

Different from the operation on the helper, the model parameter update will not be directly sent to the coordinator server due to the privacy concerns. Instead, the model update $\omega$ will be perturbed with the public update $\omega'$ returned from the helper. In detail, we will define a weight parameter $\alpha$ and perturb the local model update $\omega$ with a weighted public update $\omega'$, i.e.,

$$\omega = \omega + \alpha\omega', \tag{7}$$

where $\alpha \sim Lap(0, b)$ can be fit with the Laplace distribution following [8], and $b$ is the scale parameter. The variance of parameter $\alpha$ is $\sigma^2 = 2b^{-2}$. Note that the privacy level increases with the value of $\sigma$. However, the value of $\sigma$ cannot be increased without limit. A large value of $\sigma$ will make the coordinated model tend to that trained in the helper. Afterwards, the perturbed model update will be sent to the coordinator server.

---

**Algorithm 2** Edge server algorithm

---

1: **EdgeUpdate**$(k, \omega)$: // run on edge server $k$
2: $\mathcal{B} \leftarrow$ (split $\mathcal{D}_k$ into batches of size $B$)
3: **for** batch $b \in \mathcal{B}$ **do**
4:     $\omega \leftarrow \omega - \eta \nabla l(\omega; b)$
5: **end for**
6: randomly generate a weight $\alpha \sim Lap(0, b)$, and obtain $\omega = \omega + \alpha\omega'$
7: return $\omega$ to the coordinator server

---

### 5.1.3   Algorithm 3: Coordinator Server Algorithm

The global iteration step in the coordinator server is to take an average of the locally updated results and obtain a new global model. The model update is as below:

$$\omega_{t+1} = \sum_{k \in S_t} \frac{n_k}{n} \omega_{t+1}^k. \tag{8}$$

## 5.2   Analysis on Privacy Preservation

To quantify the level of privacy preservation provided by our *FedEVA* framework, we consider the existence of malicious attackers (e.g., eavesdroppers). According to

---

**Algorithm 3** Coordinator server algorithm

---

1: initialize $\omega_0$
2: **for** each round $t$ **do**
3:     $\mathscr{S}_t \leftarrow$ (random set of $m$ clients)
4:     **for** each client $k \in \mathscr{S}_t$ **in parallel do**
5:         $\omega_{t+1}^k \leftarrow$ **PrivateUpdate**$(k, \omega_t)$
6:     **end for**
7:     $\omega_{t+1} \leftarrow \sum_{k \in S_t} \dfrac{n_k}{n} \omega_{t+1}^k$
8: **end for**
9: return $\omega$ to the coordinator server

---

[5, 13], a privacy preservation metric, called *Privacy Preserving Distance (PPD)*, is defined as below.

**Definition 1 (Privacy Preserving Distance (PPD))** is defined as the difference between the real model parameters and the perturbed model parameters. That is,

$$\text{PPD} = \frac{\sum_{u=1}^{M} ||\mathbf{r}_u - \mathbf{E}_u||_2}{M}. \tag{9}$$

Specifically, a higher PPD is desirable for privacy protection, which can prevent the attackers from obtaining the original private model parameters.

Based on the distribution of PPDs, we further introduce another privacy preserving metric, called *Privacy Preserving Indicator (PPI)* as following.

**Definition 2 (Privacy Preserving Indicator (PPI))** For a *FedEVA* user $u$, let $\mathbf{r}_u$ denote its real model parameter vector and $\mathbf{E}_u$ denote the perturbed model parameter vector. Then, we have the privacy preservation indicator defined as:

$$I_p = \Pr \left\{ \frac{\sum_{u=1}^{M} ||\mathbf{r}_u - \mathbf{E}_u||_2}{M} \leq \epsilon_p \right\}, \tag{10}$$

where $\epsilon_p$ denotes a privacy preservation distance threshold.

Given a specific $\epsilon_p$, the value of $I_p$ indicates the per-rating distortion probability between the perturbed model parameter vector and the original model parameter vector. A lower $I_p$ indicates a better privacy preservation degree, and vice versa.

The private model parameters are perturbed by individual users before transmitting them to the coordinator server. Thus, only perturbed model parameters $\mathbf{E}_u$ can be captured by attackers. For a local *FedEVA* user $u$, the privacy preservation indicator can be rewritten as:

$$I_{p,u} = \Pr \left\{ ||\mathbf{r}_u - \check{\mathbf{r}}_u||_2 \leq \epsilon_p \right\}. \tag{11}$$

Note that the perturbed model parameter vector is not only related to real model parameters $\mathbf{r}_u$ but also related to the model parameters trained on the public dataset.

To prove the privacy level of our *FedEVA* framework, we compare it with the standard differential privacy (DP) scheme whose noise $w$ is drawn from Laplace distribution, i.e., $Lap(0, \lambda_{DP}) = \frac{1}{2\lambda_{DP}} \exp\left(-\frac{|x|}{\lambda_{DP}}\right)$, where $\lambda_{DP} = \frac{\Delta \mathbf{r}_u}{\epsilon}$ presented by Dwork and Roth [8]. We obtain the following proposition.

**Proposition 1** *For a local FedEVA user $u$, when the adding noise $\alpha$ follows the Laplace distribution $\alpha \sim Lap(0, \lambda_0)$, we can achieve*

$$I_{p,u}^{FedEVA} = I_{p,u}^{DP}, \quad \forall u \in \mathscr{U}, \tag{12}$$

*where $\lambda_0 = \frac{\lambda_{DP}}{\|\mathbf{r}'\|_2} = \frac{\Delta \mathbf{r}_u}{\epsilon \|\mathbf{r}'\|_2}$.*

**Proof** Starting from the Algorithm 3, we have

$$\|\mathbf{r}_u - \check{\mathbf{r}}_u\|_2 = |\alpha| \|\mathbf{r}'\|_2.$$

Therefore, we have

$$I_{p,u}^{FedEVA} = \Pr\{|\alpha| \leq \frac{\epsilon_p}{\|\mathbf{r}'\|_2}\} = 1 - \exp(-\frac{\epsilon_p}{\lambda_0 \|\mathbf{r}'\|_2}). \tag{13}$$

Similarly, for $\epsilon$-DP, we have

$$I_{p,u}^{DP} = \Pr\{|w| \leq \epsilon_p\} = 1 - \exp\left(-\frac{\epsilon_p}{\lambda_{DP}}\right). \tag{14}$$

Combining (13) and (14), we obtain

$$I_{p,u}^{FedEVA} = I_{p,u}^{DP}$$

$$\exp(-\frac{\epsilon_p}{\lambda_0 \|\mathbf{r}'\|_2}) = \exp\left(-\frac{\epsilon_p}{\lambda_{DP}}\right).$$

Hence to ensure same privacy level, we set $\lambda_0 = \frac{\lambda_{DP}}{\|\mathbf{r}'\|_2}$.

# 6 Performance Evaluation

In this section, we conduct field measurements to evaluate the performance of the proposed *FedEVA* scheme.

## 6.1 Experiment Settings

We reform the YOLOv3-tiny model [23] into the federated learning setting based on an open source implementation[1] and train the model on the COCO dataset [19]. To evaluate the performance of the *FedEVA* scheme, we compare it with other state-of-the-art algorithms.

- **Central YOLOv3-tiny algorithm**. YOLO is a real-time object detection algorithm that has been widely applied in video analytics.
- **Standard FedAvg algorithm**. *FegAvg* is a classical algorithm in FL which allows many clients to train a model collaboratively without sharing private data between clients or with the server, which can provide a certain level of privacy.
- **FedAvg+LDP algorithm**. Differential privacy (DP) describes the patterns of the dataset while withholding information about individuals in the dataset. Local DP (LDP) adds noise to each client's update before sharing with the server, and guarantees much stronger protection to clients' privacy.

We set the batch size as 8 and use multi-scale training. Following [16], we use Adam optimizer and gradient accumulations. In other words, the model is updated every two batches. The standard YOLOv3-tiny model is trained based on [23]. In the FL setting, we equally partition the COCO dataset into $N = 10$ parts and give each client one part. We carry out experiments on both IID and non-IID data, where the IID assumption is typically made by distributed optimization algorithms and the non-IID data is assumed by the federated optimization algorithms. On the IID case, we randomly shuffle the data before assigning to clients. On the non-IID case, we sort all data in the order of image type, which is accessible in the COCO dataset. After that, the data of the same type are aggregated together and each client will only have a few types of data.

In each global epoch, all clients participate the training process by receiving the model from the server and training it locally. We set local epoch $E = 1$. In *FedAvg*, all clients' updates are averaged with the weight of their data count. In *FedEVA*, one client is presumed as the public edge server; while other nine clients are private edge servers. The value of $\alpha$ in *FedEVA* is generated by Laplace distribution $Lap(0, b)$, where $b = 0.15$. In local DP setting, to guarantee the privacy, we clip the update with $C = 30$, which was introduced by Abadi et al. [1].

We achieve $(200, 10^{-9})$-DP using Laplace mechanism. Holohan et al. [11] proved if $b \geq \frac{C}{\epsilon - \log(1-\delta)}$, $Lap(0, b)$ satisfies $(\epsilon, \delta)$-DP. In our experiment, we have $b = \frac{30}{200 - \log(1 - 10^{-9})} \simeq 0.15$ so that we can compare *FedEVA* and *FL+LDP* fairly. The Laplace function we use is shown in Fig. 4. We generate $10^6$ random numbers with respect to $Lap(0, 0.15)$ and count the number of points falling in every interval.

The experiment results are compared with two performance metrics, including:

---

[1] https://github.com/eriklindernoren/PyTorch-YOLOv3.

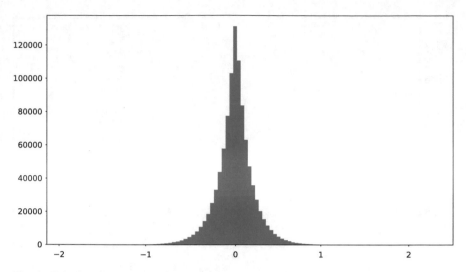

**Fig. 4** PDF of Laplace function $Lap(0, 0.15)$

- *Model accuracy.* We use the mean Average Precision (mAP) as the metric to evaluate the EVA model accuracy.
- *Privacy level.* We use the privacy preserving distance (PPD) in Proposition 1 as the metric for evaluating the privacy level. A larger PPD indicates a better privacy level, and vice versa. We assume the central training model is noise-free, i.e., the actual model.

We train YOLO model for 60 epochs and evaluate the model after each epoch.

## 6.2 Experiment Results

The experiments are conducted under both IID case and non-IID case.

### 6.2.1 IID Case

Figure 5 compares the mAP during training process with different FL models and central training. The $x$-axis represents the epoch number, while the $y$-axis represents the mAP of the aggregated model in each epoch. Note that the mAP metric of *FedAvg+LDP* scheme is 0 during 60 epochs, i.e., the *FedAvg+LDP* scheme diverges in the EVA model training. From this experiment, we can observe the mAP performance of all distributed versions of the YOLO algorithm is worse than that of the central version. We also find that the mAP metric of our *FedEVA* scheme is similar to that with the standard *FedAvg* scheme. This experiment indicates the

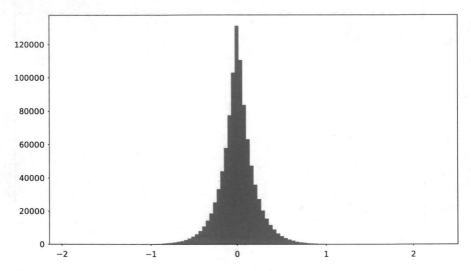

**Fig. 5** Comparison of mAP on the IID COCO dataset

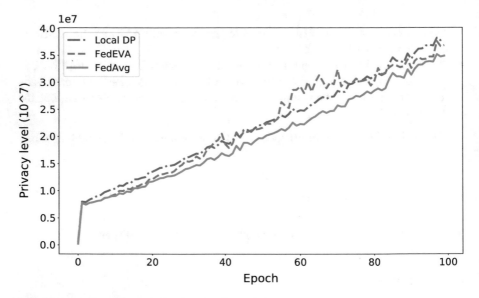

**Fig. 6** Comparison of privacy level on the IID COCO dataset

effectiveness of our *FedEVA* scheme, which greatly outperforms the *FedAvg+LDP* scheme under our experimental settings.

Figure 6 compares the privacy level of *FedAvg*, *FedEVA*, and *FedAvg+LDP*. The $x$-axis represents the epoch number, while the $y$-axis represents the privacy level. Specifically, the privacy level of the central version of YOLO is zero, i.e., the privacy issue is not considered in the standard version. By comparison, our

*FedEVA* scheme outperforms the standard *FedAvg* scheme, which implies that the design in this chapter can achieve a better privacy preservation degree. Although the *FedAvg+DLP* scheme guarantees a stronger privacy protection, the EVA model training performance under the *FedAvg+DLP* scheme is invalid.

### 6.2.2 Non-IID Case

Figure 7 compares the mAP during training process with different FL models and central training on the non-IID dataset. The $x$-axis represents the epoch number, while the $y$-axis represents the mAP of the aggregated model in each epoch. Note that the mAP metric of *FedAvg+LDP* is 0 during 60 epochs. From this experiment, we can observe the mAP performance of *FedEVA* is still similar to that of standard *FedAvg*, though both *FedEVA* and *FedAvg* perform worse on the non-IID dataset than that on the IID dataset. This experiment indicates our *FedEVA* scheme is as robust as *FedAvg* and outperforms *FedAvg+LDP* under the non-IID case.

Figure 8 compares the privacy level of *FedAvg*, *FedEVA*, and *FedAvg+LDP* versus the training epoch. The $x$-axis represents the epoch number, while the $y$-axis represents the privacy level. As the privacy issue is not considered in the standard version, the privacy level of the central version of YOLO is zero in default. Again, although the *FedAvg+DLP* scheme guarantees a stronger privacy protection, the performance of EVA model training under the *FedAvg+DLP* scheme is invalid. By comparison, the *FedEVA* scheme achieves a similar privacy level as that in the standard *FedAvg* scheme, however, our *FedEVA* scheme achieves much stronger protection on clients' privacy.

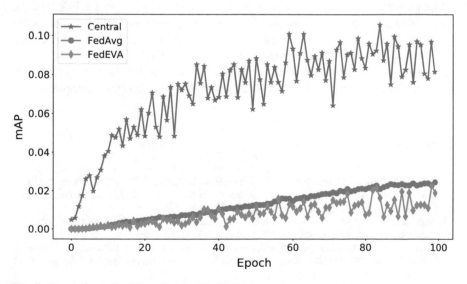

**Fig. 7** Comparison of mAP on the non-IID COCO dataset

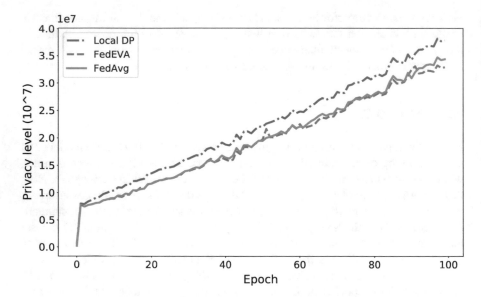

**Fig. 8** Comparison of privacy level on the non-IID COCO dataset

## 7    Conclusions

This chapter proposed a federated learning driven privacy-preserving edge video analytics model training system, named as *FedEVA*. By locally perturbing the private rating, the model parameters can still guarantee the model training accuracy, while preventing eavesdroppers from tapping clients' video contents information. The key idea lies in that the design of the perturbation method will not change the model training and updating structure. Besides, we also verify with experiments that our *FedEVA* framework outperforms the standard *FedAvg* scheme in the degree of privacy preservation. The *FedEVA* framework is practical and efficient as the perturbation operation is a linear operation with low time and space complexity.

## References

1. Abadi, M., Chu, A., Goodfellow, I., McMahan, H.B., Mironov, I., Talwar, K., Zhang, L.: Deep learning with differential privacy. In: Proceedings of the 2016 ACM SIGSAC Conference on Computer and Communications Security, pp. 308–318. (2016)
2. Bagdasaryan, E., Veit, A., Hua, Y., Estrin, D., Shmatikov, V.: How to backdoor federated learning. In: Cryptography and Security (2018). arXiv:1807.00459
3. Bilal, K., Erbad, A.: Edge computing for interactive media and video streaming. In: Proceedings of the International Conference on Fog and Mobile Edge Computing (FMEC), pp. 68–73 (2017)

4. Bonawitz, K., Ivanov, V., Kreuter, B., Marcedone, A., McMahan, H.B., Patel, S., Ramage, D., Segal, A., Seth, K.: Practical secure aggregation for privacy-preserving machine learning. In: Proceedings of the ACM SIGSAC Conference on Computer and Communications Security (CCS), pp. 1175–1191 (2017)
5. Bringer, J., Chabanne, H., Favre, M., Patey, A., Schneider, T., Zohner, M.: GSHADE: Faster privacy-preserving distance computation and biometric identification. In: Proceedings of the ACM Workshop on Information Hiding and Multimedia Security (IH&MMSec), pp. 187–198 (2014)
6. Chowdhery, A., Chiang, M.: Model predictive compression for drone video analytics. In: 2018 IEEE International Conference on Sensing, Communication and Networking (SECON Workshops), pp. 1–5. IEEE, Piscataway (2018)
7. Davies, N., Taft, N., Satyanarayanan, M., Clinch, S., Amos, B.: Privacy mediators: Helping IoT cross the chasm. In: Proceedings of the 17th International Workshop on Mobile Computing Systems and Applications, pp. 39–44. ACM, New York (2016)
8. Dwork, C., Roth, A.: The Algorithmic Foundations of Differential Privacy. Now Foundations and Trends, Hanover (2014)
9. Fan, W., He, J., Guo, M., Li, P., Han, Z., Wang, R.: Privacy preserving classification on local differential privacy in data centers. J. Parall. Distrib. Comput. **135**, 70–82 (2020)
10. Geyer, R.C., Klein, T., Nabi, M.: Differentially private federated learning: A client level perspective. In: Proceedings of NIPS Workshop: Machine Learning on the Phone and other Consumer Devices, pp. 1–7 (2017)
11. Holohan, N., Leith, D.J., Mason, O.: Differential privacy in metric spaces: numerical, categorical and functional data under the one roof. Inf. Sci. **305**, 256–268 (2015)
12. Jana, S., Narayanan, A., Shmatikov, V.: A scanner darkly: Protecting user privacy from perceptual applications. In: 2013 IEEE Symposium on Security and Privacy, pp. 349–363. IEEE, Piscataway (2013)
13. Järvinen, K., Kiss, Á., Schneider, T., Tkachenko, O., Yang, Z.: Faster privacy-preserving location proximity schemes. In: Cryptology and Network Security, pp. 3–22. Springer, Berlin (2018)
14. Jiang, L., Xu, C., Wang, X., Luo, B., Wang, H.: Secure outsourcing SIFT: efficient and privacy-preserving image feature extraction in the encrypted domain. IEEE Trans. Depend. Secure Comput. **17**(1), 179–193 (2020)
15. Khazbak, Y., Qiu, J., Tan, T., Cao, G.: Targetfinder: privacy preserving target search through IoT cameras. In: Proceedings of the International Conference on Internet of Things Design and Implementation, pp. 213–224. ACM, New York (2019)
16. Kingma, D.P., Ba, J.: Adam: A method for stochastic optimization (2014). 1412.6980
17. Li, T., Sahu, A.K., Talwalkar, A., Smith, V.: Federated learning: Challenges, methods, and future directions (2019). 1908.07873
18. Li, X., Li, J., Yiu, S., Gao, C., Xiong, J.: Privacy-preserving edge-assisted image retrieval and classification in IoT. Front. Comput. Sci. **13**(5), 1136–1147 (2019)
19. Lin, T.Y., Maire, M., Belongie, S., Hays, J., Perona, P., Ramanan, D., Dollár, P., Zitnick, C.L.: Microsoft coco: Common objects in context. In: European Conference on Computer Vision, pp. 740–755. Springer, Berlin (2014)
20. Liu, P., Qi, B., Banerjee, S.: Edgeeye: An edge service framework for real-time intelligent video analytics. In: Proceedings of the 1st International Workshop on Edge Systems, Analytics and Networking, pp. 1–6. ACM, New York (2018)
21. McMahan, B., Moore, E., Ramage, D., Hampson, S., y Arcas, B.A.: Communication-efficient learning of deep networks from decentralized data. In: Artificial Intelligence and Statistics, pp. 1273–1282 (2017)
22. Ran, X., Chen, H., Zhu, X., Liu, Z., Chen, J.: DeepDecision: A mobile deep learning framework for edge video analytics. In: Proceedings of the IEEE Conference on Computer Communications (INFOCOM), pp. 1–9 (2018)
23. Redmon, J., Farhadi, A.: Yolov3: An incremental improvement (2018). Preprint arXiv:180402767

24. Sarikaya, Y., Ercetin, O.: Motivating workers in federated learning: A stackelberg game perspective. IEEE Netw. Lett. **2**(1), 23–27 (2020)
25. Tarasov, A.V., Savchenko, A.V.: Emotion recognition of a group of people in video analytics using deep off-the-shelf image embeddings. In: International Conference on Analysis of Images, Social Networks and Texts, pp. 191–198. Springer, Berlin (2018)
26. Truex, S., Baracaldo, N., Anwar, A., Steinke, T., Ludwig, H., Zhang, R., Zhou, Y.: A hybrid approach to privacy-preserving federated learning. In: Proceedings of the 12th ACM Workshop on Artificial Intelligence and Security (AISec), pp. 1–11 (2019)
27. Uddin, M.A., Alam, A., Tu, N.A., Islam, M.S., Lee, Y.K.: Siat: a distributed video analytics framework for intelligent video surveillance. Symmetry **11**(7), 911 (2019)
28. Wang, J., Amos, B., Das, A., Pillai, P., Sadeh, N., Satyanarayanan, M.: A scalable and privacy-aware IoT service for live video analytics. In: Proceedings of the 8th ACM on Multimedia Systems Conference, pp. 38–49. ACM, New York (2017)
29. Wang, D., Peng, Y., Ma, X., Ding, W., Jiang, H., Chen, F., Liu, J.: Adaptive wireless video streaming based on edge computing: Opportunities and approaches. IEEE Trans. Serv. Comput. **12**, 1–12 (2018)
30. Wang, J., Amos, B., Das, A., Pillai, P., Sadeh, N., Satyanarayanan, M.: Enabling live video analytics with a scalable and privacy-aware framework. ACM Trans. Multimedia Comput. Commun. Appl. **14**(3), 64 (2018)
31. Wu, N., Farokhi, F., Smith, D., Kaafar, M.: The value of collaboration in convex machine learning with differential privacy. In: Proceedings of the IEEE Symposium on Security and Privacy (SP), Los Alamitos, pp. 485–498 (2020)
32. Yaseen, M.U., Anjum, A., Rana, O., Antonopoulos, N.: Deep learning hyper-parameter optimization for video analytics in clouds. IEEE Trans. Syst. Man Cyb. Syst. **49**(1), 253–264 (2018)
33. Zarepour, E., Hosseini, M., Kanhere, S.S., Sowmya, A.: A context-based privacy preserving framework for wearable visual lifeloggers. In: 2016 IEEE International Conference on Pervasive Computing and Communication Workshops (PerCom Workshops), pp. 1–4. IEEE, Piscataway (2016)
34. Zhang, T., Chowdhery, A., Bahl, P.V., Jamieson, K., Banerjee, S.: The design and implementation of a wireless video surveillance system. In: Proceedings of the 21st Annual International Conference on Mobile Computing and Networking, pp. 426–438. ACM, New York (2015)

# Part IV
# Architectural Design in Fog/Edge Computing

# Vulnerabilities in Fog/Edge Computing from Architectural Perspectives

Nhu-Ngoc Dao, Ngoc-Thanh Dinh, Quoc-Viet Pham, Trung V. Phan, Sungrae Cho, and Torsten Braun

## 1 Introduction

Nowadays, the commercial exploitation of the Internet of Things (IoT), a.k.a., IoTization, transforms traditional human life to smart life by involving a variety of daily objects and devices in Internet services. These *things* are typically equipped with sensors, independent processors, and memory as well as networking technology, such as Bluetooth, WiFi, cellular, and Ethernet connection. According to the Cisco report [3], 500 billion IoT devices are forecast to be in use by 2030, realizing the

N.-N. Dao (✉)
Department of Computer Science and Engineering, Sejong University, Seoul, South Korea
e-mail: nndao@sejong.ac.kr

N.-T. Dinh
School of Electrical and Telecommunication, Soongsil University, Seoul, South Korea
e-mail: thanhdcn@dcn.ssu.ac.kr

Q.-V. Pham
Research Institute of Computer, Information and Communication, Pusan National University, Busan, South Korea
e-mail: vietpq@pusan.ac.kr

T. V. Phan
Chair of Communication Networks, Technische Universität Chemnitz, Chemnitz, Germany
e-mail: trung.phan-van@etit.tu-chemnitz.de

S. Cho
School of Computer Science and Engineering, Chung-Ang University, Seoul, South Korea
e-mail: srcho@cau.ac.kr

T. Braun
Institute of Computer Science, University of Bern, Bern, Switzerland
e-mail: torsten.braun@inf.unibe.ch

© Springer Nature Switzerland AG 2021
W. Chang, J. Wu (eds.), *Fog/Edge Computing For Security, Privacy, and Applications*, Advances in Information Security 83,
https://doi.org/10.1007/978-3-030-57328-7_8

193

emerging trends of smartization, such as smart home, smart office, smart factory, and smart city. The connected things cover a broad range of digital devices, including home appliances, personal devices, industrial machines, and medical sensors. These IoT devices are characterized by different performances in terms of computational power, memory size, and battery capacity; they generate a variety of data [27]. Mostly, IoT data are not locally processed by IoT devices owing to resource limitations. These big IoT data are offloaded to networks [22, 27]. The heterogeneity and massiveness of the data are severe challenges for network infrastructure.

To cope with the aforementioned challenges, networks expand their computational capability from the core to the edge, resulting in a new fog/edge computing (FEC) system [9]. This computational cloudization in the whole network demonstrates a hierarchical architecture, where cloud computing is centralized on the root, fog computing is distributed in the middle, and edge computing is positioned closer to user devices [8]. In this model, service performance and latency are high at the cloud and lower from the fog to the edge. The hierarchical computing model allows networks to flexibly accommodate different users' requirements with appropriate resources on demand. For instance, real-time data offloaded from industrial machines in a smart factory are frequently prioritized to be processed by the FEC system with low latency. Meanwhile, video streamed from surveillance cameras should be offloaded to the cloud for heavy processing and huge storage [25, 31]. The supplementation of FEC has great significance in addressing issues caused by data heterogeneity and massiveness in the IoT paradigm [5].

Despite the advantages of computational flexibility and low latency, FEC systems has two major challenges: (1) power limitation owing to resource constraints and (2) security and privacy for personal user data protection. Specifically, in term of security and privacy, the supplementation of the FEC system in between the cloud and user devices makes it vulnerable against severe external attacks on north and south interfaces, as well as internally among FEC components. While the standardization of security-related functions and components in the FEC system is still in progress by the European Telecommunications Standardization Institute (ETSI) [11–13], external interfaces must deal with diverse protocols to interact with heterogeneous IoT devices and technologies.

Numerous recent researches have focused on mitigating these architectural security challenges [29, 33, 34]. State-of-the-art studies have mainly focused on protocol development, security model adaptation, and additional security component integration into the FEC system. Accordingly, they tackled security issues including authentication and access control, secure communications, trust and privacy preservation, eavesdropping prevention, attack countermeasures, and security service management. To provide a comprehensive overview of security issues in the hierarchical FEC architecture, this chapter considers FEC in multiple models, such as intrinsic architecture, the standard reference FEC architecture, FEC virtualization, and FEC integration into the 5G network. Unlike system and network architectures, these models adopt computation architectures that focus on describing and analyzing the involved computing components, their interfaces, and functional relationships.

The organization of this chapter is as follows:

- First, an introduction of computational cloudization is given with insights comparing FEC with other related platforms, such as ad hoc computing and cloud computing. Collaborations among these computing platforms in terms of security prevention are derived from the analysis as an open thought.
- Next, security and privacy issues in the intrinsic architecture of a fog/edge server are described. The analysis shows potential vulnerabilities in the main components of the fog/edge server, including buffer/memory, processor, task scheduler, input/output interfaces, and the operating platform. Consequently, recent collaborative approaches inside the fog/edge server are introduced, which have been proposed to address security issues, such as infrastructure virtualization and software containerization.
- Third, the standard reference architecture of the hierarchical FEC is analyzed to discover security and privacy issues with the involved components and their links from both operational and management/control domains. Based on this, adversary model analysis points out missing functions and procedures that may be exploited to attack the system. Finally, feasible approaches to fix these issues are recommended.
- Fourth, the FEC architecture is discussed from a network function virtualization (NFV) perspective. In this view, FEC, referred to as a virtualized network function, is controlled by the NFV management and orchestration. Hence, all common vulnerabilities of the NFV architecture have the same effect on FEC as well. This section also discusses possible integrations and implementations of security functions (as a network function) to protect the virtualized FEC system.
- Fifth, the deployment of FEC in a standard 5G mobile reference architecture is demonstrated. In this deployment, the FEC system is considered as a native function to provide computing services to mobile subscribers as well as network elements. Possible security and privacy issues related to interfaces and operational procedures are analyzed. Finally, corresponding recommendations for these issues are proposed.
- Finally, the conclusion of the chapter summarizes the security and privacy challenges in the FEC system from architectural perspectives. Future studies in this research domain are recommended.

# 2 Computational Cloudization

Computational cloudization defines the ability of networks to aggregate in-network computing resources on a pool level and flexibly allocate these resources on demand. Spreading throughout the whole network, cloudization provides computing, caching, and networking infrastructures at the core, distribution, and access tiers, i.e., cloud, fog, and edge computing, respectively [8, 29]. From a deployment perspective, a fog system located at a macro (primary) base station manages

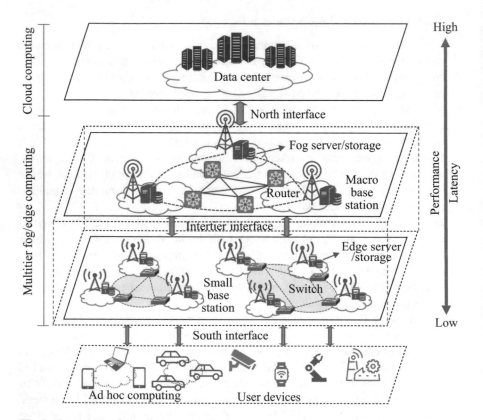

**Fig. 1** Computational cloudization overview

and connects a group of edge computing systems located at small (secondary) base stations in a local area. In this context, there exist high orchestration and harmonization between fog computing and edge computing systems to offer highly localized offloading services to the users in the local area [6, 7]. The collaboration between fog computing and edge computing systems is referred to as hierarchical FEC. Figure 1 illustrates the computational cloudization on a typical network architecture. It shows that the computing performance and latency decrease from the cloud to the edge.

On top of the cloudization model, the cloud computing offers users on-demand delivery of computing infrastructure remotely at the center of the network. This infrastructure provides a tailored amount of resources per user request in both time and space [35]. Typically, the cloud is featured by its advantages of super performance, large storage, high availability, and security; however, it faces high latency and bottle-neck bandwidth. Owing to these characteristics, cloud computing benefits offloaded services that require complex data processing and/or a large amount of storage without latency sensitivity, such as video encoding and storage, periodical data mining, ubiquitous information fusion, and machine-learning based analysis. In several contexts, the offloaded services for the cloud are the pre-processed output from FEC systems. The cloud and FEC systems interact with each other via the north interface.

In contrast, FEC systems, which consist of fog and edge computing plat-forms, offer hierarchical cloud-computing capabilities and IT infrastructure in local networks [19, 30]. This environment is characterized by low latency and high bandwidth in the proximity of user devices. In addition, the hybrid hierarchical architecture of FEC allows dynamic task sharing among fog/edge servers, enabling flexible resource allocation to accommodate various user requirements. However, because FEC systems are limited by resource constraints, they only provide users with medium/low performance and ephemeral storage [24]. Prime examples of FEC utilization include local weather forecasting, smart transportation, smart manufacturing, and healthcare services [2, 6, 25]. Because FEC is the frontier of the networks that user devices associate for offloading services, the heterogeneity of user connection technologies requires FEC to adapt to various protocols on the south interfaces.

Table 1 compares cloud and FEC systems. Based on the aforementioned analysis of computational cloudization, we observe the following security and privacy issues:

- New security and privacy issues are raised because of FEC expansion. To provide friendly services to heterogeneous user devices, FEC opens the south interface for various association protocols. This openness comes at the cost of authentication and access control vulnerabilities, which are proportional to the number of supported protocols. In particular, most user devices are lightweight IoTs that cannot adapt to robust security solutions.
- Moreover, as the frontier of in-network computing systems, FEC receives traffic directly from the user devices. Therefore, FEC suffers from denial-of-service (DoS) attacks first. This is more critical in the big IoT era, where IoT devices may be exploited by attackers to generate promiscuous flooding traffic in a distributed

**Table 1** Comparison between the cloud and FEC systems

| Characteristic | Cloud computing | Fog/edge computing |
|---|---|---|
| Deployment | Centralized at data center and core network | Distributed at local network |
| Computing capability | Super performance | Medium/low performance |
| Response latency | High | Low |
| Localization | No | Yes |
| Architectural model | Flat | Hierarchical |
| Standardization | Yes | Partial |
| Targeted services | Heavy computation | Real-time and location-aware computation |

manner. For instance, Mirai malware hijacked more than 400,000 IoT devices to generate approximately 1 Tbps of DoS traffic to an Internet host in 2016 [21].

- The communication between FEC and user devices mostly operates on open air interfaces in access networks. Therefore, secure (and possibly dedicated) channels are required to mitigate eavesdropping and jamming attacks. Spectrum efficiency and overhead should be stringently considered in any applicable solutions.
- To serve diverse user applications, security service management in FEC must ensure isolation, privacy, and anonymity of data processing and storage for tenant-users. A security-as-a-service (SaaS) platform should be deployed in FEC for this.
- On the north interface between the cloud and FEC, single points of failure and trust management are the main challenges. While the bottle-neck bandwidth is vulnerable against DoS attack, trust management is vitally important for service delegation and collaboration.

## 3 Vulnerabilities in Intrinsic FEC Architecture

A fog/edge server (FES) is a physical entity of the FEC system that performs the offloaded tasks arriving from user devices. A typical FES comprises hardware/middleware infrastructure, operating platform, and FEC services. Figure 2 depicts the intrinsic FEC architecture of an FES.

### 3.1 Hardware/Middleware Infrastructure

In an FES, the hardware/middleware infrastructure includes networking, computing, and storage resources as well as resource management. A queuing-theoretic model of the hardware/middleware infrastructure is described in Fig. 2 as follows.

- *Networking resource* represents input and output interfaces of the queue. These interfaces are identified by unique IP addresses, and each of them has a given maximum bandwidth for both user traffic arrival and departure. The output interfaces can either forward the processed user data to external parties (e.g., other FESs, the cloud, and user devices) or save the data to internal storage as a caching function. It is easily seen that the interfaces are vulnerable against two attacks. First, a DoS attack may be exploited to inject a large amount of bogus traffic to overwhelm the interfaces' bandwidth. As a result, authorized user data may not be delivered successfully. Second, a man-in-the-middle (MITM) attack an possibly spoof source addresses of the authorized user devices and destination addresses of the targeted external parties to eavesdrop the data before and after computation in front of the input and output interfaces, respectively.

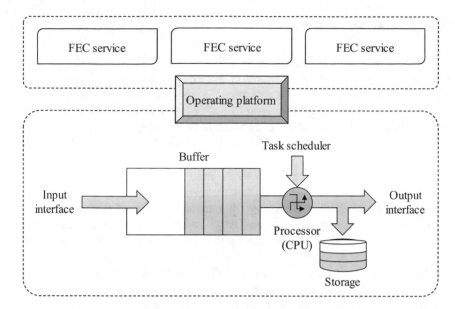

**Fig. 2** Intrinsic fog/edge computing architecture of a fog/edge server

- *Computing resource* is referred to as processor (e.g., CPU and GPU) frequencies. Tasks offloaded from different user services may have different computational complexities. In addition, the required amount of processor frequencies for task execution is usually proportional to the task complexities [7]. By exploiting this relationship, attackers can combine the IP spoofing and malfunctioned-task flooding techniques to send a massive number of bogus complex tasks aiming to exhaust the processor.
- *Storage resource* includes a queuing buffer for incoming data and a permanent storage for outgoing data. It is observed that the queuing buffer is vulnerable against DoS attack, as seen in the networking resource. Meanwhile, the permanent storage must deal with authentication and access control attacks from unauthorized users. Data privacy preservation and data loss prevention are the main responsibilities in a secure storage.
- *Resource management* covers the task scheduler and resource virtualization. The task scheduler controls task arrival and departure rates of the queuing system by adjusting the processor frequency. On the contrary, the resource virtualization function abstracts physical resource components to provide FEC services with resources on demands. Because the resource management entities are middleware, they can be hijacked by a backdoor attack and become malfunctioned.

## 3.2   Operating Platform

The operating platform running in an FES is a part of a comprehensive FEC framework that provides an environment to manage and control the operation of all FESs in the system. From a systematic perspective, an operating platform is considered as a contact agent for each FES. For internal collaboration in the FES, the operating platform handles interactions between FEC services and hardware infrastructure. The operating platform and FEC services can be either natively installed on the FES or deployed as docker containers.

Recently, several open source projects have launched for this purpose. For instance, the Linux foundation initiated the EdgeX Foundry and Akraino Edge Stack in 2017 and 2018, respectively. The integration between these projects aims at providing an orchestrated fog and edge computing service to multiple IoT applications in a practical manner. Lead by the OpenCORD projects under the Open Networking Foundation (ONF), the Central Office Rearchiteced as a Datacenter (CORD) framework combines the NFV, software defined networking (SDN), and elastic computing technologies to provide cloudization agility in the whole network. The reference implementation of CORD can be commercial off-the-shelf (COTS) servers, white-box networking switches/routers, and traffic aggregators. Other remarkable projects include Apache Edgent, StarlingX, and Eclipse Edge [32].

Because the operating platform is a software, most vulnerabilities are due to backdoor exploitation. Specific security issues depend on the selected platform software. As a contact point of FEC as well as an operation manager between FEC services and hardware infrastructure, the operating platform is considered an attractive victim for attackers. By exploiting the operating platform, attackers can perform numerous harmful actions to the system, such as modifying the FEC service policy, controlling the computing resources, stealing the user data, and destroying the computing services.

## 3.3   FEC Services

FEC services are applications dedicated to executing offloaded tasks from user devices. Depending on the class of the computing tasks, FEC services are different from each other in terms of software configuration, resource requirement, library dependency, and content restriction. For optimal deployment, FEC services typically adopt the containerization mechanism. If the FES is a dedicated hardware, each FEC service is within a container. Otherwise, if a common hardware is partly shared for the FES, the FEC operating platform and involved services are encapsulated in a container with its own running environment.

Facilitated by the advantages of the containerization mechanism, key security concerns for FEC services are related to trustworthiness and authentication/access control issues. Trust management is performed by a mutual attestation between the contained entity and the supported environment. Meanwhile, authentication and access control ensure secure software operation by using various mechanisms, such as identification (ID), hardware root of trust (HW-RoT), and Authentication/Authorization/Accounting (AAA).

## 4 Vulnerabilities in Hierarchical FEC Reference Architecture

The mission of FEC standardization has attracted considerable attention owing to the exponential growth of in-network computing demands. To supplement each other, individual organizations participating in developing the FEC reference architecture have their own focuses. For instance, the OpenFog consortium (recently merged with the Industrial Internet Consortium on January 2019) introduced the OpenFog reference model [26], which specifies major attributes that a system should embody to offer advanced features satisfying various user services. The proposed OpenFog attributes include security, scalability, openness, autonomy, reliability/availability/serviceability (RAS), agility, hierarchy, and programmability. The OpenFog model has been released in the technical document OPFRA001.020817 and approved in the Institute of Electrical and Electronics Engineers (IEEE) 1934–2018 standard [16]. Moreover, the Telecommunication Standardization Sector of the International Telecommunication Union (ITU-T) has issued Recommendation ITU-T Q.5001 [17], which specifies user cases, signalling requirements, and interaction procedures among computing components in an intelligent edge computing framework. The application of this recommendation ensures interoperability in FEC system. From a computational perspective, ETSI proposed a multi-access edge computing (MEC) framework and reference architecture in standard ETSI GS MEC 003 and related specification documents [11–14, 19]. The reference architecture defines functions and components of a hierarchical FEC framework as well as their interactive interfaces. Referring to the NFV models and mobile network architectures, the standards describe how the proposed FEC architecture integrates into these environments adaptively and efficiently. The interoperability of these integrations is validated through collaborations with related groups, including but not limited to, the Open Connectivity Foundation (OCF), the Open Network function virtualization (OpenNFV), and the third Generation Partnership Project (3GPP). The following sections consider security and privacy issues in a hierarchical FEC reference architecture adopting these aforementioned standards.

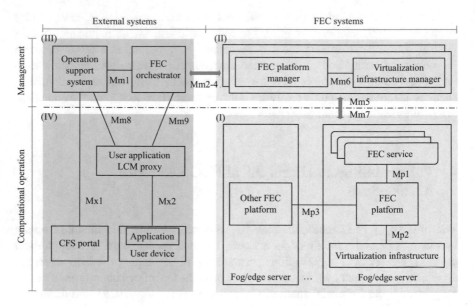

**Fig. 3** Fog/edge computing reference architecture

## 4.1 Reference Architecture

Figure 3 illustrates internal/external functional components and interfaces in a comprehensive hierarchical FEC reference architecture. The reference architecture is described in both computational operation and management planes on the basis of the ETSI MEC standards [13]. Typically, all functional components (a.k.a. FEC nodes—FECNs) in the FEC architecture can be divided into four functional blocks labelled as (I)–(IV). A combination of blocks (I) and (II) represents a single FEC system architecture (i.e., a cluster of FESs), and blocks (III) and (IV) are considered external systems. From a functional perspective, blocks (I) and (IV) are the task responsor and interrogator in the operational domain, respectively. Meanwhile, blocks (II) and (III) are low/high-level controllers in the management domain, respectively.

- *Block (I)* includes FESs and their peer interface Mp3. The Mp3 defines reference points between two FEC platforms for control communication. Optionally, control messages can be exchanged between FEC platforms belonging to different FEC systems to facilitate intertier/intercluster features of FEC system coordination, such as computing handover, service redundancy, and load balancing. Note that the intrinsic FEC architecture of an FES as well as its vulnerabilities were in Sect. 3.
- *Block (II)* contains FEC platform and virtualization infrastructure managers for each FEC system. The FEC platform manager is responsible for the life cycle of FEC services and FEC service rules and requirements, such as authorization,

traffic engineering, and service conflict handling. The FEC platform manager contacts multiple FEC platforms in a single FEC system via the Mm5 interface. In addition, the virtualization infrastructure manager configures and provisions the virtualization infrastructures in FESs to prepare sufficient resources to run FEC services on demand. Interaction between the FEC platform manager and the virtualization infrastructure manager is to exchange service requirements and virtualized resource states.

- *Block (III)* comprises a central FEC orchestrator for resource orchestration and FEC service harmonization in a whole hierarchical FEC system. The FEC orchestrator is connected to the operation support system (OSS) of the network to obtain service requests from user devices. The interfaces between components in blocks (II) and (III) are Mm2 (OSS–FEC platform managers), Mm3 (FEC orchestrator–FEC platform managers), and Mm4 (FEC orchestrator–virtualization infrastructure managers).
- *Block (IV)* includes frontend components, which gather computational requests from user devices. The requests are aggregated at the user application lifecycle management (LCM) proxy and customer facing service (CFS) portal, then delivered to the OSS and FEC orchestrator.

## 4.2   Adversary Models

Figure 4 illustrates an FEC security stack derived from the above description of the function and interaction relationships among components in the hierarchical FEC architecture [15, 26, 28]. The FEC security stack complies with the ITU-T X.800 recommendation and covers five viewing levels, including communication level security (communication view), application level security (software view), system software security (system view), FEC platform security (platform view), and node hardware security (node view). Because four lower security levels were discussed

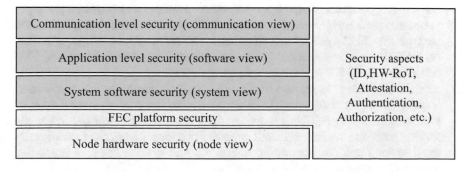

**Fig. 4** Fog/edge computing security stack

as vulnerabilities of intrinsic architecture for each individual FECN in Sect. 3, this section focuses on analyzing common adversary models at the communication level.

### 4.2.1 Eavesdropping

- *Objective:* Eavesdropping attacks aim at confidentiality violation by capturing information delivered via communication channels, such as internal links (Mm1, Mm6, and Mp3), intertier/intercluster links (Mm5 and Mm7), and external links (Mm2–4, Mm8, and Mm9).
- *Initial capabilities:* To execute the attacks, the adversary is assumed to feature the following capabilities. First, a system penetration is exploited to reach the reference points on communication channels. For this purpose, the adversary may scan addresses and/or identifications of the channels actively/passively. Second, the adversary must obtain knowledge about the configuration parameters of the transaction protocols used among FECNs for protocol cracking. Third, decryption capability is requested to decrypt the captured messages on the channels to reveal clear-text information. Eavesdropping activities can also be performed as a hidden data crawler after successfully taking control of the FECNs by other attacks, such as malware and hijacking.
- *Tools:* Sniffer, malware, social engineering, cryptographic attack, etc.
- *Attack process:* Typically, the attack process includes the following steps. First, the adversary must specify the point of sniffing to copy the transmission message. The sniffing points can be a switching node operating on the communication channel or act as a listener in a wireless environment. Once the sniffing point is attached to the channel, the next step is to crack the transmission protocol used among FECNs if possible. Otherwise, the transmission messages are copied and then decrypted offline using a powerful computer.

### 4.2.2 Denial of Service (DoS)

- *Objective:* DoS attacks target a serviceability violation, which causes resources to become unreachable/unavailable at the FEC systems. Two typical DoS attack strategies on FEC systems are volumetric and task-complex traffic injections (as well as their combination). In this circumstance, external interfaces, FEC orchestrator, and FEC services are attractive targets.
- *Initial capabilities:* To prepare for an effective DoS attack, the adversary must identify the attack destination. In other words, the input interface addresses of FEC systems and/or FECNs are needed for a volumetric traffic injection, while service ports and service authentication/authorization permits are additionally required for a task-complex traffic injection. In all cases, a chain of supportive tools may be used to boost the attack impact.
- *Tools:* Sniffer, malware, social engineering, botnets, IP spoofing, amplifiers, faked data generator, etc.

- *Attack process:* Initially, the attack destination identification can be achieved using several methods, such as sniffer and social engineering. Next, to generate a large malicious traffic targeting the destination, vulnerable user devices may be hijacked using a malware or software crack. Then, a grid of botnets, command and control infrastructure, and amplifiers is used to enforce the hijacked devices flooding bogus traffic to the destination. In addition, an IP spoofing technique can be applied to produce diverse source addresses. In particular, authorized hijacked devices may be recruited to create task-complex traffic to severely exhaust the resources at the destination. While a volumetric attack consumes network resources and session port pools, a task-complex attack causes the depletion of computing resource. Note that a hybrid attack may be utilized to achieve significantly stronger attack effects.

### 4.2.3 Man in the Middle

- *Objective:* MITM attacks cause data integrity and privacy violation where sensitive information is leaked and/or modified without legal authorization. As a result, service authentication violation and data confidentiality violation are additional effects.
- *Initial capabilities:* To be a "hidden man in the middle", the adversary secretly positions himself in the communication path between user devices and the FECNs. For this purpose, the adversary must impersonate or take control of one of the parties participating in the communication. In other words, at least the middle devices and/or the victims (i.e., user devices and the FECNs) must be identified. Moreover, the adversary must have knowledge about configuration parameters of the transaction protocols as well as authentication/authorization materials of the service session for a data modification attack.
- *Tools:* Sniffer, malware, social engineering, jammer, cryptographic attack, etc.
- *Attack process:* The adversary does the same actions to obtain the identification of the targeted devices and/or take control of them. Once adversary is in between the user devices and the FECNs, and depending on the role of the adversary in the communication, the adversary can have different effects on the data by using various attacks, such as message/transaction replay, spoofing, and traffic pattern collection.

## 4.3 Security Recommendations

To deal with the aforementioned threats, OpenFog recommends appropriate transaction and security protocols using the communication channels among FECNs. In particular, a client-server paradigm is applied for the transaction between FECNs in internal and interior communications. Meanwhile, the event-based publish-subscribe messaging patterns are implemented for information exchanges.

**Table 2** Secure FECN-to-FECN communications

| Application models | Transaction protocols | Security protocols |
|---|---|---|
| Client-Server | SOAP over HTTP | WSS, TLS/DTLS |
|  | RESTful HTTP/COAP |  |
| Publish-Subscribe | MQTT, AMQP, RTPS | TLS/DTLS |

Table 2 lists the protocol suites recommended in OpenFog OPFRA001.020817 [26]. Depending on the deployment infrastructure and technologies, the following security standards should be additionally utilized:

- 802.1AR—Secure device identity
- 802.1AE—Media access control (MAC) security
- 802.1X—Port-based (authenticated) media access control
- IPsec AH & ESP, Tunnel/Transport modes
- (D)TLS—(Datagram) Transport layer security

## 5   Vulnerabilities in FEC Virtualization

Softwarization has been considered as one of the key foundations of next-generation networks [8, 9], where SDN and NFV technologies are implemented to replace dedicated devices and network services with elastic virtualized resources and software solutions. In this context, the FEC virtualization architecture standardized by ETSI [11, 12] allows FEC services to be instantiated on NFV infrastructure. In this virtualized environment, ETSI NFV management and orchestration (MANO) components are utilized to mitigate a part of the FEC management and control operations. From the NFV architectural perspective, FECNs are considered as virtualized network functions (VNFs) of the networks. Figure 5 describes the FEC virtualization model. In vertical management, FEC systems are monitored by the OSS via the interface Mm1, as analyzed in Sect. 4. In horizontal management, the NFV MANO manages FECNs from a network virtualization perspective.

For virtualization management, NFV MANO architecture comprises three major components: virtualization infrastructure manager (VIM), VNF manager (VNFM), and NFV orchestrator (NFVO) [10]. The VIM performs resource management and allocation regarding the computing, storage, and network resources as well as monitors virtualization operations, such as capacity states, fault collection, and issue analysis. On the contrary, the VNFM is in charge of VNF lifecycle management such as service installation, update, query, and termination. A couple of VIMs and VNFMs are deployed for each VNF tier or cluster. In a hierarchical NFV system, a central NFVO exists to manage multiple couples of VNFM and VIM at NFV tiers and/or clusters as well as external interaction with other systems.

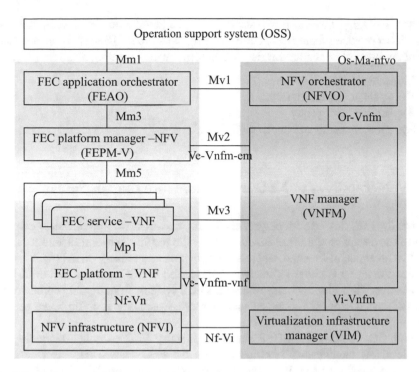

**Fig. 5** Fog/edge computing function virtualization

Matching the FEC components and services into the NFV architecture, the VIM is responsible for managing FES resources as an NFV infrastructure (NFVI) via interface Nf-Vi, the VNFM is responsible for managing FEC platforms, FEC services, and FEC platform manager (FEPM-V) as three types of VNF services via three corresponding interfaces: Ve-Vnfm-vnf, Mv3, and Ve-Vnfm-em/Mv2, respectively. At the center, the NFVO considers and manages the FEC orchestrator as a VNF application manager (referred to as FEAO) via interface Mv1. Note that in this integration model, the FEC system does not need its own virtualization infrastructure managers for FEC tiers and clusters. These works are covered by the VIMs of the NFV MANO.

Equipped with additional external interfaces, the FEC virtualization architecture exposes several vulnerabilities. Major threats are protocol attacks and NFV MANO hijacking. In particular, the protocols used on interfaces Mv1, Mv2/Ve-Vnfm-em, Mv3, Ve-Vnfm-vnf, and Nf-Vi are targets for sniffing, corrupting, and modifying through eavesdropping, jamming, MITM, and protocol cracking attacks. Meanwhile, any security exploitation in the NFV MANO may cause unexpectable effects on the FEC functions and operations via spurious management message exchanges.

Despite of these security issues, the NFV MANO offers undeniable advantages to FEC systems in terms of outside security support. Various comprehensive security solutions as VNF applications can be deployed to effectively protect the FEC system against emerging threats. For instance, a blockchain-based framework has been utilized to provide secure transactions for FEC operations [18, 23]. Typical defense solutions are recommended for the FEC system by the OpenFog standard [26]:

- Intrusion prevention systems (IPS)
- Deep packet inspection (DPI)
- Application layer proxy
- System monitoring and audit trail

# 6 Vulnerabilities of FEC Integration into the 5G Network

Currently, FEC has been deployed in existing networks (4G mobile networks and below) as an optional add-on function to handle traffic in several different modes, such as breakout, in-line, tap, and independent modes, as described in [14]. Via these modes, the traffic is either redirected or duplicated to FEC systems as a third-party application. These deployments restrict the FEC systems from flexibly collaborating with advanced network functions, such as service handover, mobility management, and unified threat management.

With 5G, because the 3GPP standards identify FEC as a key technology to enable a low latency and intelligent edge, the FEC system has been designed to be integrated into the 5G network as a native application function (AF) [19]. This integration allows the FEC system to exploit a wide range of services and information offered by other standard network functions. Figure 6 illustrates FEC integration into the 5G network.

To map the FEC system onto a network AF, two reference interfaces have to be considered: N6 and Naf. The N6 interface provides connection between

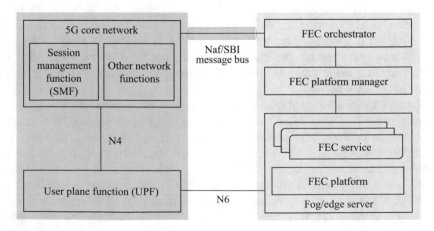

**Fig. 6** Fog/edge computing integration into the 5G network

**Table 3** Vulnerabilities in FEC via other 5G network elements

| Threat | Description |
| --- | --- |
| DoS attack | DoS attacks aim to interrupt operation of 5G network elements serving FEC systems, such as subscriber information management, mobility management, and handover management |
| Hijacking attack | The adversary aims to take control of network elements. Consequently, these elements are utilized to generate fake information and requests, modify legal information, and ignore message exchange with the FEC system |
| MITM attack | MITM attacks impersonate legal network elements to eavesdrop information exchanged with the FEC system. In high-level attacks, the attackers can modify the information |
| IP spoofing attack | The adversary generates and advertises spoofed addresses of authorized network elements by attacking directory service functions, e.g., domain name system (DNS) servers. This action leads to unstable communication between the FEC system and other network elements |
| Misconfiguration attack | The adversary prevents configuration transfer from the OSS to the FEC orchestrator, leading to malfunctioned FEC services. In high-level attacks, the configuration may be modified to accommodate other attacks |

an FES and a user plane function (UPF). In the 5G network, the UPF is a fundamental component that aggregates local traffic and performs data packet engineering, as controlled by the session management function (SMF). Via the N6 interface, the UPF encapsulates and exchanges user data to/from the FES. For management purposes, the FEC orchestrator collaborates with other 5G network functions through the common service based interface (SBI) message bus. The FEC orchestrator supports the Naf interface to provide standard application programming interfaces (APIs) for the others. Joining the SBI message bus demonstrates a high-level integration of the FEC into the 5G network.

Regarding security and privacy, for ease of imagination, FEC integration into the 5G core ecosystem can be thought to be similar to the case of a candidate country becoming a member state of the European Union. Definitely, a member state enjoys many advantages shared by the others in the union. However, the openness among member states also opens for possible threats from the remaining countries. Similarly, after integration, the FEC system obtains many advanced features supported by other network functions. However, the vulnerabilities in security and privacy of the FEC system increase along with risks from other network functions [1]. Table 3 summarizes the threats possibly affecting the FEC system via other 5G network elements.

As a countermeasure against the abovementioned attacks, a two-level protection should to be established. The first level of protection is provided by other 5G network elements and dedicated security service system for DoS prevention, traffic isolation, and configuration verification. The second level of protection is locally equipped at the FEC system that mainly aims at providing AAA security and privacy for identity verification and service access control [4, 20]. In addition, self management and orchestration should be configured to maintain FEC functionalities in cases of control missynchronization among network elements.

# 7 Summary

This chapter provided a comprehensive overview of vulnerabilities in FEC systems within multiple architectural levels, including intrinsic FEC, hierarchical FEC system, FEC virtualization, and FEC integration into the 5G networks. Adversary analysis of each FEC architecture has exposed particular security and privacy issues in the computing components and their interfaces. Based on the analysis, feasible countermeasure strategies were recommended. This chapter presents the readers with state-of-the-art knowledge to develop, modify, customize, and redesign secure FEC architectures adapting to various applied scenarios.

# References

1. Ahmad, I., Kumar, T., Liyanage, M., Okwuibe, J., Ylianttila, M., Gurtov, A.: Overview of 5G security challenges and solutions. IEEE Commun. Stand. Mag. 2(1), 36–43 (2018)
2. Al Hamid, H.A., Rahman, S.M.M., Hossain, M.S., Almogren, A., Alamri, A.: A security model for preserving the privacy of medical big data in a healthcare cloud using a fog computing facility with pairing-based cryptography. IEEE Access 5, 22313–22328 (2017)
3. Cisco Inc: Internet of things at a glance. https://www.cisco.com/c/dam/en/us/products/collateral/se/internet-of-things/at-a-glance-c45-731471.pdf (2020). Accessed 20 Feb 2020
4. Dao, N.N., Kim, J., Park, M., Cho, S.: Adaptive suspicious prevention for defending DoS attacks in SDN-based convergent networks. PLOS ONE 11(8), e0160375 (2016)
5. Dao, N.N., Lee, Y., Cho, S., Kim, E., Chung, K.S., Keum, C.: Multi-tier multi-access edge computing: The role for the fourth industrial revolution. In: Proceeding of the 2017 International Conference on Information and Communication Technology Convergence (ICTC). pp. 1280–1282. IEEE, Jeju (2017)
6. Dao, N.N., Vu, D.N., Lee, Y., Cho, S., Cho, C., Kim, H.: Pattern-identified online task scheduling in multitier edge computing for industrial IoT services. Mob. Inf. Syst. **2018**, 2101206 (2018)
7. Dao, N.N., Vu, D.N., Na, W., Kim, J., Cho, S.: SGCO: Stabilized green crosshaul orchestration for dense IoT offloading services. IEEE J. Sel. Areas Commun. 36(11), 2538–2548 (2018)
8. Dao, N.N., Na, W., Cho, S.: Mobile cloudization storytelling: Current issues from an optimization perspective. IEEE Internet Comput. 24(1), 39–47 (2020)
9. Dao, N.N., Tran, Q.D., Dinh, N.T., Cho, S., Braun, T.: Edge computing architectures. In: Taheri, J., Deng, S. (eds.) Edge Computing: Models, Technologies and Applications. The Institution of Engineering and Technology (IET), London (2020)
10. ETSI GS NFV 002 V1.2.1: Network functions virtualisation (NFV); Architectural framework. Technical Report, European Telecommunications Standards Institute, Sophia Antipolis CEDEX (2014)
11. ETSI GS NFV-MAN 001 V1.1.1: Network functions virtualisation (NFV); Management and orchestration. Technical Report, European Telecommunications Standards Institute, Sophia Antipolis CEDEX (2014)

12. ETSI GR MEC 017 V1.1.1: Multi-access edge computing (MEC); Deployment of mobile edge computing in an NFV environment. Technical Report, European Telecommunications Standards Institute, Sophia Antipolis CEDEX (2018)

13. ETSI GS MEC 003 V2.1.1: Multi-access edge computing (MEC); Framework and reference architecture. Technical Report, European Telecommunications Standards Institute, Sophia Antipolis CEDEX (2019)

14. Giust, F., et al.: ETSI White paper No. 28 - MEC deployments in 4G and evolution towards 5G. Technical Report, European Telecommunications Standards Institute, Sophia Antipolis CEDEX (2018)

15. Hao, Z., Novak, E., Yi, S., Li, Q.: Challenges and software architecture for fog computing. IEEE Internet Comput. 21(2), 44–53 (2017)

16. IEEE 1934-2018: IEEE standard for adoption of OpenFog reference architecture for fog computing. Technical Report, Institute of Electrical and Electronics Engineers (IEEE), Piscataway (2018)

17. ITU-T Q.5001: Signalling requirements and architecture of intelligent edge computing. Technical Report, International Telecommunication Union (ITU), Geneva (2018)

18. Kang, J., Yu, R., Huang, X., Wu, M., Maharjan, S., Xie, S., Zhang, Y.: Blockchain for secure and efficient data sharing in vehicular edge computing and networks. IEEE Internet Things J. 6(3), 4660–4670 (2018)

19. Kekki, S., et al.: ETSI White paper No. 28 - MEC in 5G networks. Technical Report, European Telecommunications Standards Institute, Sophia Antipolis CEDEX (2018)

20. Khan, R., Kumar, P., Jayakody, D.N.K., Liyanage, M.: A survey on security and privacy of 5G technologies: potential solutions, recent advancements and future directions. IEEE Commun. Surv. Tutor. 22, 196–248 (2020)

21. Kolias, C., Kambourakis, G., Stavrou, A., Voas, J.: DDoS in the IoT: Mirai and other botnets. IEEE Comput. 50(7), 80–84 (2017)

22. Li, S., Da Xu, L., Zhao, S.: 5G Internet of things: a survey. J. Ind. Inf. Integr. 10, 1–9 (2018)

23. Liu, H., Zhang, Y., Yang, T.: Blockchain-enabled security in electric vehicles cloud and edge computing. IEEE Netw. 32(3), 78–83 (2018)

24. Mai, L., Dao, N.N., Park, M.: Real-time task assignment approach leveraging reinforcement learning with evolution strategies for long-term latency minimization in fog computing. Sensors 18(9), 2830 (2018)

25. Na, W., Jang, S., Lee, Y., Park, L., Dao, N.N., Cho, S.: Frequency resource allocation and interference management in mobile edge computing for an Internet of things system. IEEE Internet Things J. 6(3), 4910–4920 (2019)

26. OPFRA001.020817: OpenFog reference architecture for fog computing. Technical Report, OpenFog Consortium, Fremont, California (2017)

27. Pham, Q.V., Fang, F., Ha, V.N., Piran, M.J., Le, M., Le, L.B., Hwang, W.J., Ding, Z.: A survey of multi-access edge computing in 5G and beyond: Fundamentals, technology integration, and state-of-the-art. IEEE Access. 8, 116974–117017 (2020)

28. Roman, R., Lopez, J., Mambo, M.: Mobile edge computing, fog et al.: a survey and analysis of security threats and challenges. Future Gener. Comp. Syst. 78, 680–698 (2018)

29. Sittón-Candanedo, I., Alonso, R.S., Corchado, J.M., Rodríguez-González, S., Casado-Vara, R.: A review of edge computing reference architectures and a new global edge proposal. Future Gener. Comp. Syst. 99, 278–294 (2019)

30. Taleb, T., Samdanis, K., Mada, B., Flinck, H., Dutta, S., Sabella, D.: On multi-access edge computing: a survey of the emerging 5G network edge cloud architecture and orchestration. IEEE Commun. Surv. Tutor. 19(3), 1657–1681 (2017)

31. Vu, D.N., Dao, N.N., Jang, Y., Na, W., Kwon, Y.B., Kang, H., Jung, J.J., Cho, S.: Joint energy and latency optimization for upstream IoT offloading services in fog radio access networks. Trans. Emerg. Telecommun. Technol. 30(4), e3497 (2019)

32. Yousefpour, A., Fung, C., Nguyen, T., Kadiyala, K., Jalali, F., Niakanlahiji, A., Kong, J., Jue, J.P.: All one needs to know about fog computing and related edge computing paradigms: a complete survey. J. Syst. Archit. 98, 289–330 (2019)

33. Zhang, J., Chen, B., Zhao, Y., Cheng, X., Hu, F.: Data security and privacy-preserving in edge computing paradigm: Survey and open issues. IEEE Access **6**, 18209–18237 (2018)
34. Zhang, P., Zhou, M., Fortino, G.: Security and trust issues in fog computing: a survey. Future Gener. Comput. Syst. **88**, 16–27 (2018)
35. Zhang, Q., Cheng, L., Boutaba, R.: Cloud computing: State-of-the-art and research challenges. J. Internet Serv. Appl. **1**(1), 7–18 (2010)

# Security and Intelligent Management for Fog/Edge Computing Resources

Jun Wu

## 1  Introduction

With the evolutionary growth of Internet of Things (IoT), it is estimated that almost 50 billion devices will be interconnected by 2020, and the generated data traffic will grow by another 1000 times. Sensing data from huge number of heterogeneous sensors will generate big data at the edge of IoT. With the emergence of diverse IoT applications (e.g., environment monitoring, e-health, industrial control), it becomes challenging for fog/edge computing to deal with these heterogeneous IoT environments with edge big data. Driving by this trend, fog/edge computing, an emerging computing paradigm, has received a tremendous amount of interest. By pushing data storage, computing, analysis and controls closer to the network edge, fog/edge computing has been widely recognized as a promising solution to meet the requirements of low latency, high scalability and energy efficiency, as well as mitigate the network traffic burdens.

Currently, there are four novel development trends for fog/edge computing. First, motivated by the success of artificial intelligence (AI) in a wide spectrum of fields, it is envisaged that AI powered fog/edge computing could enhance the intelligent processing and analysis capabilities at the edge of the networks. Based on the fog/edge computing, edge AI or edge intelligence, is beginning to receive a tremendous amount of interest. Second, as next generation networking technologies, software-defined networks (SDN) and information-centric networks (ICN) has been introduced into networked fog/edge computing. These deep integration technologies provide evolutionary networking approach for fog/edge computing, which can support the reconfigurable fog/edge computing architecture and contents process-

J. Wu (✉)
Institute of Cyber Science and Technology, Shanghai Jiao Tong University, Shanghai, China
e-mail: junwuhn@sjtu.edu.cn

© Springer Nature Switzerland AG 2021
W. Chang, J. Wu (eds.), *Fog/Edge Computing For Security, Privacy, and Applications*, Advances in Information Security 83,
https://doi.org/10.1007/978-3-030-57328-7_9

ing/analysis capabilities at the communication layer for 5G/6G. Third, optimized big data architecture is a must for edge big data analysis. Improved architectures of traditional big data, such as Hadoop and MapReduce, have attracted a lot of attentions.

Aforementioned novel development trends introduce a lot of benefits into fog/edge computing. However, the security and intelligent management problems of fog/edge computing resources are meanwhile introduced, which are still open issues. First, because fog/edge computing is usually deployed in large-scale IoT, it faces various threats from untrusted distributed geographic multi-sources and differentiated layer of the networks. Traditional security approaches cannot be used in fog/edge computing due to the limited computing and storage resources at edge. Second, content threats will be generated at the communication layer, because SDN/ICN technologies has been introduced into networked fog/edge computing nodes. Third, at the edge of the networks, there is unbalance between the users and providers of fog/edge computing resources. On-demand resource scheduling and balance are the must for fog/edge computing. Based on aforementioned motivations, this chapter aims to study the lightweight security and intelligent scheduling approaches for fog/edge computing resources. Novel technologies, such as blockchain, edge learning and semantic reasoning, will integrated seamlessly in the proposed architecture. To resolve aforementioned problems, this chapter studies the collaborative trust, content intrusion detection and security isolation, storage resource intelligent orchestration, smart resources partitioning technologies for fog/edge computing. This work is significant to promote the highly secure and efficient fog/edge computing for next generation networks.

## 2  Related Works

Currently, the security and smart scheduling of fog/edge computing resources have been attracted a lot of attentions. Basically, existing works focus on the security of communications, storage, data analysis of fog/edge computing. First, because context-aware capability is a novel and special feature of the fog/edge computing, content security protection should be provided at the communication layer. Second, big data at the edge raise up the requirements of efficient and dynamic edge storage resource scheduling. Third, the computing and processing resources configurations should be considerations. Related works are presented as follows.

Considering content awareness and edge distribution, edge/fog computing can provide benefits to defense in the performance of threat-aware filtering and semantic reasoning to construct edge defense isolation. In order to satisfy the secure automation control requirements of contents in IoT environments, exploiting edge/fog computing to achieve adaptive operations platform has been perceived as a promising approach, which enabled high manageability of IoT [1]. Moreover, content and data analysis technologies of fog/edge computing has been widely applied in next generation networks [2], security defense [3, 4], and optimizing computation resources of the application layer [5].

Edge storage related technologies have attracted a lot of attentions. Most of the recent research works about edge storage mainly focuses on the optimization and enrichment of the storage algorithms. In [6], Guanlin Wu et al. proposed a multiplier cooperative storage algorithm based on alternating directions. It minimizes the latency of task implementation and the total cost of the entire operation while maximizing node utilization of local information and system reliability. Besides, considering fairness metrics, the authors proposed an approximation algorithm to achieve caching load balance based on an integer linear programming problem in [7]. In [8], a three-layer architecture model for data storage management is proposed. It provides an adaptive algorithm that dynamically increases the high predictive precision required to provide efficient real-time decisions making and minimizes the amount of data stored in limited storage space. Cloud-edge collaborative work mode makes information interaction more convenient. Recently, most researches focus on resource scheduling and delay optimization. A vehicle control framework coordinated an Upper Edge Server is proposed in [9]. It enables more flexible scheduling of edge servers in view of the autonomous control of the vehicle. It also addresses the balance among the size of required edge servers, the capacity, and the ratio of dominated time. In [10], by integrating the advantages of edge platform and cloud platform, a new framework is proposed for the joint processing between the edge and cloud. It leverages the full network cognition and recorded information provided by the cloud, guiding edge computing components to meet the multiple performance demands of heterogeneous IoT networks. In addition, in [11], three scheduling algorithms (static, dynamic and batch synchronization) are able to solve problems when edge and cloud work collectively. The emergence of Hadoop allows data storage to be distributed, while HDFS, the bottom of Hadoop, is not suitable for use at the edge. Its framework is built on a specific collection of nodes. Specifically, the NameNode (only one) that provides metadata services and the DataNode that provides storage blocks are the units of the HDFS. However, HDFS has a drawback, which is a single point of failure, because there is only one NameNode. The work in [12] investigates the performance of the Hadoop benchmark suite, which runs on both physical and virtual infrastructure on the test platform for edge computing deployment. Moreover, most of the work focuses on analyzing and compressing data with MapReduce. Compared with cloud computing, edge computing lacks research on data storage architecture.

Many existing works focus on the processing and computing resources management of fog/edge computing. Different from the cloud-based system that aggregates all edge data into a remote data center, edge/fog computing provides a more efficient and scalable platform that enables context-awareness, low latency, energy efficiency, and big data analytics [13, 14]. Resource partitioning is a hot topic in wireless communication filed [15–17]. Existing studies always focused on the management of radio and frequency resources in femtocell and small cells. Due to the heterogeneity nature of wireless communication, the most popular approach for radio resource partitioning was frequency reuse. Singh and Andrews [18] provided a joint analytical framework for users offloading and resource partitioning in co-channel heterogeneous networks. Recently, another study [19] exploited the Stackelberg game model to optimize cooperatively the resource partitioning and data offloading in co-channel two-tier heterogeneous networks. However, different from resource partitioning in wireless communication, edge/fog computing pay more attention on recognizing which is the most popular delay sensitive services regardless of the data scale or user number in a domain so that the existing studies on resource partitioning in wireless communication cannot be applied in fog/edge computing systems. Moreover, the existing resources partitioning approaches designed for cloud-enabled IIoT also cannot be applied to fog-enabled IoT directly. For example, Mach and Becvar [20] formulated a load balancing problem between multiple fog servers as the cooperative resource sharing. However, the existing load balancing scheme required all data traffic to pass through an additional load balancer. To improve the efficiency of big data analysis, the literature [21] proposed a computation partitioning model for mobile cloud computing. However, this method only can improve the data processing efficiency in data center, but not adapt to fog computing paradigm due to the decentralization nature of fog computing [22]. Expect for the studies on flows shunting in IIoT, some early proposals in [23, 24] also tried to develop the autonomous resources allocation platforms for IoT to reduce the service response time under the fog environment. Recently, advocating the underlying edge/fog computing infrastructures to share their resources was also very insightful [25, 26]. However, it was not easy to observe the computing states of all heterogeneous edge devices in realtime [27].

## 3   Collaborative Trusted Edge/Fog Computing

A typical collaborative trusted service discovery system comprises three categories of entities: trust evidence providers, fog nodes, and trusted service discovery users, which are shown in Fig. 1. (1) Trust Evidence Providers are IoT devices that have cooperated with fog nodes. When they work with fog nodes, each IoT device will record trust evidence of fog nodes according to performance of fog nodes. The trust evidence can be recorded based on diverse trust properties in terms of Quality-of-Service (QoS) trust and social trust, where trust properties indicate the variables employed to measure the trustworthiness. By aggregating the trust evidence, one can obtain the trust values of fog nodes and block untrusted nodes. (2) Fog Nodes run a cross-blockchain structure consisting of multiple parallel blockchains. Each parallel blockchain stores encrypted data of fog nodes that serve in a specific application. The encrypted data includes the encrypted location information of fog nodes and the corresponding encrypted trust evidence collected by edge/fog devices. Fog nodes search for trusted fog nodes for users using encrypted data in the blockchain. After

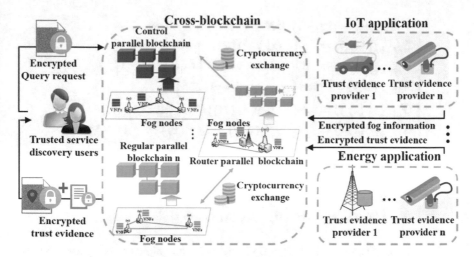

**Fig. 1** Collaborative trusted service discovery architecture for fog computing

that, fog nodes send back the encrypted trust evidence of fog nodes which are in the preset areas. (3) Trusted Service Discovery Users require to locate trusted fog nodes that can provide specific service in predefined areas. They send encrypted query request to fog nodes and ask them to send back encrypted trust evidence belonging to the fog nodes who can provide that service in the search areas. Then the user will purchase the decryption key from the trust evidence provider and calculate the trust values of these nodes.

## 3.1 Fog Nodes Information Encryption

Step. 1: The generation of $\{l[i]\}$: Set the initial values of the scaled Zhongtang chaotic system as $x[0]$, $y[0]$, and $z[0]$. Three pseudorandom sequences $\{x[i]\}$, $\{y[i]\}$, $\{z[i]\}$ are constructed by iterating the chaotic system $2 \times N^2 \times 3 \times r$ times, where $N^2$ and $r$ denote the image size, the frequency of color information encryption scheme, respectively. Then $l[i]$ equals to $\max(x[i], y[i], z[i])$, The element of pseudorandom sequence $\{L[i]\}$ is calculated as $\mathrm{mod}(floor(l[i] \times 10^{14}), 256)$.

Step. 2: The construction of $M_2$: Present fog nodes as icons on blank image $M_2$ using rendering rules. The rules are given as follows. The location of one icon on the image is determined by the coordinate information of one fog node. The color of the icon is determined by the service type of that fog node. Performing $r$ rounds of color information encryption operations on $M_1$ to obtain intermediate image $M_2$.

$$
\begin{cases}
f(x, y, k) = (y-1) \times N + (k-1) \times N^2 + x, \\
R'_{M_1,k-1}(x, y) = L[f(x, y, k)] \oplus R_{M_1,k-1}(x, y), \\
R''_{M_1,k-1}(x, y) = R'_{M_1,k-1}(x, y) + L[f(x, y, k) + 1], \\
R_{M_1,k}(x, y) = \mathrm{mod}\left(R''_{M_1,k-1}(x, y), 256\right), \\
G'_{M_1,k-1}(x, y) = L[f(x, y, k+1)] \oplus G_{M_1,k-1}(x, y), \\
G''_{M_1,k-1}(x, y) = G'_{M_1,k-1}(x, y) + L[f(x, y, k+1) + 1], \\
G_{M_1,k}(x, y) = \mathrm{mod}\left(G''_{M_1,k-1}(x, y), 256\right), \\
B'_{M_1,k-1}(x, y) = L[f(x, y, k+2)] \oplus B_{M_1,k-1}(x, y), \\
B''_{M_1,k-1}(x, y) = B'_{M_1,k-1}(x, y) + L[f(x, y, k+2) + 1], \\
B_{M_1,k}(x, y) = \mathrm{mod}\left(B''_{M_1,k-1}(x, y), 256\right),
\end{cases}
\tag{1}
$$

$R_{M1,k}(x, y)$, $G_{M1,k}(x, y)$, $B_{M1,k}(x, y)$ denote the ciphertexts generated by performing $k$ rounds of color information encryption operations on $R_{M1}(x, y)$, $G_{M1}(x, y)$, and $B_{M1}(x, y)$. $N^2$ is the total number of pixels in $M_1$.

Step. 3: The creation of $M_3$: Final encrypted image $M_3$ is gained by performing $t$ times of generalized Arnold transform on $M_2$. $m$ and $n$ are set as the control parameters of the generalized Arnold transform. Then the process of scrambling the coordinates of pixel $(x, y)$ with one round of generalized Arnold transform can be described as

$$
\begin{cases}
R_{M_3}(x, y) = R_{M_2}(\mathrm{mod}(x + n \times y, N), \mathrm{mod}(m \times x + (n \times m + 1) \times y, N)), \\
G_{M_3}(x, y) = G_{M_2}(\mathrm{mod}(x + n \times y, N), \mathrm{mod}(m \times x + (n \times m + 1) \times y, N)), \\
B_{M_3}(x, y) = B_{M_2}(\mathrm{mod}(x + n \times y, N), \mathrm{mod}(m \times x + (n \times m + 1) \times y, N)).
\end{cases}
\tag{2}
$$

And the final encrypted image $M_3$ is obtained by encrypting intermediate image $M_2$ with $t$ rounds of generalized Arnold transform. Moreover, the trust evidence of each fog node is also encrypted by a symmetric key algorithm with a specific encryption key. After that, $M_3$ is stored in the parallel blockchain along with the encrypted trust evidence.

## 3.2 Trusted Service Request Generation

Trusted service request generation includes following two steps. Step. 1: The construction of the trusted service request: In the trusted service request, the trusted service discovery user defines the service type of fog nodes and the area where it should provide service. Then the user renders the search area on an image $M_4$ using the rendering rules. Step. 2: The encryption of the trusted service request: Encrypt $M_4$ to obtain final encryption image $M_5$, where the encryption algorithm and key set are the same with that performed on $M_1$. Then the query request $M_5$ is sent to the fog node of the parallel blockchain.

## 3.3 Privacy-Preserving Range Query and Response

Step. 1: The construction of query criteria: Query criteria are used to judge whether $M_3$ meets the demand of the trusted service discovery user. Then the query criteria of pixel $M_5(x, y)$ can be calculated as

$$
\begin{cases}
U(x, y) = R_{M_3}(x, y) \oplus R_{M_5}(x, y), \\
V(x, y) = G_{M_3}(x, y) \oplus G_{M_5}(x, y), \\
W(x, y) = B_{M_3}(x, y) \oplus B_{M_5}(x, y).
\end{cases}
\tag{3}
$$

Step. 2: Results list $(RL)$ creation: For $M_5(x, y)$, the coordinates information and color information of the pixel will be stored into $RL$ if $U(x, y)$, $V(x, y)$, $W(x, y)$ are all zero. Then, the query response $RL$ and the corresponding encrypted trust evidence of the fog nodes are stored into the router parallel blockchain and sent to the corresponding trusted service discovery user in an off-chain manner by the fog node.

## 3.4 Trusted Evidence Aggregation

For $R_{M5}(x, y)$, $G_{M5}(x, y)$, $B_{M5}(x, y)$ in the results list, its original coordinate values $(X, Y)$ can be obtained by performing $t$ times of inverse generalized Arnold transform on $(x, y)$. After that, the color component values of the corresponding original pixel can be retrieved by decrypting $R_{M5}(x, y)$, $G_{M5}(x, y)$, $B_{M5}(x, y)$ with $r$ rounds of color information decryption operations. And the details on performing $k$ - th round of color information decryption operations on $M_5(x, y)$ can be expressed as in (4).

$$
\begin{cases}
f(X, Y, k) = N \times (Y - 1) + X + (k - 1) \times N^2, \\
DR_{M5,k-1}(x, y) = DR_{M5,k-1}(x, y) - L[f(X, Y, k) + 1], \\
DR''_{M5,k-1}(x, y) = \mathrm{mod}\left(DR'_{M5,k-1}(x, y), \ 256\right) \\
DR_{M5,k}(x, y) = L[f(X, Y, k)] \oplus DR''_{M5,k-1}(x, y), \\
DG'_{M5,k-1}(x, y) = DG_{M5,k-1}(x, y) - L[f(X, Y, k + 1) + 1], \\
DG'_{M5,k-1}(x, y) = \mathrm{mod}\left(DG'_{M5,k-1}(x, y), \ 256\right), \\
DG_{M5,k}(x, y) = L[f(X, Y, k + 1)] \oplus DG''_{M5,k-1}(x, y), \\
DB'_{M5,k-1}(x, y) = DB_{M5,k-1}(x, y) - L[f(X, Y, k + 2) + 1], \\
DB''_{M5,k-1}(x, y) = \mathrm{mod}\left(DB'_{M5,k-1}(x, y), \ 256\right), \\
DB_{M5,k}(x, y) = L[f(X, Y, k + 2)] \oplus DB''_{M5,k-1}(x, y).
\end{cases}
\tag{4}
$$

The location information of fog nodes that fulfill the requirement of the trusted service discovery user are stored in the decrypted *RL*. Then the user sends some cryptocurrency to the trust evidence provider to obtain the decryption key. The user can purchase encrypted trust evidence from other parallel blockchain using cryptocurrency exchange. The trusted service discovery user can evaluate trustworthiness of fogs by aggregating the obtained trust evidence.

The proposed Collaborative Trusted Service Discovery (CTSD) can evaluate the credibility of fog nodes by collaboratively aggregating trust evidence using cross-blockchain-enabled fog computing. A cross-blockchain structure is firstly proposed to ensure the encrypted location information and trust evidence of fog nodes can be propagated in a tamper-proofing and eavesdropping-resistance manner. And novel privacy-preserving range query based collaborative trust evidence aggregation is proposed to aggregate encrypted trust evidence using encrypted location information. The proposed CTSD improves the trustworthiness of fog computing.

# 4  Intrusion Detection and Security Isolation for Edge/Fog Computing

Currently, intrusion detection and security isolation are very important issues for edge/fog computing in content centric environments. Host Defense Fog Nodes (HDFNs) are constructed between host nodes and network nodes in edge/fog computing networks. Host defense is deployed logically between hosts and next hop of edge/fog computing nodes to prevent malicious data from entering contents. It is placed on fog nodes that achieve seamless coverage of host geographically. Each host is in a jurisdiction of one corresponding fog node. Moreover, due to the computation of fog nodes, the burden for hosts of configuring defense mechanism is sharply reduced.

## 4.1  Basic Idea of Edge Defense Mechanism with Content Semantic Awareness

The proposed edge defense for edge/fog computing to provide protection with semantic reasoning and smart content threat-aware. The proposed system utilizes fog computing for boundary isolation. Fog computing, offloading intelligence and recourses from cloud center to edge network, is introduced to provide edge computation and distribution required by the proposed defense mechanism. Fog computing provides context and content awareness for semantic analysis and customized configuration of intrusion detection and security isolation. Without being placed on routers or hosts, the requirements of infrastructure performance are greatly reduced. Moreover, we proposed a semantic reasoning approach based on Knowledge Graph (KG). It is designed to collect security knowledge and mine the illegal information relationships that may exist between the requested content and the blacklist. To protect against potential and ongoing attacks, our algorithm collects contextual traffic related to the pending packets, emphasizing the relevant semantic dimensions to guide reasoning. Potential threats of pending packets and their response data can be predicted by analyzing content attributes by interest names.

Smart reasoning algorithms with semantic knowledge to mine potential content threats are proposed with KG. Firstly, communication context was selected as weights based on semantics to guide the inference direction. Moreover, weighted semantic inference was designed to reason the threatening relations and knowledge with interest packets and then limit the content of response packets. Our proposed semantic inference mechanism can perceive penetrated and obfuscating content threats, and configured customized knowledge policies with distinct interests according to the security knowledge constructed from inference.

## 4.2 Architecture of Smart Reasoning Based Content Threat Fog-Defense

Host Defense Fog Nodes (HDFNs) are constructed between hosts and networks of. Host defense is deployed logically between hosts and next hop of edge/fog nodes to prevent malicious data from entering edge/fog systems. It is placed on fog nodes that achieve seamless coverage of host geographically. Each host is in a jurisdiction of one corresponding edge/fog computing node. Moreover, due to the computation of fog nodes, the burden for hosts of configuring defense mechanism is sharply reduced. The basic architecture is shown in Fig. 2.

The monitoring layer receives packets from the source and record communication histories of covered edge/fog nodes including location, terminals, activities and resources. Parsing and detecting packets including both requests and responses, and device attributes of each edge/fog computing node, the monitoring layer can help to perceive host behavior and contextual traffic.

The context analysis layer parses the pending packet and extracts necessary content. Content including packet names, publisher keys and excluding information for names of responses is extracted from interest packets. For data packets, components of name, signature, sighed information and content are extracted into the database. The extracted content is analyzed to compute the relevance of history communication to select the most related context packets.

The strategy layer implements semantic reasoning with Knowledge Graph on packet names and content to mine inherent threats. For interest packets, the policy layer firstly selects the relevant context traffic and calculate a correlation weight matrix to guide the direction of reasoning. It then mines implicit threatening entities and underlying threatening relations between content and blacklists. By combining

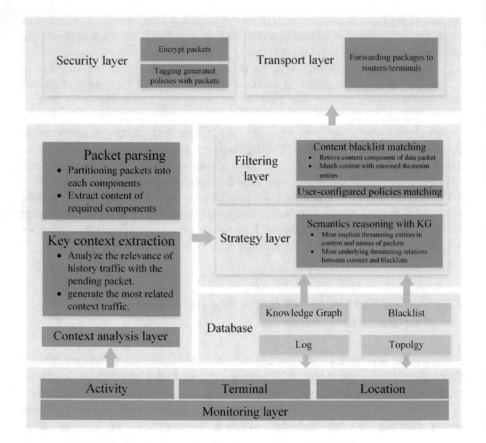

**Fig. 2** Structure design of proposed HDFNs

the interest packet components to limit producer identity and exclude package names, the user-configured strategy is generated and are called interest-configured strategy (ICSs).

In the filtering layer, blacklist-based content matching is performed to prevent exception requests. In addition, the data integrity and validity are checked to prevent requests and ICS from being tampered with by malicious nodes for interest packets. For received data packets, ICS matching and content filters, integrity and validity permission checking are implemented in an orderly fashion. In data packets, the whole content is filtered to prevent illegal and falsity content attacking whereas in interest packets the content names are filtered for malicious request detection.

In the security layer, in order to add up the blacklist with reasoned threats, the ICS tag is added by the tagging module on interest packets. Therefore, the extra edge/fog computing nodes save the time and energy of processing packets and acquire ICSs when caching the interest. The encryption module conducts encryption for a tagged interest packet. The transport layer then forwards new packets to the next hop.

## 4.3   Smart Semantic Reasoning for Defense Knowledge Policy

The PTransE model is applied to find the potential sensitive entities to composite a relation between requests and blacklists, which are likely to be the obfuscating objects in the responded data packets. One captures its meaning, such as head entity $e_h$, relation r and tail entity $e_t$, and the other are projection vectors $e_{hp}$, $e_{tp}$, constructing two mapping matrices, $\beta_{reh}$ and $\beta_{ret}$. PTransE takes multiple-step relation paths into consideration for representation learning, the score function of multiple steps is defined in.

$$F_r\left(O^\beta\right) = \frac{\sum_{p^m \in O^\beta} \Gamma(p^m|e_h,e_t)F_r(p^m)}{\sum_{p^m \in O^\beta(e_h,e_t)} \Gamma(p^m|e_h,e_t)} \tag{5}$$

During the training, we took KG triples as learning samples to optimize parameters in loss function as is proposed in. We construct false triples as negative samples by replacing random elements in KG triples. The loss function value continuously reduced through learning, and the entity vector and relation matrix can better reflect the semantic information of entities and relations. Relations with the similar semantics with context and requests were learnt as a result. We trained the model to optimize the loss metric thus obtaining multistep relations of entity pairs $O^\beta$ $(e_h, e_t) = \{p^m{}_1,\cdots p^m\alpha\}$, where each relation path is $p^m = (\tau_1,\cdots\tau_1)$.

To guide the relation path direction, we modified the resource function $\Gamma(p^m|e_h, e_t)$ which measures the resource flowed from $e_h$ to $e_t$ as the path reliability. When the middle entity $e_{i-1} \in W_e$, the resource allocated to the next entity is weakened by the relevance weight resulting in lower loss function in learning. And $\theta$ is defined as below.

$$\Gamma\left(p^m|e_{i-1},e_i\right) = \sum_{e_{i-1}\in E_{i-1}} \theta \frac{\Gamma(p^m|e_h,e_{i-1})}{|E_i|}$$

$$where\ \theta = \begin{cases} e^{-\tau_i^c}, if\ e_h\ (') \ or\ e_t\ \binom{'}{} \in W_c \\ 1, others \end{cases} \tag{6}$$

As a result, the blacklists are expanded with the reasoned middle entities for each request. A response will be blocked when the middle entities are detected in the packet content.

To build edge defense against potential content threats, we proposed a fog based content threat defense scheme with content-oriented semantic reasoning. The proposed mechanism realized edge defense against content threats by blocking illegal content and unexpected access. Smart reasoning models for semantics with context awareness were proposed to mine potential threatening knowledge from packet. The simulation results showed the proposed fog based ICN defense mechanism could provide valid and efficient isolation defense. This work is significant to improve ICN security.

# 5 Storage Resource Intelligent Orchestration for Edge/Fog Computing

## 5.1 Basic Idea of Proposed Edge Unified Big Data Intelligent Storage Architecture

The proposed architecture divides edge nodes into two types based on their functions, namely edge servers, and edge data prosumers. We call the edge nodes as edge data prosumers, which are the data producers and consumers simultaneously. While each edge server is composed of a Master and several data containers. The Master here is equivalent to an agent with control and management capabilities, and the data container performs the storage tasks assigned by the Master.

There are three main functional modules in the Master, synchronization communication management, dynamic storage, as well as multi-user data write and mapping. The most distinctive feature is storing data dynamically to meet the requirements of edge computing. It can use intelligent recommendation algorithms to decide the stored location of the data, by learning data checking and labeling. As for communication management, the Master manages the request and reception of data, which improves the liquidity of data between edge servers. We also set up a communication protocol pool to enhance the portability and scalability of the edge server. Similar to HDFS, the data mapping table can facilitate data lookup. The difference is that it supports arbitrarily modify and delete data, and a file can have multiple writers. While the data container reports the remaining storage space to the Master in time. It stores edge fragmented data and user private data, such as the ID number of a personal medical record in medical scenes, as well as public shared data that was previously stored in the cloud. Data popularity is the criterion for a data storage location and is, therefore, an important component of data tags.

The workflow of the proposed architecture is shown in Fig. 3, achieving cloud-edge collaborative mechanism. Edge data prosumers can do data pre-classification and preprocess for edge service. The initial popularity of the data is marked by them and initialized based on the data being called. They make the data more valuable and easy to analyze for some tasks, such as machine learning. Data uploading and downloading are the basic functions for the data prosumers. In this paper, we have enhanced the characteristics of the edge servers. Edge servers can store data sent by the cloud and the edge. The stored data is dynamic and circulated in the edge servers. This architecture pays more attention to the interactivity between edge servers. Besides, the cloud can audit messages. It meanwhile provides computing service and data storage. Both data sharing and message sharing happen in the cloud.

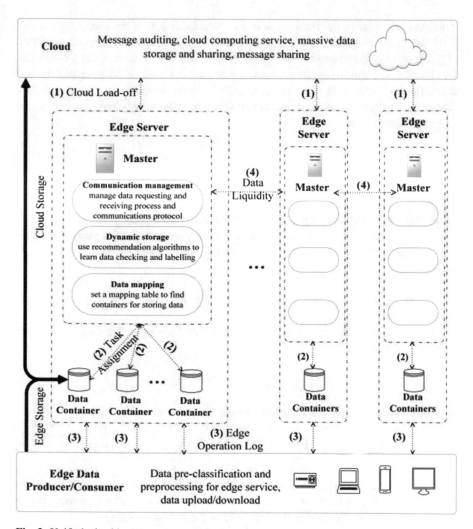

**Fig. 3** Unified edge big data storage architecture

## 5.2 *Machine Learning Based Dynamic Data Storage Strategy*

The dynamic storage area in Master ensures the real-time characteristics of edge computing. Firstly, the Master identifies the data popularity on the data tag, with popularity as an important indicator, and Q-learning to determine where the data is stored. This ensures data that is frequently used can be stored in the edge server, and data that is not used frequently and that is not user privacy can be stored in the cloud.

This process reduces the time for data recall and the cost of resources consumed. And the Master updates the data label according to scalable recommendation algorithms (e.g. knowledge-based recommendation, deep learning, etc.). Secondly, by checking the timestamp of the data, the Master determines whether expires and drops the expired data in time. Thirdly, when the data container reaches its capacity threshold, the Master cleans its redundant storage contents until there is extra space to assign tasks again.

## 5.3 Dynamic Storage Model

Let $a_e(t_e)$ denote that the storage action vector in slot $t_e$. $|a_e(t_e)|_{dji} = 1$ indicates that data $d_{ji}$ is stored in a container of Master $y_j$, and $|a_e(t_e)|_{dji} = 0$ otherwise. We update the popularity of the data depending on the received requests from the prosumers, defined as

$$DP_{d_{ji}}(t_e) = \alpha \cdot DP_{d_{ji}}(t_e - 1) + (1 - \alpha) \cdot N_{d_{ji}} / \frac{1}{D} \sum_{q=1}^{D} N_{q_{ji}} \qquad (7)$$

Having observed the prosumer requests at the end of slot $t_e$, our edge server state is expressed as

$$s_e(t_e) = \left[a_e^T(t_e), DP^T(t_e)\right]^T \qquad (8)$$

Storage performance can be estimated via the state value function

$$V_\pi(s_e(t_e)) = \lim_{T \to \infty} E\left[\sum_{\tau=t_e}^{T} \gamma^{\tau-T} C(s_e[\tau], \pi(s_e[\tau]))\right] \qquad (9)$$

which is the overall average cost generated by the infinite time range, with the future discount parameter $\gamma$ between 0 and 1. The discount factor $\gamma$ tunes balances current versus future costs. The best policy $\pi_*$ making the minimal cost is

$$\pi^* = argmin_{\pi \in \Pi} V_\pi(s_e), \forall s_e \in S$$

To give a clear overview of how Q-learning works, we define the state-action value function based on the policy $\pi$, namely. We use $\varepsilon_t -$ greedy algorithm to tend to a best policy. Algorithm 1 shows the dynamic storage mathematics model.

---

**Algorithm 1** Dynamic storage at Edge Server

---

1: Initialize state $s_e(0)$ randomly, $Q_0(s_e, a_e) = 0 \quad \forall s_e, a_e$
2: Initialize $\epsilon_t \in (0, 1)$, step size $\lambda$, $\lambda_e = 1 - \lambda$
3: **for** $t_e = 1, 2, \cdots$ **do**
4:     Take storage action $a_e(t_e)$ in a probabilistic manner
5:

$$a_e(t_e) = \begin{cases} \underset{a_e}{argmin} \, Q(s_e(t_e - 1), a_e) & w.p. \quad 1 - \epsilon_t \\ random \quad a_e \in A & w.p. \quad \epsilon_t \end{cases}$$

6:     $DPU(t_e)$ is updated based on prosumer requests
7:     Set $s_e(t_e) = [a_e(t_e)^T, DP^T(t_e)]^T$
8:     Calculate cost $C(s(t_e - 1), a_e(t_e)|DP(t_e))$
9:     Update $Q(s_e(t_e-1), a_e(t_e)) = \lambda_e Q(s_e(t_e-1), a_e(t_e))$
10:     $+\lambda[C(s_e(t_e-1), a_e(t_e)|DPU(t_e)) + \gamma \underset{a}{Q_{min}}(s_e(t_e), \alpha)]$
11: **end for**

---

# 6 Service Popularity-Based Smart Resources Partitioning for Edge/Fog Computing

## 6.1 Basic Idea

To resolve the unbalance problem between resources providers and consumers in the network edge, a smart resources partitioning scheme, SRPS, is proposed, which is shown in Fig. 4.

SRPS has three key components: (1) Global fog identifier (GFID): The SRPS exploits the GFID to name each of fog node. The separation between the GFID and the global service identifier (GSID) provides support for global observability. (2) SRPS controller: The SRPS controller is utilized to monitor and control the computing states of all fog nodes in F-edge/fog. The SRPS controller maintains the mapping from GFID to GSID. If a fog node moves from GSID to GSID' or the computing resource of fog node is exhausted, service providers can redirect the service requests to a new address to find computing resources without any service interrupts. (3) Computing task stream list (CTSL): To realize automatic resources partitioning, the CTSL is presented, which includes three basic tuples: *MatchField*, *ActionField*, and *Counter*. The parameters in each tuple can be pre-customized by system designers of F-edge/fog. Since the cloud usually aggregates large-scale computing, storage, and network resources, a SRPS controller can be implemented in cloud to monitor the global states of geo-distributed fog entities. And also, edge/fog users can optimize the resources allocation of each fog node by

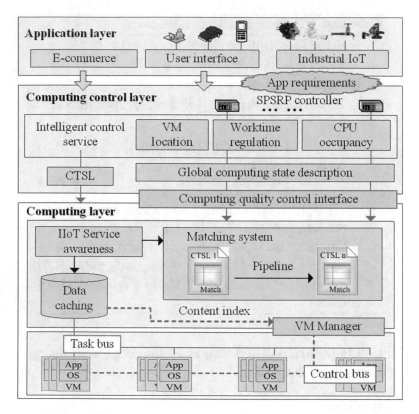

**Fig. 4** Architecture of SRPS

adding their own scheduling algorithms into SRPS controller. The states of each underlying infrastructure (e.g., GPS, camera, liquid meter, and the mometer) are identified by geo-distributed fog nodes. We emphasize that all of edge devices are equipped with SDN protocols. Especially, all of the computing tasks on different edge devices are labeled as record items and added into the defined CTSL.

## 6.2 Service Popularity Model

Consider there are many different types of edge/fog services to be processed in edge/fog. An edge/fog service is denoted as $E = \{\alpha_{type}, \beta_{task}, \gamma_{SLA}\}$, where $\alpha_{type}$, $\beta_{task}$, $\gamma_{SLA}$ denote application type, computing task and computing quality contract (CQC). It is common to see that a fog node simultaneously serves for multiple edge/fog service sessions. Similar to the content caching problem in edge/fog computing systems, the edge/fog service $E$ on the $i_{th}$ fog node is modeled through a generalized Zipf function.

$$Z_i^E(k_t) = \frac{\Omega}{k_t^\gamma}, k_t = 1, 2, \ldots, K,$$

$$k_{t+\Delta t} = Z_i^{E-1}\left(Z_i^E(k_t) + \lambda_{\Delta t}\right). \tag{10}$$

where $\Omega = \left(\sum\limits_{k=1}^{K} \frac{1}{k^\gamma}\right)^{-1}$ and $0 \le \gamma \le 1$ is the exponent and $k_t$ denotes the popularity ranking of edge/fog service $E$ on the $i_{th}$ fog node at time $t$. $\lambda$ is the number of arrival $E$ type of edge/fog services on fog node $i_{th}$ during $\Delta t$ spot. And also, the $Z_i^{E-1}(*)$ is the inverse function of $Z_i^E(*)$.

Originally, Zipf's law was found by observing and analyzing the word frequency distribution. About 20 years ago, the distribution of many Internet services was proven to follow Zipf's law and many existing web caching strategies used Zipf's law to model Internet users' service requests. Recently, popularity-based smart caching for information-centric networking (ICN) has utilized Zipf' law to model the content distribution. Now, Zipf's law is being applied in many fields such as linguistics, geography, economics, and broadcast TV. Similar to Internet services, the distribution of edge/fog services also follows Zipf's law. This paper exploits Zipf's law to predict the computing cost of edge/fog services by calculating their popularity rankings. Fog node gets popularity rankings of edge/fog services by analyzing the statistics of past and current logs in real time.

## 6.3 Computing Cost

To improve the resources utilization and computing quality, fog nodes are more willing to locally process popular edge/fog services and work with fewer remote-control operations (e.g., wake up, sleep, and migration).

For multiple types of edge/fog services at time $t$, the computing cost for one example edge/fog service on fog node is defined as the following function:

$$C_i^{Ej} = \frac{C_i^{E0}}{Z_i^{Ej}(k_t)} \tag{11}$$

Where $C_i^{E0}$ is the fixed original computing cost on fog node $j_{th}$ when $Z_i^{Ej}(k_t) = 1$.

By combining the Eqs. (1)–(3), the relationship between computing cost and service popularity is a convex function when $\gamma < 1$, while the relationship between computing cost and service popularity is a concave function when $\gamma > 1$. Moreover, for a fixed $\Delta R$, $\Delta C_2$ is larger than $\Delta C_1$ and $\Delta C_2'$ is smaller than $\Delta C_1'$. In the other word, for $\gamma < 1$, the change of service popularity when $k_t < 7$ has a greater impact on the computing cost than the change of service popularity when $k_t > 16$. For $\gamma > 1$, the change of service popularity has a greater impact on the computing cost when

$k_t > 13$ than when $k_t < 20$. In this paper, the SRPS shifts the less popular services on $i_{th}$ fog nodes into the other fog nodes to minimize their computing costs under $\gamma > 1$.

## 6.4   Popularity-Aware Computation Partitioning Algorithm

The working flow of algorithm 2 illustrated is described step by step as follows. The input parameters of algorithm 2 contain $\lambda_{\delta_t}$, $f_h$, $k_i$, $Th$, $R_{ij}$, $L_f^R$, $\gamma_{CQC}$. Therein, $\lambda_{\delta_t}$, $f_h$, $R_{ij}$, $L_f^R$ can be calculated by fog server based on the edge/fog service requests in a real system. $Th$ and $\gamma_{CQC}$ are two constants, which are configured by the edge/fog engineer according to the engineering experience in the applied edge/fog scenario. $k_i$ is a statistical variable that can be calculated. When the data flows of edge/fog services arrive at the fog node, the service type of these data flows will be identified and then the edge/fog service popularity rank on this fog node will be updated. If the rank of an arriving edge/fog service is less than $Th$, it will be pushed into the pending list. Otherwise, it will be pushed into the forwarding list ($FW_{List}$). For the edge/fog service on the pending list, fog node will calculate the computing cost of providing this edge/fog service and observe if the computing quality is in the scope of $\gamma_{CQC}$. The fog node will select a policy (it may be an identity of a virtual machine). For the edge/fog service on the $FW_{List}$, the fog node will send it to the SRPS controller for deeper analysis.

The SRPS scheme modeled the relationship between service popularity and computing cost with Zipf's law. Moreover, the SRPS scheme decoupled the computing control from data processing and support mobile and heterogeneous computing resource scheduling.

## 7   Analysis

In this section, we give the main contributions and cost analysis of the proposed approach for security and intelligent management for fog/edge computing resources.

## 7.1   Main Contributions

The contributions of aforementioned approaches are as follows:

Firstly, collaborative trust and security protection scheme was proposed for edge/fog computing systems. The trust evidence can be recorded based on diverse trust properties in terms of Quality-of-Service (QoS) trust and social trust, where trust properties indicate the variables employed to measure the trustworthiness. Fog

---

**Algorithm 2:** Popularity-aware computation partitioning algorithm intra fog node.

---

**Input:** $\lambda_{\Delta_t}, f_h, k_i, Th, R_{ij}, L_f^R, \gamma_{CQC}$
**Output:** $VM_{ID}, Hop_{next}$
1: Identify the service type of $\lambda_{\Delta_t}$
2: Upgrading the popularity of all IIoT services on $f_h$
3: **for** $(i = 1, i + +, i <= n)$
4:    **if** $k_i < Th$, $k_i \in PendingList$; **else** $k_i \in FWList$
5: **end**
6: **if** $PendingList \neq NULL$
7:    **for** j=1:length($PendingList$)
8:       1) Calculate the computing cost $C_{f_h}^{E_x}$
9:       2) Find $R_{ij}^*$ the $E_x$ belongs to; 3) Find which $L_f^R = 1$
10:       **for** $l = TotalNumberofVMs$
11:          Select a $VM$ to process the arrival data flows
12:          Calculate the $V_{R_{ij}^*}^{\gamma_{CQC}}$ and add it to $Array[l]$
13:       **end**
14:       Find the minimization value $min[V_{R_{ij}^*}^{\gamma_{CQC}}]$ in $Array[l]$
          and output the parameters of corresponding VM
15:    **end**
16: **end**
17: **if** $FWList \neq NULL$
18:    **for** j=1:length($FWList$)
19:       **if** $FWList(j) == CTSL$, {Forwarding to $Hop_{next}$}
20:       **else** {Add $FWList(j)$ into $PendingFlow$,
21:       Forwarding to SPSRP controller to get $Hop_{next}$}
22:    **end**
23: **end**
24: **end**

---

Nodes run a cross-blockchain structure consisting of multiple parallel blockchains. The encrypted data includes the encrypted location information of fog nodes and the corresponding encrypted trust evidence collected by edge/fog devices. Semantic based security detection and isolation scheme are proposed for edge/fog computing system to defense against content threat.

Secondly, a unified data storage architecture that is dedicated to managing data at the edge is proposed. The characteristic of our proposed architecture is to migrate the advantages of Hadoop Distributed File System (HDFS) in Cloud Computing to the edge to ensure that edge services provide better QoS. Moreover, to maximize the capability of edge nodes, we devise a dynamic storage policy-making mechanism based on Q-learning, which can recommend data with high invoked popularity for edge servers and updating data in time accordingly. To achieve a high-level linkage, we also propose a communication model for edge-cloud and edge-edge communication. Edge nodes can share their storage information with neighbors by the synchronous communication.

Thirdly, a scheme is proposed for service popularity-based smart resources partitioning. The Zipf's law is used to calculate the popularity rank of the IIoT service and predicted the computing cost of arriving IoT services on edge/fog computing. We provided a solving method of threshold value for forwarding edge/fog services, and applied it to decide whether the arriving IIoT service should be locally handled. The work first decoupled the computing control layer from the computing layer, and provided a programmable interface for edge/fog computing operators.

## 7.2   Main Cost

In the proposed approach, each edge/fog computing node can adaptively pick up and process the most popular IIoT services and smartly partition its resources based according to the popularity rankings of picked IIoT services. Unpopular IoT services on an edge/fog node will be forwarded to the other FN for efficient processing. In other words, it is no need for each edge/fog node to ask for the states of other edge/fog nodes. Thus, the complexity of proposed algorithms is $\Theta(n)$. The function of proposed algorithms was not to copy the load balancing and VM migration in cloud data to distributed edge/fog nodes. By using Algorithms 1 and 2 to partition the resources of edge/fog nodes, we can obtain minimized computing cost and minimized CQC validation. All the performance improvements of the proposed scheme were directly beneficial to edge/fog computing users because the service popularity reflected the real demands of edge/fog computing users. In terms of whether it will cause additional computing cost, the answer is inevitable. However, compare to the improvements of proposed approach, the additional computing cost caused by complexity of proposed algorithms is minor. Moreover, the additional computing cost can be handled by resources offloaded from cloud. Besides, the edge/fog nodes selectively deals with the local delay-sensitive services rather than all of the arriving edge/fog computing services.

## 8   Conclusion

In this chapter, the methods and technologies of security and intelligent management for fog/edge computing resources were studied. Blockchain and semantic are introduced to enhance the trust and security protection capabilities of the edge/fog computing systems. Moreover, we are dedicated to the complex application scenarios and massive data generated by edge nodes, which takes challenges to the edge-cloud collaboration. By taking the advantages of Hadoop, a unified edge-cloud intelligent storage architecture is proposed to improve the performance of edge services. Finally, the proposed resource partitioning scheme modeled the relationship between service popularity and computing cost with Zipf's law, which decoupled the computing control from data processing and support mobile and heterogeneous computing resource scheduling. Future work is the artificial intelligence collaborations technologies for edge/fog computing systems.

# References

1. Steiner, W., Poledna, S.: Fog computing as enabler for the industrial internet of things. e i Elektrotechnik und Informationstechnik. **133**(7), 310–314 (2016)
2. Zeng, D., Gu, L., et al.: Joint optimization of task scheduling and image placement in fog computing supported software-defined embedded system. IEEE Trans. Comput. **65**(12), 3702–3712 (2016)
3. Wu, J., Dong, M., et al.: Big data analysis based security cluster Management for Optimized Control Plane in software-defined networks. IEEE Trans. Netw. Serv. Manag. **15**(1), 27–38 (2018)
4. Wu, J., Dong, M., et al.: FCSS: fog-computing-based content-aware filtering for security Services in Information-Centric Social Networks. IEEE Trans. Emerg. Top. Comput. https://doi.org/10.1109/TETC.2017.2747158
5. Dong, M., Ota, K., et al.: Preserving source-location privacy through redundant fog loop for wireless sensor networks. Proc. of IEEE DASC. **2015**, 1835–1842 (2015)
6. Wu, G., Chen, J., Bao, W., Zhu, X., Xiao, W., Wang, J., Liu, L.: MECCAS: collaborative storage algorithm based on alternating direction method of multipliers on Mobile edge cloud. In: IEEE International Conference on Edge Computing, pp. 40–46 (2017)
7. Huang, Y., Song, X., Ye, F., Yang, Y., Li, X.: Fair caching algorithms for peer data sharing in pervasive edge computing environments. In: IEEE 37th International Conference on Distributed Computing Systems, pp. 605–614 (2017)
8. Zhao, C., Dong, M., Ota, K., Li, J., Wu, J.: Edge-MapReduce based intelligent information-centric IoV: cognitive route planning. IEEE Access. **7**, 50549–50560 (2019)
9. Sasaki, K., Makido, S., Nakao, A.: Vehicle control system for cooperative driving coordinated multi-layered edge servers. In: IEEE 7th International Conference on Cloud Networking, pp. 1–7 (2018)
10. Sharma, S.K., Wang, X.: Live data analytics with collaborative edge and cloud processing in wireless IoT networks. IEEE Access. **5**, 4621–4635 (2017)
11. Olaniyan, R., Maheswaran, M.: Synchronous scheduling algorithms for edge coordinated internet of things. In: IEEE 2nd International Conference on Fog and Edge Computing, pp. 1–10 (2018)
12. Femminella, M., Pergolesi, M., Reali, G.: Performance evaluation of edge cloud computing system for big data applications. IEEE International Conference on Cloud Networking. 170–175 (2016)
13. Tang, B., et al.: Incorporating intelligence in fog computing for big data analysis in smart cities. IEEE Trans. Ind. Informat. **13**(5), 2140–2150 (Oct. 2017)
14. Zeydan, E., et al.: Big data caching for networking: moving from cloud to edge. IEEE Commun. Mag. **54**(9), 36–42 (Sep. 2016)
15. Jeon, W.S., Kim, J., Jeong, D.G.: Downlink radio resource partitioning with fractional frequency reuse in femtocell networks. IEEE Trans. Veh. Technol. **63**(1), 308–321 (Jan. 2014)
16. Dhungana, Y., Tellambura, C.: Multichannel analysis of cell range expansion and resource partitioning in two-tier heterogeneous cellular networks. IEEE Trans. Wirel. Commun. **15**(3), 2394–2406 (Mar. 2016)
17. Tefek, U., Lim, T.J.: Relaying and radio resource partitioning for machine-type communications in cellular networks. IEEE Trans. Wirel. Commun. **16**(2), 1344–1356 (Feb. 2017)

18. Singh, S., Andrews, J.G.: Joint resource partitioning and offloading in heterogeneous cellular networks. IEEE Trans. Wirel. Commun. **13**(2), 888–901 (Feb. 2014)
19. Tai, M.H., Tran, N.H., Le, L.B., Saad, W., Kazmi, S.M.A., Hong, C.S.: Coordinated resource partitioning and data offloading in wireless heterogeneous networks. IEEE Commun. Lett. **20**(5), 974–977 (May 2016)
20. Mach, P., Becvar, Z.: Mobile edge computing: a survey on architecture and computation offloading. IEEE Commun. Surv. Tuts. **19**(3), 1628–1656 (Jul.–Sep. 2017)
21. Pu, L., Chen, X., Xu, J., Fu, X.: D2D fogging: an energy-efficient and incentive-aware task offloading framework via network-assisted D2D collaboration. IEEE J. Sel. Areas Commun. **34**(12), 3887–3901 (Dec. 2016)
22. Mubeen, S., Nikolaidis, P., Didic, A., Pei-Breivold, H., Sandstrom, K., Behnam, M.: Delay mitigation in offloaded cloud controllers in industrial IoT. IEEE Access. **5**, 4418–4430 (2017)
23. Li, J., Huang, L., Zhou, Y., He, S., Ming, Z.: Computation partitioning for mobile cloud computing in a big data environment. IEEE Trans. Ind. Informat. **13**(4), 2009–2018 (Aug. 2017)
24. Vasconcelos, D.R.D., Andrade, R.M.D.C., Souza, J.N.D.: Smart shadow–an autonomous availability computation resource allocation platform for internet of things in the fog computing environment. In: Proc. Int. Conf. Distrib. Comput. Sens. Syst, pp. 216–217 (2015)
25. Zhang, H., Xiao, Y., Bu, S., Niyato, D., Yu, F.R., Han, Z.: Computing resource allocation in three-tier IoT fog networks: a joint optimization approach combining stackelberg game and matching. IEEE Int. Things J. **4**(5), 1204–1215 (Oct. 2017)
26. Yin, B., Shen, W., Cheng, Y., Cai, L.X., Li, Q.: Distributed resource sharing in fog-assisted big data streaming. In: Proc. IEEE Int. Conf. Commun, pp. 1–6 (2017)
27. Nishio, T., Shinkuma, R., Takahashi, T., Mandayam, N.B.: Service oriented heterogeneous resource sharing for optimizing service latency in mobile cloud. In: Proc. IEEE Int. Conf. Cloud Comput. Technol. Sci, pp. 19–26 (2013)

# Algorithms for NFV-Enabled Multicasting in Mobile Edge Computing

Zichuan Xu and Weifa Liang

## 1 Introduction

Mobile devices, including smart phones and tablets, gain increasing popularity as communication tools of users for their business, social networking, and personal entertainment. However, the computing, storage and battery capacities of each mobile device are very limited, due to its portal size. Leveraging by rich computing and storage resources in clouds, mobile devices can offload some of their tasks to clouds for processing and storage, while the clouds usually are remote located from their end users. Thus, the response delay to user requests may not be tolerable for some real-time applications. Instead, a new network service paradigm, Mobile Edge Computing (MEC), is emerged, which can provide cloud-computing capability at the edge of core networks in the proximity of mobile users [1]. MEC can significantly shorten the response delay to user applications, ensure highly efficient network operation and service delivery, and improve user experience of using the services, which is an ideal platform to meet ever-growing resource demands of mobile users for their applications, by enhancing mobile device capabilities in a real-time manner [34].

Considering that the computing resource in an MEC network is highly distributed in base stations, edge servers and mobile devices, providing security and privacy guarantees for multicasting services is fundamentally challenging. Unlike

Z. Xu (✉)
School of Software, Dalian University of Technology, Dalian, Liaoning Province, China
e-mail: z.xu@dlut.edu.cn

W. Liang
Research School of Computer Science, The Australian National University, Canberra, ACT, Australia
e-mail: wliang@cs.anu.edu.au

© Springer Nature Switzerland AG 2021
W. Chang, J. Wu (eds.), *Fog/Edge Computing For Security, Privacy, and Applications*, Advances in Information Security 83,
https://doi.org/10.1007/978-3-030-57328-7_10

conventional core networks or data center networks, various security and privacy solutions are available. For example, various hardware-oriented middleboxes are deployed in data center networks to enforce security and privacy functions of various services in the data center networks. Examples of such network functions typically include Intrusion Protection/Detection System (IPS/IDS), firewalls, web filtering, flow filtering, deep packet inspection, or pattern matching and remediation. However, the deployment of dedicated hardware middleboxes in MEC networks will increase the operational and maintenance costs, considering the service and user dynamics of mobile users. Specifically, a deployment hardware-based flow filter in an edge server means a static placement and configuration of the function in the network. Security and privacy events can happen in any node of the MEC network dynamically. The static placement of hardware network functions is no longer meeting the security and privacy requirements of users. Network Function Virtualization (NFV) [25] which runs the network function in generic servers as virtual machines have been envisioned as a promising technique to enable high security and privacy guarantees in MEC networking. It can also enable fast, agile service deployment and cost-effective yet error-free service provisioning in future communication networks. It replaces security network functions from expensive, dedicated hardware-middleboxes, by software implementation, where each network function is virtualized as a virtualized network function instance that runs in a virtual machine in an edge server in MEC.

Although implementing network functions as VNF instances is a promising technology to guarantee security and privacy of MEC networks, admitting NFV-enabled multicast requests in an MEC poses several challenges. Firstly, both computing and storage resources at edge servers and communication resources at links are not unlimited, in comparison with its counterpart - the powerful centralized data center network (a cloud). It is of paramount importance to optimize the performance of the MEC network through judicious allocating its limited resources to meet user resource demands. Secondly, each NFV-enabled multicast request has a service function chain requirement, how to steer the data traffic of the request to go through each network function in its service function chain correctly? Thirdly, the service chain implementation may either share some existing network function instances with the other requests or instantiate new VNF instances. How to make such a decision to minimize the admission cost of the request? Finally, how to maximize the network throughput by admitting or rejecting each arrived request immediately if requests arrive one by one without the knowledge of future request arrivals? In this chapter, we will address the aforementioned challenges.

The novelties of the study in this chapter are as follows. We study NFV-enabled multicast request admissions in MEC by formulating three novel optimization problems that explore VNF instance placement and sharing among different NFV-enabled multicast requests. We aim to maximize the network throughput while minimizing the accumulative admission cost of admitted requests through striving for fine tradeoffs between the usages of computing and bandwidth resources. We devise the very first approximation algorithms for a single NFV-enabled multicast request admission with the objective to minimize its admission cost, with the assumption that the VNFs of a service chain may or may not be consolidated into a single edge server. We also consider dynamic admissions of NFV-enabled multicast requests, by developing an online algorithm with a provable competitive ratio. The key ingredients in the development of these proposed algorithms lie in (1) dynamically determining the use of existing VNF instances or instantiating new VNF instances for each request admission; and (2) determining the admission order of a given set of requests as admitted requests will heavily impact the admissions of future requests, due to the availability of the demanded resources and whether existing VNF instances can be shared by future requests.

The main contributions of this chapter are summarized as follows. We study the NFV-enabled multicast request admissions in an MEC network with the aim to either minimize the request admission cost, or maximize the network throughput for a set of requests or a sequence of requests arriving one by one without the knowledge of future arrivals, subject to both computing and bandwidth resource capacities on edge servers and links in the network, respectively. We first propose an approximation algorithm for the cost minimization problem of a single NFV-enabled multicast request admission. We then develop an efficient heuristic for a set of NFV-enabled multicast request admissions, by reducing the problem to the single NFV-enabled multicast request admission. We thirdly consider dynamic NFV-enabled multicast request admissions by devising an online algorithm with a provable competitive ratio. We finally evaluate the performance of the proposed algorithms through experimental simulations. Simulation results reveal that the proposed algorithms are very promising.

The rest of this chapter is organized as follows. Section 2 conducts literature review. Section 4 introduces notions, notations, and problem definitions. Section 5 devises an approximation algorithm for the cost minimization problem with consolidated VNFs. Section 6 devises an approximation algorithm for the cost minimization problem of a single NFV-enabled multicast request admission where different VNF instances of a service function chain can be deployed to different edge servers. Section 7 devises an online algorithm for dynamic NFV-enabled multicast request admissions. Section 8 evaluates the proposed algorithms empirically, and Sect. 9 concludes the chapter.

# 2 Related Work

As a key-enabling technology of 5G, MEC networks have gained tremendous attentions by the research community recently. There are extensive studies of user unicast and multicast request admissions through resource provisioning in MEC networks [3, 6, 7, 9, 10, 12, 13, 19, 21, 24, 32, 33]. For example, Jia et al. [11] considered the assignment of user requests to different edge servers in a Wireless Metropolitan Area Network with the aim to minimize the maximum delay among offloaded tasks, by developing heuristics for the problem. Ceselli et al. [3] focused on the design optimization such as the VM placement and migration, and user request assignment, by formulating a Mixed Integer Linear Programming (MILP) solution and heuristic algorithms for the problem. Xia et al. [26] investigated opportunistic task offloading under link bandwidth, residual energy in mobile devices, and computing capacity constraints of edge servers.

All the aforementioned studies assumed that each task will be allocated with dedicated computing resource, and there is no consideration of utilizing existing VNF instances to serve new tasks. However, many requests usually demand the same type of services. If the VNF instance of a specified service has already been instantiated with sufficient residual processing capacity, the other tasks that request for the service can make use of the VNF instance. Several recent studies explored the placement and sharing of VNF instances [9, 13, 30]. For example, Jia et al. [13, 30] studied a novel task offloading problem in an MEC network, where each offloading task requests a network function service with a maximum tolerable delay requirement. They aimed at maximizing the number of requests admitted while minimizing their admission cost, for which they proposed an efficient online algorithm. He et al. [9] studied the joint service placement and request scheduling in order to optimally provision edge services while taking into account the demands of both sharable and non-sharable resources. They aim to maximize the network throughput, for which they showed that this joint optimization problem is NP-hard and then developed heuristic algorithms.

There are several studies of NFV-enabled multicasting in MEC environments [2, 13, 29]. For example, Zhang et al. [35] investigated the NFV-enabled multicasting problem in SDNs. They assumed that there are sufficient computing and bandwidth resources to accommodate all multicast requests, for which they provided a 2-approximation algorithm if only one server is deployed. In reality, it is not uncommon that both computing and bandwidth resources in MEC are limited, which need to be carefully allocated. Furthermore, they did not consider dynamic admissions of NFV-enabled multicast requests, which is much complicated compared with the problem of admitting a single or a set of given requests. Xu et al. [29] studied the cost minimization problem of admitting a single NFV-enabled multicast request, where the implementation of the service chain of each request will be consolidated into a single edge server. Xu et al. [31] recently considered the admissions of NFV-enabled multicast requests with QoS constraints in MEC by proposing approximation and heuristic algorithms for the problem. Ma et al. [18, 20] considered the profit maximization problem in MEC by dynamically admitting NFV-enabled unicast requests with QoS requirements, for which they developed an efficient heuristic, and an online algorithm with a provable competitive ratio if the QoS requirement can be ignored. Although they considered the sharing of existing VNF instances among different unicast requests, the problem of NFV-enabled unicast request admissions in [14, 18, 20] is a special case of the problem of NFV-enabled multicast request admissions where the destination set contains only one node. The essential differences of the study in this chapter from these mentioned studies [13, 29, 30, 35] are (1) the VNF instances of the service chain of each NFV-enabled multicast request in this chapter can be placed to multiple edge servers, not just one edge server in the previous studies; and (2) the sharing of existing VNF instances among different multicast requests has not been explored, this exploration makes the problem become more challenging.

## 3 Security Network Functions and Chaining

In this section, we given the definition of security service functions, and describe their virtualization and chaining.

### 3.1 Security Network Functions and Their Virtualization

Traditionally, network operators deploy various hardware-based security switches to ensure secure data transfers within the network. With network function virtualization, these hardware-based switches are replaced by softwarelization, i.e., the security functions are implemented in pieces of software in virtual machines (VMs) in servers. Such security functions include firewalls, Intrusion Prevention/Detection System (IPS/IDS), Deep Packet Inspection (DPI), Application Visibility and Control (AVC), network virus and malware scanning, sandbox, Data Loss Prevention (DLP), Distributed Denial of Service (DDoS) mitigation and TLS proxy. For example, one typical security attack is the DDOS attack, which can easily be detected via an IDS, by identifying malicious system activities and violation of system policies. In general, a security network function is a type of network function that carries out specific security tasks. In addition to packet forwarding that makes use of security functions, security network functions can be used for buffering, injecting or blocking malware packets, as well as proxy connections, and most network security network functions maintain states at the connection, session or transaction levels.

It is well known that the deployment and purchase cost of hardware-based switches of security network functions are usually very high. and, their maintenance cost is also expensive, thus the operational costs of network operators for secure data transfer within its network usually is not cheap at all. To reduce the operational cost of provisioning security services, network function virtualization enables running virtualized security function instances to become agile and non-expensive. In particular, by leveraging the technique of NFV, security functions can be implemented in VMs or containers. Security network functions then can be dynamically instantiated, automatically deployed, and transparently inserted into the traffic flow to address different security needs for various applications.

## 3.2 Security Network Function Chaining

In case a packet flow needs to be processed by a sequence of security network functions, we term this sequence as a security network function chain for the data traffic. Specifically, security network function chaining is a technique for selecting and steering data traffic flows through a sequence of security network functions by leveraging both Network Function Virtualization (NFV) and Software Defined Networking (SDN) technologies.

Consider an application scenario where a set of users is going to hold a tele-conference. To ensure all messages in the tele-conference multicast to all participants securely, there is a specified security network function chain for this tele-conference which includes (1) each multicast packet should be encrypted before reaching other destinations in the untrusted core network; and (2) all traffic should be inspected by an IDS. Meeting those security requirements need careful chaining of instances of network encryption functions and IDSs.

## 4 Optimization Problems for Security Function Chaining

In this section, we first introduce the system model, notions and notations, and then define the problems precisely.

### 4.1 System Model

We consider a mobile edge cloud (computing) network (MEC) in a metropolitan region that is modelled by an undirected graph $G = (V, E)$, where $V$ is a set of *access points* (APs) located at different locations in a metropolitan region, e.g., shopping centers, airports, restaurants, bus stations, and hospitals. An edge server is co-located with each AP node $v \in V$ via a high-speed optical cable. This implies that the communication delay between them is negligible due to plenty of bandwidth on the cable. For simplicity, each AP node and its co-located edge server will be used interchangeably if no confusion arises. Each edge server has computing capacity $C_v$ for implementing various virtualized network functions (VNFs). $E$ is the set of links between APs. Each link $e \in E$ has a bandwidth capacity $B_e$. The available bandwidth resource of each link is time-varying, because requests arrive in and depart from the system dynamically [4]. We assume that each AP node covers a certain area, in which each mobile user can access the MEC service wirelessly through the AP. In case a mobile user located at an overlapping coverage region of multiple APs, the mobile user can register itself to an appropriate AP and tune itself to the parameters of the AP [23]. Figure 1 is an example of an MEC network.

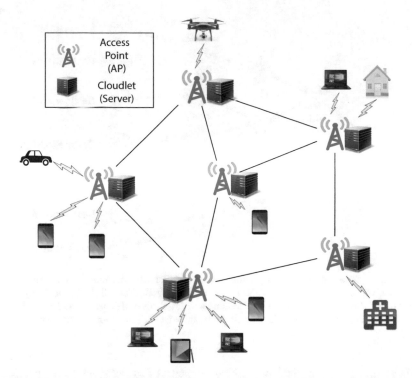

**Fig. 1** An illustrative example of an MEC network consisting of 6 APs with each co-located with an edge server

## 4.2 NFV-Enabled Multicast Requests with Service Function Chain Requirements

Consider *an NFV-enabled multicast request* $r_j = (s_j, D_j, \rho_j, SFC_j)$ that transmits its data traffic from the source node $s_j \in V$ to the given set $D_j \subseteq V$ of destination nodes with a specified packet rate $\rho_j$. Each packet in the data traffic stream must pass through the sequence of network functions of its *service function chain* $SFC_j = \langle f_{j,1}, \ldots, f_{j,l}, \ldots, f_{j,L_j} \rangle$ before reaching each of the destinations, where $L_j$ is the length of $SFC_j$. We assume that a unit packet rate of $r_j$ requires bandwidth resource $b_e$ in a link $e \in E$, thus, the total amount $\rho_j \cdot b_e$ of bandwidth required for $r_j$ in $e$.

We assume that resources in edge servers are virtualized, using container-based lightweight virtualization technologies, and thus can be allocated and shared flexibly. Each instance of a *virtualized network function* (VNF) is a virtual machine in an edge server. Without loss of generality, we assume that different types of VNFs among all service function chains of requests can be classified into $K$ types. Denote by $f^{(k)}$ and $C(f^{(k)})$ the VNF of type $k$ and the amount of computing resource consumed for its implementation in an edge server, respectively, $1 \leq k \leq K$. Suppose each VNF instance of $f^{(k)}$ has *a maximum processing capacity* $\mu^{(k)}$.

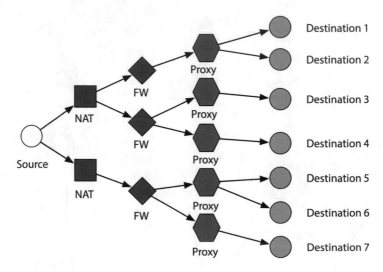

**Fig. 2** An example of an NFV-enabled multicast request with a service function chain that consists of three network functions: Network Address Translation (NAT), Firewall (FW), and Proxy. Its data packet traffic flows from the source node *Source* to a set of seven destination nodes. Each packet must pass through one VNF instance of each of the three network functions in the service function chain

Furthermore, if the residual processing capacity of an existing VNF instance is sufficient to process the data traffic of a newly admitted request, this VNF instance can be shared by the request. Otherwise, a new VNF instance for the request needs to be instantiated in an edge server with sufficient residual computing resource in order to admit the request.

To admit an NFV-enabled multicast request $r_j$, each packet of its data traffic is enforced to go through a VNF instance of each network function in its $SFC_j$ prior to reaching each of the destinations in $D_j$. Denote by $T(j)$ *the pseudo-multicast tree* that transmits the data traffic of request $r_j$ from the source $s_j$ to the destinations in $D_j$, where *a pseudo-multicast tree* [28] in fact may be a graph, not a tree. A pseudo-multicast tree is a directed pseudo-steiner tree which starts from a source node and reaches each node in a destination set. However, due to the availability of some edge servers in the MEC (i.e., be able to accommodate the VNF instances with sufficient resources), each edge server node and physical link of the network may appear multiple times in the pseudo-multicast tree. Figure 2 is an example to illustrate the admission of an NFV-enabled multicast request, where for each network function $f_{j,l}$ in the service function chain $SFC_j$, either an existing VNF instance (with sufficient residual processing capacity) is selected or a new VNF instance is instantiated in an edge server $f_{j,l}$ in each path from the source node $s_j$ to each destination node in $D_j$, and these VNF instances can be placed at different edge servers.

An example of the pseudo-multicast tree and its relationship with a multicast tree $T$ is shown in Fig. 3. Notice that given an NFV-enabled multicast request,

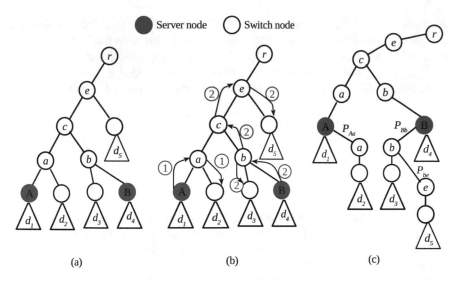

**Fig. 3** A pseudo-multicast tree $G_T$ derived from a multicast tree $T$ for an NFV-Enabled multicast request $r_k$, and another tree $T'$ derived from $G_T$ which has the identical cost as $G_T$. (**a**) The multicast tree $T$. (**b**) The pseudo-multicast tree $G_T$ from $T$. (**c**) The cost-identical multicast tree $T'$ from $G_T$.

its pseudo-multicast tree may not be unique, because its packet can be directed to different destinations via different paths. However, the determination of a pseudo-multicast tree can be tailored to fit different optimization objectives. For example, if the objective is to minimize the cost of implementing the request, the found pseudo-multicast tree should be able to achieve the lowest cost.

## 4.3 Admission Costs of an NFV-Enabled Multicast Request

The admission cost of an NFV-enabled multicast request in an MEC network is the sum of three constituent costs: the VNF instance processing cost for processing its data packets, the VNF instance instantiation cost for instantiating new VNF instances in edge servers, and the bandwidth cost for routing its data traffic along links in its pseudo-multicast tree. Instantiating VNF instances at edge servers consumes both computing and storage resources of the edge servers, thus incurs the VNF instantiation cost. Denote by $c_{ins}(f^{(k)}, v)$ the instantiation cost a VNF instance of network function $f^{(k)}$ in an edge server $v$, and $\rho_j \cdot c_{proc}(f^{(k)}, v)$ the processing cost of data traffic of a request $r_j$ at a VNF instance of $f^{(k)}$ at edge server $v$, where $c_{proc}(f^{(k)}, v)$ is the cost of processing a packet by a VNF instance $f^{(k)}$ at edge server $v$ and $\rho_j$ is the packet rate of $r_j$. Notice that the processing cost $c_{proc}(f^{(k)}, v)$ of a data packet of different VNF instances at different edge servers may be significantly different, since different VNF instances consume different amounts of computing resources and different edge servers have different amounts of energy consumptions. In addition, each packet of the data traffic of request $r_j$ is routed along a pseudo-multicast tree $T(j)$ that incurs the communication cost $\rho_j \cdot \sum_{e \in T(j)} c_e$, where $c_e$ is the unit transmission cost on link $e \in E$, and $\sum_{e \in T(j)} c_e$ is the cost of transferring a packet along the pseudo-multicast tree $T(j)$.

## 4.4   Problem Definitions

In practice, different network operators may have different settings of their MEC networks. Therefore, different network performance indicators can be taken into account when optimizing the multicasting service delivery process. In particular, the choice of the objective and resource settings depends on the characteristics of the provider, e.g., types of available resources, geographical scope of the infrastructure, nature of the offered services etc. We here define different versions of the NFV-enabled multicast request admission problem in an MEC network $G$ that cover the requirements of a wide range of network service providers.

**Definition 1** For large-scale network operators that build their own infrastructures, they want to minimize the cost of operating the MEC networks while guaranteeing the performance of multicast services. They thus have sufficient computing and bandwidth resources, and consolidate all VNFs of a service chain into a single edge server to avoid performance degradation due to inter-VNF data transmission. Assuming that the MEC network $G = (V, E)$ has sufficient computing and bandwidth resources to meet the resource demands of a single NFV-enabled multicast request and the VNFs of each $SFC_j$ of a NFV-enabled request $r_j$ can be consolidated into a single edge server, the *the cost minimization problem with consolidated VNFs* in $G$ for an NFV-enabled multicast request $r_j$ is to find a pseudo-multicast tree such that its implementation cost is minimized, if no more than $M$ edge servers are used for implementing its service chain $SFC_j$, assuming that $G$ has sufficient computing and bandwidth resources for the request.

**Definition 2** For medium and small-scale network operators that lease certain amounts of computing and bandwidth resources from infrastructure providers, they want to maximize the utilization of their leased resources while minimizing the cost of resource usages. We thus have the following optimization problem. Given an MEC network $G = (V, E)$ with a set $V$ of edge servers (or APs), each $v \in V$ has computing capacity $C_v$, let $B_e$ be the bandwidth capacity of each link $e \in E$, assuming that the previous $j - 1$ NFV-enabled multicast requests have been responded (admitted or rejected), consider an incoming NFV-enabled multicast request $r_j = (s_j, D_j, \rho_j, SFC_j)$, *the cost minimization problem* of admitting request $r_j$ is to find a pseudo-multicast tree $T(j)$ in $G$ to route its data traffic from the source node $s_j$ to each destination node in $D_j$ while each packet in the data traffic must pass through each VNF instance in the service function chain $SFC_j$, such that its admission cost is minimized, subject to computing and bandwidth capacities on both edge servers and links of $G$.

**Definition 3** In practical MEC networks, multicast requests arrive into the system dynamically. Such dynamic requests need to be admitted on their arrivals, such that the limited resources of an MEC network is maximally utilized. We then have the following online optimization problem. Given an MEC network $G = (V, E)$ with a set $V$ of edge servers, each $v \in V$ has computing capacity $C_v$, and each link $e \in E$ has bandwidth capacity $B_e$. Let $r_1, r_2, \ldots, r_j$ be a sequence of NFV-enabled multicast requests that arrive one by one without the knowledge of future request arrivals, *the online throughput maximization problem* in $G$ is to maximize the number of requests admitted, subject to computing and bandwidth capacities on both edge servers and links of $G$.

# 5 Approximation Algorithms for the Cost Minimization Problem with Consolidated VNFs

In this section we deal with the cost minimization problem with consolidated VNFs.

## 5.1 Algorithm Overview

The basic idea of the proposed approximation algorithms is to find a pseudo-multicast tree rooted at the source and spanning all destinations, and each packet from the source to destinations passes through an edge server in the tree, such that the cost of the tree is minimized. To this end, the finest trade-off between the computing and communication costs needs to be explored. Specifically, if an edge server $v$ with a lower computing cost is included in the pseudo-multicast tree for multicast request $r_j$, the processing cost of NFV-enabled $r_j$ may be reduced. This however will increase the communication cost if the edge server $v$ is far from the destinations of $r_j$. On the other hand, if there are multiple edge servers located at different branches of the multicast tree, then the packet can pass through each of the edge servers to reach its destinations in $D_k$. This will lead to less bandwidth usages from the source to the destinations, which is achieved at the expense of high computing cost by employing multiple edge servers. We thus need to identify a set of edge servers with each implementing the service chain $SFC_j$ of $r_j$ and find a pseudo-multicast tree including the identified edge server(s) on the path from the source $s_j$ to each destination $u \in D_j$. As $M$ is a constant, we aim to find a pseudo-multicast tree in $G$ that contains no more than $M$ edge servers and the path in the tree from $s_j$ to each destination $u \in D_j$ must pass through one of the identified edge servers such that the cost of the tree is minimized.

Recall that there are $|V|$ APs that are attached with edge servers, clearly $M \leq |V|$. As a pseudo-multicast tree for any NFV-enabled multicast request can contain at least one but no more than $M$ edge servers, there are at most $\binom{|V|}{M}$ combinations of edge servers that can meet the computing resource demand of service chain $SFC_j$ of request $r_j$. For each combination of edge servers, a pseudo-multicast tree in $G$ can be identified, and the tree with the minimum cost is then used to implement $r_k$. We thus reduce the NFV-enabled multicast problem to a Steiner tree problem in an auxiliary undirected graph. An approximate solution to the latter returns an approximate solution to the former.

## 5.2 Approximation Algorithm

Given an NFV-enabled multicast request $r_j$, we now devise an approximation algorithm for the cost minimization problem with consolidated VNFs in $G$, by reducing it to the Steiner tree problem in an auxiliary undirected graph $G_j^i = (V_j^i, E_j^i; c)$ with an edge weight function $c$ for all $i$ with $1 \leq i \leq \binom{|V|}{M}$, where $V_j^i = V \cup \{s_j'\}$, $E_j^i = E \cup \{(s_j', v) \mid v \in V^i\}$, $V^i$ ($\subseteq V$) is the $i$th combination of edge servers in $V$, and $s_j'$ is a *virtual source* of request $r_j$. For each $v \in V^i$, if edge $(s_j, v) \in E$ in $G$, the cost of edge $(s_j, v) \in E_j^i$ is assigned zero. $s_j'$ is the new source in $G_j^i$, replacing the original source $s_j$. Notice that the original source $s_j$ is still contained in $G_j^i$ serving as a 'regular' switch node without an attached edge server. To guarantee that the traffic of $r_j$ passes through its service chain $SFC_j$ that is implemented in one or multiple edge servers in $V^i$ ($\subseteq V_S$), we connect $s_j'$ with all edge server nodes in $V^i$, where the edge between $s_j'$ and each edge server node $v \in V^i$ in $G_j^i$ represents a shortest path $p_{s_j,v}$ in $G$ between nodes $s_j$ and $v$. The weight of edge $(s_j', v)$ is the cost sum of the edges in path $p_{s_j,v}$ plus the cost of implementing $SFC_j$ in edge server $v$, i.e., $c_{(s_j',v)} = \sum_{e \in p_{s_j',v}} c_e \cdot b_k + c_v(SFC_j)$, where $c_v(SFC_j)$ is the cost of the amount $C_v(SFC_j)$ of computing resource consumption for implementing $SC_k$. In addition, the weight $c_e$ of each edge $e \in E_k^i \cap E$ is the cost $c_e \cdot \rho_j$ of processing packet rate $\rho_j$ of $r_j$ on edge $e \in E$. An example of the constructed auxiliary graph $G_j^i$ is shown in Fig. 4.

For the sake of convenience, in the rest of this chapter we assume that the APs with attached edge servers is $V = \{v_1, v_2, \ldots, v_{|V|}\}$. Having constructed the auxiliary graph $G_j^i$, we now find a Steiner tree in $G_j^i$ for request $r_j$. We first find a minimum spanning tree (MST) $T_{mst}^i$ in a complete graph consisting of nodes in $\{s_j'\} \cup D_j$, in which each edge is assigned a weight that is equal to the length of the shortest path in $G_j^i$ between its two endpoints. Let $H_j^i$ be a subgraph of $G_j^i$ derived from $T_{mst}^i$ by replacing each edge of $T_{mst}^i$ with its corresponding shortest path in $G_j^i$.

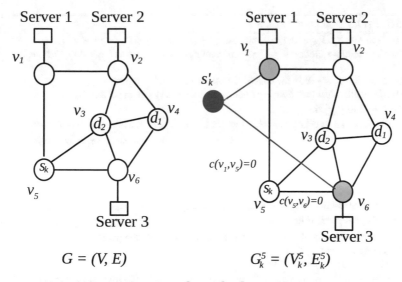

**Fig. 4** An example of the auxiliary graph $G_k^5 = (V_k^5, E_k^5)$ constructed from an SDN $G = (V, E)$ with $V^5 = \{v_1, v_6\}$, assuming that $M = 2$

We then find an approximate Steiner tree $T_j^i$ in $H_j^i$, by applying the approximation algorithm due to Kou et al. [16], which will serve as the pseudo-multicast tree for $r_j$. The detailed algorithm for the cost minimization problem with consolidated VNFs is given in Algorithm 1.

## 5.3   Algorithm Analysis

The rest is to show the correctness of Algorithm `Appro_Multi`, and analyze its time complexity and the approximation ratio.

**Lemma 1** *Algorithm* `Appro_Multi` *delivers a feasible solution for the NFV-enabled multicasting problem with and without SDN resource capacity constraints.*

**Proof** We here show that the solution delivered by Algorithm 2 is feasible. Each path $p$ in $T_j$ from $s_j'$ to one destination $d \in D_j$ corresponds to a path in $G$ from $s_j$ to $d$. This is evidenced by the fact that any path in subgraph $H_j^i$ of $G_j^i$, starting from $s_j'$ must use one of its incident edges in $E_j^i$, and the another endpoint $v$ of the edge must be an edge server in $V^i \subseteq V$, following the construction of $G_j^i$. This implies that each path in $T_j$ from $s_j$ to any destination $d \in D_j$ must use one of the edge servers in $V^i$. And $T_j$ includes all nodes in $D_j$. Thus, the tree obtained is a feasible pseudo-multicast tree for multicast request $r_j$.                                                                      □

---

**Algorithm 1** `Appro_Multi`

---

**Input:** $G = (V, E)$, $V_S$, a multicast request $r_k = (s_k, D_k; b_k, SC_k)$, and $M \geq 1$.
**Output:** A pseudo-multicast tree $T_j$ for implementing the multicast request $r_j$ with the minimum
      cost.

1:  $cost_j \leftarrow \infty$; $T_j \leftarrow \emptyset$; /* the cost of the pseudo-multicast tree */
2:  /* each combination of choosing $i$ edge servers from $|V|$ edge servers */;
3:  **for** $i \leftarrow 1$ to $\binom{|V|}{M}$ **do**
4:     Construct an auxiliary undirected graph $G^i_j = (V^i_j, E^i_j)$, as illustrated in Fig. 4;
5:     Find an MST $T^i_{mst}$ in a complete graph induced by the nodes in $\{s'_j\} \cup D_j$ with the weight
       of each edge being the length of the shortest path in $G^i_j$ between its two endpoints;
6:     Let $H^i_j$ be a subgraph of $G^i_j$ derived from $T^i_{mst}$, by replacing each edge of $T^i_{mst}$ with the
       corresponding shortest path in $G^i_j$; find an approximate Steiner tree $T^i_j$ in $H^i_j$ rooted at $s'_j$
       and spanning nodes in $D_j$, by invoking the approximation algorithm due to Kou et al. [16];
7:     **if** $c(T^i_j) < cost_j$ **then**
8:        $cost_j \leftarrow c(T^i_j)$, $T_j \leftarrow T^i_j$; /* a candidate solution to the problem */
9:  **if** $T_j$ contains node $s_j$ **then**
10:    Merge nodes $s_j$ and $s'_j$ into $s'_j$;
11: Rename $s'_j$ in $T_j$ as $s_j$, and let $T_j$ be the resulting graph (the pseudo-multicast tree) for data
    traffic routing of request $r_j$;
12: **return** $T_j$ and its cost $c(T_j)$.

---

**Theorem 1** *Given an MEC network $G = (V, E)$, a set $V$ of APs with each having an attached edge server, and an NFV-enabled multicast request $r_j = (s_j, D_j, \rho_j, SFC_j)$, there is a $2M$ approximation algorithm, Algorithm 2, for the cost minimization problem with consolidated VNFs, assuming no more than $M$ edge servers will be employed for its service chain implementation, where the approximation ratio $2M$ is the best. The time complexity of the algorithm is $O(|V|^3 \cdot |V|^M)$, where $M \geq 1$ is a small integer.*

**Proof** We first analyze the approximation ratio of Algorithm 2. Let $G^*_T$ be the optimal pseudo-multicast tree for the NFV-enabled multicast request $r_k$ in $G$. If $G^*_T$ is not a multicast tree, there is a corresponding tree $T'$ with the identical cost as $G^*_T$, following the definition of pseudo-multicast tree; otherwise, $G_T$ itself is a multicast tree. From now on, we denote by $T^*$ either the optimal multicast tree $G^*_T$ or its corresponding cost-identical tree $T'$. We assume that there are $l$ edge servers in $T^*$ with each implementing $SFC_j$ with $1 \leq l \leq M$. Without loss of generality, we assume that these $l$ nodes are $v_1, v_2, \ldots, v_l$, respectively. Clearly, it can be shown that none of pairs of these nodes in $T^*$ has the ancestor and descendant relationship in terms of a node being used as an edge server, otherwise the node in $V$ will be treated as a regular AP without the use of its edge server. Each subtree $T^*_{v_i}$ of $T^*$ rooted at $v_i$ contains some destinations, and all of the $l$ subtrees will contain all the destinations in $D_j$, following its definition. We construct another tree $T^*_c = (V', E')$ which is derived from $T^*$ by compressing the path in $T^*$ from $s_j$ to each node $v_i$ as follows. We replace the source node $s_j$ by a node $s'_j$ and the path in $T^*$ from $s_j$ to $v_i$ by an edge $(s'_j, v_i)$, and assign the edge a weight that is the sum of all edge costs in the path plus the cost of using edge server $v_i$. In the worst case, each of such compressions can increase the cost of the optimal tree, and there are in total $l$ compressions. Furthermore, for each compression, if the cost sum of all edges from the source $s_j$ to each $v_i$ dominates the cost of the tree, each of such compressions can increase the total cost by a value that equals the cost of $T^*$. We thus claim that the cost of tree $T_c^*$ is no greater than $l$ times the cost of tree $T^*$, i.e., $c(T_c^*) \le l \cdot c(T^*)$.

It can be seen that there is a multicast tree $T_j^i$ in $G_j^i$ rooted at source $s_j'$ and spanning all destinations in $D_j$, which has the same topological structure as $T_c^*$, however, it has a lower cost compared with that of $T_c^*$, i.e., $c(T_j^i) \le c(T_c^*)$. This is because the weight of each edge in $G_j^i$ between $s_j'$ and $v_i$ is the length of the shortest path in $G$ between the two nodes plus the cost of using edge server $v_i$, while the corresponding edge weight in $T_c^*$ is the sum of all edge weights in the path in $T^*$ between $s_j$ and $v_i$ plus the cost of using edge server $v_i$.

Let $T_j^{OPT,i}$ be an optimal multicast tree in $G_j^i$ rooted at $s_j'$ and spanning all destinations in $D_j$ and each path in the tree from $s_j'$ to a destination goes through one of the edge servers in $V'$. Then, $c(T_j^{OPT,i}) \le c(T_j^i)$ as $T_j^i$ is one of the multicast trees for NFV-enabled multicast request $r_j$. Let $T_j^{app,i}$ be an approximate multicast tree in $G_j^i$ for multicast request $r_j$, by the approximation algorithm with an approximation ratio of 2 due to Kou et al. [16]. We then have $c(T_j^{app,i}) \le 2c(T_j^{OPT,i})$. Since $c(T_j^i) \le c(T_c^*)$ and $c(T_c^*) \le l \cdot c(T^*)$, we have $c(T_j^{app,i}) \le 2c(T_j^{OPT,i}) \le 2c(T_j^i) \le 2c(T_c^*) \le 2 \cdot l \cdot c(T^*)$. Since a pseudo-multicast tree $T_j$ with the minimum cost from the $\binom{|V|}{M}$ auxiliary undirected graphs $G_j^i$ for all $i$ with $1 \le i \le \binom{|V|}{M}$ will be found and the value of $l$ is within $[1, M]$, the cost of the pseudo-multicast tree $T_j$ for $r_j$ is no greater than $2M \cdot c(T^*)$.

We finally analyze the time complexity of Algorithm 2 as follows. The algorithm proceeds iteratively. Within each iteration, it first constructs an auxiliary graph $G_j^i$, and then finds an approximate Steiner tree $T_j^{app,i}$ in $G_j^i$ for each multicast request $r_j$. It takes $O(|E| + |V| \log |V|)$ time to find a single-source shortest path tree in $G_j$ by Dijkstra's algorithm, while it takes $O(|V_j^i|^3) = O(|V|^3)$ time to find an approximate Steiner tree $T_j^{app,i}$ [16]. There are $O(\binom{|V|}{M})$ ($= O(|V|^M)$) iterations. The algorithm thus takes $O(|V|^3 \cdot |V|^M)$ time. For example, if $M = 3$, then the time complexity of Algorithm 2 is $O(|V|^3 \cdot \log^3 |V|)$. □

# 6 An Approximation Algorithm for the Cost Minimization Problem

In this section, we deal with the cost minimization problem of a single NFV-enabled multicast request admission. We first devise an approximation algorithm for the problem, and then analyze its performance.

## 6.1    Algorithm Overview

Given an MEC network $G = (V, E)$ and an NFV-enabled multicast request $r_j$, we aim to minimize the admission cost of the request by steering its data traffic from the source $s_j$ to the set of destinations in $D_j$ while each packet of the data traffic must pass through a sequence of network functions in its specified service function chain $SFC_j$. To tackle the problem, it poses three challenges. One is the resource availability in MEC. Whether request $r_j$ should be admitted or not is determined by the availability of its demanded resources in $G$; the other is which edge servers should be identified to implement which network functions of its $SFC_j$; and finally, whether new VNF instances will be instantiated or existing VNF instances can be shared for the implementation of $SFC_j$ must be made dynamically. It is essential to address the aforementioned challenges in order to deliver a cost-efficient solution to the problem.

The basic idea behind the proposed approximation algorithm for the problem is reducing it to the directed multicast tree problem in an auxiliary, directed acyclic graph. If there is a multicast tree in the auxiliary graph rooted at the source $s_j$ and spanning all destinations in $D_j$, then, request $r_j$ can be admitted, otherwise, $r_j$ should be rejected due to lack of sufficient resources to meet its resource demands. This claim will be shown later in algorithm analysis. A pseudo-multicast tree $T(j)$ in $G$ [28] finally can be derived from the multicast tree $T'(j)$ in the auxiliary graph for the implementation of request $r_j$.

## 6.2    Approximation Algorithm

Given an NFV-enabled multicast request $r_j$, we can either make use of existing network function instances as long as their residual processing capacities are sufficient to admit the request. Or if there is sufficient available computing resource in an edge server, a new instance for the requested type of network function can be instantiated in the edge server. Thus, there are multiple candidate instances for each network function $f_{j,l}$ in its service function chain $SFC_j$ in $G$ to be dynamically determined with $1 \leq l \leq L_j$.

Denote by $\lambda(j, l) = k$ the type of network function which is the $l$th network function $f_{j,l}$ in $SFC_j$ of request $r_j$ with $1 \leq k \leq K$ and $1 \leq l \leq |SFC_j|$, and denote by $F_v^{(k)}$ the set of VNF instances of type $k$ instantiated in edge server $v$. Let $\mu_i^{re}$ be the residual processing capacity of VNF instance $i \in F_v^{(k)}$. Let $C_v^{re}$ be the residual computing capacity of edge server $v \in V$. Denote by $N_{l,v}$ the set of VNF instances that can be employed as the $l$th network function $f_{j,l}$ in $SFC_j$ in edge server $v$, including both existing network function instances with sufficient residual processing capacities, i.e., $\mu_i^{re} \geq \rho_j$ with $i \in F_v^{(\lambda(j,l))}$, as well as a new VNF instance $i'$ to be created providing sufficient computing resource in edge server $v$, i.e., $C_v^{re} \geq C(f^{(\lambda(j,l))})$. Then, $N_l$ is the set of VNF instances that can be employed as the $l$th network function $f_{j,l}$ in $SFC_j$ among all edge servers in $V$, i.e., $N_l = \cup_{v \in V} N_{l,v}$. We assume that the number of VNF instances of the same type in each edge server is a small constant. To this end, we construct an auxiliary, directed acyclic graph $G'_j = (V'_j, E'_j)$ for request $r_j$ from $G$ as follows.

Let $G'$ be a subgraph graph of $G$ after removing each link from $G$ if the residual bandwidth of the link is less than $\rho_j \cdot b_e$. The node set $V'_j$ of $G'_j$ is the union on sets $N_l$ of VNF instances with $1 \le l \le L_j$, with the source node $s_j$, the destination node set $D_j$ of multicast request $r_j$, i.e., $V'_j = \cup_{l=1}^{L_j} N_l \cup \{s_j\} \cup D_j$. To ensure that the network functions of $SFC_j = \langle f_{j,1}, \ldots, f_{j,l}, \ldots, f_{j,L_j} \rangle$ are traversed in order, we add a directed edge from a node $x \in N_{l-1}$ to each node $y \in N_l$ with $2 \le l \le L_j$ if there is a shortest path in graph $G'$ between $x$ and $y$, and the weight $w(x, y)$ assigned to the directed edge is the sum of *the communication cost* along the shortest path in $G'$ between the edge servers implementing VNF instances $x$ and $y$ and *the processing and VNF instance instantiation cost* of network function $y$. Notice that if the VNF instance is an existing one, its instantiation cost is 0; and if the two network functions $x$ and $y$ reside in the same edge server, their communication cost is 0. We then add a directed edge from $s_j$ to each node $y \in N_1$ if such a shortest path in $G'$ exists, and the weight assigned to the edge is the sum of the communication cost along the shortest path and the processing and VNF instance instantiation cost of network function $y$. Also, we add a directed edge from each node $x \in N_L$ to a node $y \in D_j$, and set the communication cost along the shortest path from an edge server that implements network function $x$ to the AP node $y$ as its weight if such a shortest path in $G'$ exists. Thus, $E'_j = \cup_{l=2}^{L_j} \{\langle x, y \rangle \mid x \in N_{l-1}, y \in N_l\} \cup \{\langle s_j, y \rangle \mid y \in N_1\} \cup \{\langle x, y \rangle \mid x \in N_{L_j}, y \in D_j\}$.

To ensure that a multicast request can be admitted without violating computing capacity of any edge server, it must be mentioned that we here adopt a conservative request admission strategy. That is, only if the residual computing capacity of an edge server is sufficient to accommodate all necessary VNF instance instantiating (any VNF instances in its $SFC_j$), it can be allowed to create new VNF instances for request $r_j$. Figure 5 shows the construction of graph $G'_j$ for request $r_j$ after the first $j - 1$ NFV-enabled multicast requests have been considered. Notice that if there is a shortest path between a VNF instance hosted in edge server $u \in V$ and another VNF instance hosted in edge server $v \in V$, there will be a shortest path between any pair of VNF instances hosted in these two edge servers, respectively. For simplicity, we use an edge in the graph to represent a set of edges between each pair of VNF instances residing in the two edge servers, respectively.

Having constructed graph $G'_j$, the cost minimization problem of the admission of request $r_j$ is reduced to find a directed multicast tree $T'(j)$ in $G'_j$ rooted at $s_j$ and spanning all nodes in $D_j$, such that the weighted sum of the edges in $T'(j)$ is minimized. Notice that the cost $c(T'(j))$ is *the minimum admission cost* of request $r_j$ in $G$. This is the classic directed Steiner tree problem, which is NP-hard. There is an approximate solution within $|D_j|^\epsilon$ times of the optimal one [5], where $\epsilon$ is a constant with $0 < \epsilon \le 1$. The value choice of $\epsilon$ reflects a tradeoff between the solution accuracy and the running time to obtain the solution. If the multicast tree

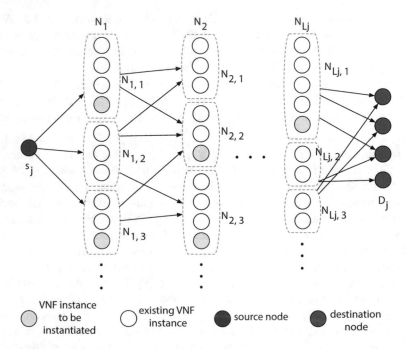

**Fig. 5** The auxiliary directed acyclic graph $G'_j$ for NFV-enabled multicast request $r_j$ consists of $L_j + 2$ layers from left to right, where layer 0 is the source node $s_j$ and layer $L_j + 1$ contains all destination nodes in $D_j$. Each layer $l$ with $1 \leq l \leq L_j$, consists of the VNF instances of type $\lambda(j, l)$ that can be deployed to process the data traffic of request $r_j$ in some of the edge servers $v \in V$, and if there is sufficient residual computing resource in an edge server, a new VNF instance of that type can be instantiated in edge server $v$ as well

$T'(j)$ in $G'_j$ rooted at $s_j$ and spanning all destinations in $D_j$ does exist, a pseudo-multicast tree $T(j)$ in $G$ rooted at $s_j$ and spanning all nodes in $D_j$ can then be derived. Specifically, we replace each directed edge in the multicast tree $T'(j)$ by a set of edges in its corresponding shortest path in $G$. The detailed description of the algorithm for the cost minimization problem is given in Algorithm 2, which is referred to as algorithm `Appro_Multi_CH`.

## 6.3 Algorithm Analysis

In the following, we show the correctness of the proposed algorithm, **Algorithm** `Appro_Multi_CH`, and analyze its approximation ratio and time complexity.

---

**Algorithm 2** Finding a minimum-cost pseudo-multicast tree in $G$ for request $r_j$, i.e., `Appro_Multi_CH`

---

**Input:** An MEC network $G = (V, E)$ with a set $V$ of edge servers. Assume that the first $j - 1$ NFV-enabled multicast requests have been considered, and some VNF instances have been instantiated for the admissions of requests. Now consider an NFV-enabled multicast request $r_j = (s_j, D_j, \rho_j, SFC_j)$.

**Output:** Admit or reject request $r_j$, and if $r_j$ is admitted, a pseudo-multicast tree $T(j)$ in $G$ will be delivered.

1: A subgraph $G'$ is obtained by removing all edges from $G$ whose residual bandwidth is strictly less than $\rho_j \cdot b_e$;

2: Compute all pairs shortest paths in $G'$ between each pair of AP nodes;

3: Construct the auxiliary directed acyclic graph $G'_j = (V'_j, E'_j)$ from $G$, and assign a weight on each edge in $E'_j$;

4: Find an approximate multicast tree $T'(j)$ in $G'_j$ rooted at $s_j$ and spanning all nodes in $D_j$, by applying the approximation algorithm on $G'_j$ due to Charikar et al. [5];

5: **if** $T'(j)$ in $G'_j$ exists **then**

6:   A pseudo-multicast tree $T(j)$ in $G$ is derived, by replacing each edge in $T'(j)$ with the edges of its corresponding shortest path in $G$;

7:   If a selected VNF instance is to be instantiated, create a new VNF instance in its edge server;

8:   Update residual resource capacities of links, edge servers, and VNF instances in $G$;

9: **else**

10:   Reject request $r_j$.

---

**Lemma 2** *An NFV-enabled multicast request $r_j$ is admissible in $G$ if and only if there is a multicast tree $T'(j)$ in graph $G'_j$ rooted at $s_j$ and spanning all nodes in $D_j$.*

**Proof** We first show that if there is a multicast tree $T'(j)$ in $G'$ rooted at $s_j$ and spanning all nodes in $D_j$, there is a feasible solution to the cost minimization problem for request $r_j$ in $G$. It can be seen that $G'_j$ contains $L_j + 2$ layers with source $s_j$ in layer 0 and all destination nodes in $D_j$ in layer $L_j + 1$. Thus, for each destination node $d \in D_j$, there is a directed path in $G'_j$ from $s_j$ to $d$ that goes through a node in each layer, which implies that each packet of request $r_j$ will be processed by either an existing VNF or a newly instantiated VNF of the network function in that layer, and the segment of the routing path in $G$ that corresponds a directed edge in $G'_j$ has sufficient communication bandwidth to meet the requirement of data traffic of $r_j$ in $G$. Thus, the solution delivered is a feasible solution.

We then show that if there does not exist a multicast tree $T'(j)$ in $G'_j$ rooted at $s_j$ and spanning all nodes in $D_j$, then request $r_j$ is inadmissible and should be rejected, i.e., there is not sufficient resources in $G$ to admit the request. Assume that a destination node $d \in D_j$ is not reachable from $s_j$ (or $d$ is not contained in $T'(j)$). Assume that node $v_l$ in layer $l$ is the smallest layer from which $d$ is reachable in $G'_j$ with $1 \le l \le L_j$, this implies that there is not any directed edge from any node in layer $l - 1$ to node $v_l$ in layer $l$. Following the construction of $G'_j$, there are three possibilities for absence of any such an edge: (1) among all edge servers in $G$, either none of them has sufficient computing resource to instantiate a new VNF for $f_{j,l} \in SFC_j$; or (2) all existing VNF instances of $f_{l,j}$ in these edge servers have less residual processing capacities for $f_{j,l}$; or (3) there is not any path in $G'$ from any node in layer $l - 1$ to a node (edge server) in layer $l$ due to the lack of communication bandwidth to meet the bandwidth requirement of $r_j$. In other words, it is lack of sufficient resources in $G$ to meet the resource demands of $r_j$, thus, it will be rejected.                                                                                          □

**Theorem 2** *Given an MEC network $G = (V, E)$ with a set $V$ of APs that each is attached an edge server, and an NFV-enabled multicast request $r_j = (s_j, D_j, \rho_j, SFC_j)$, there is an approximation algorithm, Algorithm Appro_ Multi_CH, for the cost minimization problem with an approximation ratio of $|D_j|^\epsilon$. The algorithm takes $O((L_j \cdot |V|)^{\frac{1}{\epsilon}} |D_j|^{\frac{2}{\epsilon}} + |V|^3)$ time, where $L_j (= |SFC_j|)$ is the length of $SFC_j$ of request $r_j$, and $\epsilon$ is a constant with $0 < \epsilon \leq 1$.*

**Proof** The solution obtained by the proposed algorithm Algorithm Appro_ Multi_CH is feasible, which has been shown by Lemma 2. In the following, we analyze the approximation ratio of the proposed algorithm. The admission cost of multicast request $r_j$ is the sum of (1) the VNF instance processing cost; (2) the VNF instance instantiation cost, and (3) the communication bandwidth usage cost. Each packet of the data traffic of request $r_j$ is transferred from the source node $s_j$ to each destination node in $D_j$ while passing through each VNF instance in its service function chain $SFC_j$. The sum of these three costs is assigned to each directed edge in $E'_j$. Thus, the cost of the minimum Steiner tree $T'(j)$ found in $G'_j$ rooted at $s_j$ and spanning all nodes in $D_j$, is the minimum admission cost of $r_j$ in $G$. Following [5], the approximation ratio of the proposed algorithm for the cost minimization problem for a single multicast request admission is $|D_j|^\epsilon$, where $\epsilon$ is a constant with $0 < \epsilon \leq 1$.

We finally analyze the time complexity of Algorithm Appro_Multi_CH as follows. Finding all pairs shortest paths in $G'$ between each pair of AP nodes takes $O(|V|^3)$ time, by invoking the well-known Floyd-Warshall algorithm. The construction of the auxiliary directed acyclic graph $G'_j$ for each request $r_j$ takes $O(L_j \cdot |V|)$ time, since there are $L_j + 2$ layers in $G'_j$ and each layer contains $O(|V|)$ nodes, assuming that the number of VNF instances of the same type in each edge server is a small constant, i.e., $O(1)$. Recall that $L_j = |SFC_j|$. Finding an approximate multicast tree in $G'_j$ for request $r_j$ takes time $O(|V'_j|^{\frac{1}{\epsilon}} |D_j|^{\frac{2}{\epsilon}})$, by applying the $(|D_j|^\epsilon)$-approximation algorithm due to Charikar et al. [5]. Thus, the running time of Algorithm Appro_Multi_CH is $O((L_j \cdot |V|)^{\frac{1}{\epsilon}} |D_j|^{\frac{2}{\epsilon}} + |V|^3)$ where $\epsilon$ is a constant with $0 < \epsilon \leq 1$.                                                                                          □

# 7 An Online Algorithm for the Online Throughput Maximization Problem

In this section, we study the online throughput maximization problem, where NFV-enabled multicast requests arrive one by one without the knowledge of future request arrivals. We first propose an online algorithm for the problem, through building a novel cost model to capture dynamic resource consumptions in $G$ and performing resource allocations for request admissions based on the built cost model. We then analyze the competitive ratio and time complexity of the proposed online algorithm.

## 7.1 The Usage Cost Model of Resources

The basic idea behind the proposed online algorithm is to regulate an online admission control policy to respond to each arrived request by either admitting or rejecting it, depending on the availability of its demanded resources and a given admission control policy. We still make use of the auxiliary directed acyclic graph $G'_j$ as an important data structure for the online throughput maximization problem. The weight assigned to each edge in $G'_j$ here however is different from its weight in the previous section that is defined as follows.

We here introduce a resource usage cost model to measure all different types of resource consumptions of each VNF instance (processing capacity), each edge server (computing resource), and each link (bandwidth resource) when admitting requests. Given the dynamics of resource demands of user requests and occupied resource releasing in the network, there is a need of a cost model to capture the dynamic consumptions of various resources in the network in order to assist the admissions of future requests and better utilize the resources. Intuitively, overloaded resources usually have higher probabilities to be violated by the resource demands of currently admitted requests, due to the high dynamics of resource consumptions. This eventually will affect the admissions of future requests. Therefore, if a type of resource has been highly utilized, it should be assigned a higher usage cost to reduce its usage in future; otherwise, it should be assigned a lower usage cost to encourage its usage in future.

The proposed online algorithm examines each incoming NFV-enabled multicast request one by one. When request $r_j$ arrives, the resource availabilities of the VNF instances of network functions in its service chain, computing resources in edge servers, and bandwidth resources in links will determine whether it is admissible. Recall that $F_v^{(k)}$ is the set of existing VNF instances of type $k$ in edge server $v$. If there is sufficient computing resource in edge server $v$, a new VNF instance of type $k$ can be instantiated at it. For the sake of convenience, assume that set $F_v^{(k)}$ contains the newly instantiated VNF instance of type $k$ as well.

Denote by $\mu_{v,i}^{(k)}(j)$ the residual processing capacity of the VNF instance $i \in F_v^{(k)}$ of type $k$ in edge server $v$ when request $r_j$ arrives with $\mu_{v,i}^{(k)}(0) = \mu^{(k)}$ initially. If request $r_j$ is admitted and its packets is processed by VNF instance $i$, then $\mu_{v,i}^{(k)}(j) = \mu_{v,i}^{(k)}(j-1) - \rho_j$, otherwise, its residual computing capacity does not change.

As for each network function in service function chain $SFC_j$ of $r_j$, a new VNF instance of it can be instantiated or an existing VNF instance of it can be shared, a binary variable $x_v^{(\lambda(j,l))}$ is introduced for each network function $f_{j,l}$ in $SFC_j$ with $1 \leq l \leq |SFC_j| = L_j$, where $x_v^{(\lambda(j,l))}$ is 1 if the $l$th VNF instance is newly instantiated in edge server $v$; otherwise 0. Then, denote by $C_v(j)$ the residual computing capacity at edge server $v \in V$ when request $r_j$ arrives with $C_v(0) = C_v$ initially. If request $r_j$ is admitted and some VNF instances are instantiated in edge server $v$, then $C_v(j) = C_v(j-1) - \sum_{l=1}^{L_j} C(f^{(\lambda(j,l))}) \cdot x_v^{(\lambda(j,l))}$. Similarly, denote by $B_e(j)$ the residual bandwidth in link $e \in E$ when request $r_j$ arrives with $B_e(j) = B_e(j-1) - \rho_j \cdot b_e$ if request $r_j$ is admitted and $B_e(0) = B_e$.

To capture the resource usage of request $r_j$, we use an exponential function to model the cost $W_{v,i}^{(k)}(j)$ of processing packets of $r_j$ by the VNF instance $i \in F_v^{(k)}$ as follows,

$$W_{v,i}^{(k)}(j) = \mu^{(k)}(\alpha^{1 - \frac{\mu_{v,i}^{(k)}(j)}{\mu^{(k)}}} - 1), \tag{1}$$

where $\alpha \ (> 1)$ is a tuning parameter to be decided later, and $1 - \frac{\mu_{v,i}^{(k)}(j)}{\mu^{(k)}}$ is the processing capacity utilization ratio in the VNF instance $i$ when request $r_j$ is considered. Similarly, the cost $W_v(j)$ of instantiating new VNF instances for request $r_j$ at edge server $v \in V$ and the cost $W_e(j)$ of using bandwidth resource at link $e \in B$ are defined, respectively,

$$W_v(j) = C_v(\beta^{1 - \frac{C_v(j)}{C_v}} - 1), \tag{2}$$

$$W_e(j) = B_e(\gamma^{1 - \frac{B_e(j)}{B_e}} - 1), \tag{3}$$

where $\beta \ (> 1)$ and $\gamma \ (> 1)$ are tuning parameters to be decided later, and $1 - \frac{C_v(j)}{C_v}$ and $1 - \frac{B_e(j)}{B_e}$ are the resource utilization ratios in edge server $v$ and link $e$, respectively, when request $r_j$ is considered. In order to encourage the sharing of VNF instances among multicast requests, we assume that the cost of creating a new VNF instance is much higher than the cost of processing capacity usage, i.e., $\beta \gg \alpha$.

We then define *the normalized usage cost* $\omega_{v,i}^{(k)}(j)$ of each VNF instance $i \in F_v^{(k)}$ in edge server $v$ for request $r_j$ as follows,

$$\omega_{v,i}^{(k)}(j) = W_{v,i}^{(j)}(j)/\mu^{(k)} = \alpha^{1 - \frac{\mu_{v,i}^{(k)}(j)}{\mu^{(k)}}} - 1. \tag{4}$$

Similarly, the normalized usage costs $\omega_v(j)$ at each edge server $v \in V$ and $\omega_e(j)$ at each link $e \in E$ for request $r_j$ are defined as follows,

$$\omega_v(j) = W_v(j)/C_v = \beta^{1-\frac{C_v(j)}{C_v}} - 1, \tag{5}$$

$$\omega_e(j) = W_e(j)/B_e = \gamma^{1-\frac{B_e(j)}{B_e}} - 1. \tag{6}$$

Having defined the usage costs of different resources in $G$, now consider the current incoming NFV-enabled multicast request $r_j$, we construct an auxiliary graph $G'_j = (V'_j, E'_j)$ which is almost identical to the one for the cost minimization problem. The difference lies in the weight assignment of edges in $G'_j$. Specifically, here the weight assigned to each directed edge in $E'_j$ is the sum of the three normalized constituent usage costs defined in (4), (5), and (6), respectively. That is, each edge $(x, y) \in E'_j$ has a weight

$$w(x, y) = \omega_{v,y}^{(\lambda(j,l))}(j) + \omega_v(j) + \sum_{e \in P(u,v)} \omega_e(j), \tag{7}$$

assuming that $x$ is a VNF instance in level $l - 1$ deployed in edge server $u$, $y$ is a VNF instance in level $l$ deployed in edge server $v$, $P(u, v)$ is a shortest path in $G$ between edge servers $u$ and $v$.

To avoid admitting requests that consume too much resources, thereby undermining the performance of the MEC, we adopt the following admission control policy. If (1) the sum of normalized usage costs of the VNF instances in its service function chain is greater than a given threshold $\sigma_1$, i.e., $\sum_{v \in V} \sum_{l=1}^{L_j} \sum_{i \in F_v^{(\lambda(j,l))}} \omega_{v,i}^{(\lambda(j,l))}(j) > \sigma_1$, where $L_j = |SFC_j|$; or (2) the sum of normalized usage costs of its VNF instantiations is greater than another given threshold $\sigma_2$, $\sum_{v \in V} \omega_v(j) > \sigma_2$; or (3) the sum of normalized usage costs of its bandwidth in links is greater than the third threshold $\sigma_3$, $\sum_{e \in E} \omega_e(j) > \sigma_3$, request $r_j$ will be rejected, where $\sigma_1 = \sigma_2 = \sigma_3 = n$, and $n = |V|$. The detailed algorithm for the online throughput maximization problem is given in Algorithm 3, which is referred to as algorithm `Online_Multi` in the rest of this chapter.

## 7.2   Algorithm Analysis

We now analyze the competitive ratio and time complexity of the proposed online algorithm, Algorithm `Online_Multi`. We first show the upper bound on the total cost of admitted requests. We then provide a lower bound on the cost of a rejected request by Algorithm `Online_Multi` but admitted by an optimal offline algorithm. We finally derive the competitive ratio of Algorithm `Online_Multi`.

---

**Algorithm 3** Online algorithm for the online throughput maximization problem, i.e., Online_Multi

---

**Input:** An MEC network $G = (V, E)$ with a set $V$ of APs, each $v \in V$ is attached an edge server with computing capacity $C_v$, a sequence of NFV-enabled multicast requests $r_j = (s_j, D_j, \rho_j, SFC_j)$ arriving one by one without the knowledge of future arrivals.

**Output:** Maximize the network throughput by admitting or rejecting each arrived request $r_j$ immediately. If $r_j$ admitted, a pseudo-multicast tree $T(j)$ for $r_j$ in $G$ from source node $s_j$ to a set of destination nodes in $D_j$ will be delivered.

1: **while** request $r_j$ arrives **do**
2:    A subgraph $G'$ of $G$ is constructed by removing each edge with residual bandwidth capacity less than $\rho_j \cdot b_e$;
3:    Construct the auxiliary graph $G'_j = (V'_j, E'_j)$ for request $r_j$, assign a weight to each edge in $E'_j$ according to Eq. (7);
4:    Find an approximate multicast tree $T'(j)$ in $G'_j$ rooted at $s_j$ and spanning all nodes in $D_j$, by applying the approximation algorithm on $G'_j$ due to Charikar et al. [5];
5:    **if** $T'(j)$ does not exist **then**
6:        Reject request $r_j$;
7:    **else**
8:        Determine whether $r_j$ will be accepted by the admission control policy;
9:        **if** $r_j$ is admissible **then**
10:           A pseudo-multicast tree $T(j)$ in $G$ is derived from $T'(j)$, by replacing each edge in $T'(j)$ by the edges in its corresponding shortest path in $G$;
11:           If a VNF instance in an edge server is to be instantiated, create the new VNF instance;
12:           Update residual resource capacities of VNF instances, links and edge servers in $G$;

---

**Lemma 3** *Given an MEC network $G = (V, E)$, with each edge server $v \in V$ has computing capacity $C_v$ and a set $E$ of links that each link $e \in E$ has bandwidth capacity $B_e$, denote by $\mathcal{A}(j)$ the set of NFV-enabled multicast requests admitted by the algorithm, Algorithm* Online_Multi*, until the arrival of request $r_j$. Then, the cost sums of VNF instances, edge servers, and links when multicast request $r_j$ arrives are*

$$\sum_{v \in V} \sum_{l=1}^{L_j} \sum_{i \in F_v^{(\lambda(j,l))}} W_{v,i}^{(\lambda(j,l))}(j) \leq 2n \log \alpha \cdot \mathbb{B}(j), \tag{8}$$

$$\sum_{v \in V} W_v(j) \leq 2nL_{max} \log \beta \cdot |\mathcal{A}(j)| \cdot C(f_{max}), \tag{9}$$

$$\sum_{e \in E} W_e(j) \leq 2n \log \gamma \cdot \mathbb{B}(j), \tag{10}$$

*respectively, provided that the maximum length $L_{max}$ of any service function chain is no greater than $n$, i.e., $L_{max} = \max_{1 \leq j' \leq j}\{|SFC_{j'}|\} \leq n$, and $\rho_{j'} \leq \frac{\min_{1 \leq l \leq L_{j'}}\{\mu^{(\lambda(j',l))}\}}{\log \alpha}$, $\sum_{l=1}^{L_{j'}} C(f^{(\lambda(j',l))}) \cdot x_v^{(\lambda(j',l))} \leq \frac{\min_{v \in V} C_v}{\log \beta}$, $\rho_{j'} \cdot b_e \leq \frac{\min_{e \in E} B_e}{\log \gamma}$*

with $1 \leq j' \leq j$, where $\sum_{l=1}^{L_{j'}} C(f^{(\lambda(j',l))}) \cdot x_v^{(\lambda(j',l))}$ is the computing resource being occupied by newly instantiated VNF instances in edge server $v$ for request $r_{j'}$, $\mathbb{B}(j)$ is the accumulative bandwidth resource being occupied by the admitted requests, i.e., $\mathbb{B}(j) = \sum_{r_{j'} \in \mathcal{A}(j)} \rho_{j'} \cdot b_e$, and $C(f_{max})$ is the maximum computing resource required among all VNF instance types, i.e., $C(f_{max}) = \max_{1 \leq k \leq K} \{C(f^{(k)})\}$.

**Proof** Consider a request $r_{j'} \in \mathcal{A}(j)$ admitted by Algorithm Online_Multi. For any VNF instance $i \in F_v^{(k)}$, we have

$$W_{v,i}^{(k)}(j'+1) - W_{v,i}^{(k)}(j') = \mu^{(k)}(\alpha^{1-\frac{\mu_{v,i}^{(k)}(j'+1)}{\mu^{(k)}}} - 1) - \mu^{(k)}(\alpha^{1-\frac{\mu_{v,i}^{(k)}(j')}{\mu^{(k)}}} - 1)$$

$$= \mu^{(k)} \alpha^{1-\frac{\mu_{v,i}^{(k)}(j')}{\mu^{(k)}}} (\alpha^{\frac{\mu_{v,i}^{(k)}(j') - \mu_{v,i}^{(k)}(j'+1)}{\mu^{(k)}}} - 1) = \mu^{(k)} \alpha^{1-\frac{\mu_{v,i}^{(k)}(j')}{\mu^{(k)}}} (\alpha^{\frac{\rho_{j'}}{\mu^{(k)}}} - 1)$$

$$= \mu^{(k)} \alpha^{1-\frac{\mu_{v,i}^{(k)}(j')}{\mu^{(k)}}} (2^{\frac{\rho_{j'}}{\mu^{(k)}} \log \alpha} - 1) \leq \mu^{(k)} \alpha^{1-\frac{\mu_{v,i}^{(k)}(j')}{\mu^{(k)}}} \cdot \frac{\rho_{j'}}{\mu^{(k)}} \cdot \log \alpha \qquad (11)$$

$$= \alpha^{1-\frac{\mu_{v,i}^{(k)}(j')}{\mu^{(k)}}} \cdot \rho_{j'} \cdot \log \alpha, \qquad (12)$$

where Ineq. (11) holds due to that $2^a - 1 \leq a$ for $0 \leq a \leq 1$.

Similarly, for any edge server $v \in V$, we have $W_v(j'+1) - W_v(j') \leq \beta^{1-\frac{C_v(j')}{C_v}} (\sum_{l=1}^{L_{j'}} C(f^{(\lambda(j',l))}) \cdot x_v^{(\lambda(j',l))}) \log \beta$ and for any link $e \in E$, we have $W_e(j'+1) - W_e(j') \leq \gamma^{1-\frac{B_e(j')}{B_e}} \cdot \rho_{j'} \cdot b_e \cdot \log \gamma$.

We then calculate the cost sum of all VNF instances when admitting request $r_{j'}$. The difference of the cost sum of VNF instances before and after admitting request $r_{j'}$ is

$$\sum_{v \in V} \sum_{l=1}^{L_{j'}} \sum_{i \in F_v^{(\lambda(j',l))}} W_{v,i}^{(k)}(j'+1) - W_{v,i}^{(k)}(j') = \sum_{l=1}^{L_{j'}} W_{v,i}^{(k)}(j'+1) - W_{v,i}^{(k)}(j')$$

$$\qquad (13)$$

$$\leq \sum_{l=1}^{L_{j'}} \alpha^{1-\frac{\mu_{v,i}^{(\lambda(j',l))}(j')}{\mu^{(\lambda(j',l))}}} \cdot \rho_{j'} \cdot \log \alpha, \quad \text{by Ineq. (12)}$$

$$= \rho_{j'} \cdot \log \alpha \sum_{l=1}^{L_{j'}} \alpha^{1-\frac{\mu_{v,i}^{(\lambda(j',l))}(j')}{\mu^{(\lambda(j',l))}}} = \rho_{j'} \cdot \log \alpha \left( \sum_{l=1}^{L_{j'}} (\alpha^{1-\frac{\mu_{v,i}^{(\lambda(j',l))}(j')}{\mu^{(\lambda(j',l))}}} - 1) + \sum_{l=1}^{L_{j'}} 1 \right)$$

$$= \rho_{j'} \cdot \log \alpha \left( \sum_{l=1}^{L_{j'}} \omega_{v,i}^{(k)}(j') + L_{j'} \right) \leq 2n\rho_{j'} \cdot \log \alpha \qquad (14)$$

Ineq. (12) holds since $\sum_{i=1}^{n} A_i \cdot B_i \leq \sum_{i=1}^{n} A_i \cdot \sum_{i=1}^{n} B_i$, for all $A_i \geq 0$ and $B_i \geq 0$. Equation (13) holds due to that for each network function $f_{j',l}$, only one VNF instance is employed to process data traffic of request $r_{j'}$. Ineq. (14) holds due to the fact that if request $r_{j'}$ is admitted, the admission control policy is met, i.e., $\sum_{l=1}^{L_{j'}} \omega_{v,i}^{(\lambda(j',l))}(j') = \sum_{l=1}^{L_{j'}} \alpha^{1 - \frac{\mu_{v,i}^{(\lambda(j',l))}(j')}{\mu^{(\lambda(j',l))}}} - 1 \leq \sigma_1 = n$, and the length of service function chain of request $r_{j'}$ is less than the number of APs, i.e., $|SFC_{j'}| = L_{j'} \leq L_{max} \leq n$.

Similarly, the difference of the cost sum of edge servers before and after admitting request $r_{j'}$ is $\sum_{v \in V} W_v(j'+1) - W_v(j') \leq 2nL_{j'} \cdot C(f_{max}) \cdot \log \beta$, where $C(f_{max})$ is the maximum computing resource consumption of any VNF instance $f^{(k)}$, $1 \leq k \leq K$ in the MEC. And the difference of the cost sum of links before and after admitting request $r_{j'}$ is $\sum_{e \in E} W_e(j'+1) - W_e(j') \leq 2n\rho_{j'} \cdot b_e \cdot \log \gamma$.

The cost sum of VNF instances for request admissions when $r_j$ arrives thus is

$$
\sum_{v \in V} \sum_{l=1}^{L_j} \sum_{i \in F_v^{(k)}} W_{v,i}^{(k)}(j) = \sum_{j'=1}^{j-1} \sum_{v \in V} \sum_{l=1}^{L_{j'}} \sum_{i \in F_v^{(k)}} W_{v,i}^{(k)}(j'+1) - W_{v,i}^{(k)}(j') \tag{15}
$$

$$
= \sum_{r_{j'} \in \mathcal{A}(j)} \sum_{v \in V} \sum_{l=1}^{L_{j'}} \sum_{i \in F_v^{(\lambda(j',l))}} (W_{v,i}^{(k)}(j'+1) - W_{v,i}^{(k)}(j'))
$$

$$
\leq \sum_{r_{j'} \in \mathcal{A}(j)} 2n\rho_{j'} \cdot \log \alpha, \quad \text{by Ineq. (14)}
$$

$$
= 2n \log \alpha \sum_{r_{j'} \in \mathcal{A}(j)} \rho_{j'} = 2n \log \alpha \cdot \mathbb{B}(j),
$$

where Eq. (15) follows from the fact that if a request is not admitted, none of the processing capacity of any VNF instance will be consumed.

Similarly, the cost sum of edge servers for request admissions when $r_j$ arrives is $\sum_{v \in V} W_v(j) \leq 2nL_{max} \log \beta \cdot |\mathcal{A}(j)| \cdot C(f_{max})$, and the cost sum of links for request admissions when $r_j$ arrives is $\sum_{e \in E} W_e(j) \leq 2n \log \gamma \cdot \mathbb{B}(j)$. $\qquad \square$

We now provide a lower bound on the weight of a rejected request by Algorithm Online_Multi but admitted by an optimal offline algorithm denoted by $OPT$. Before we proceed, we choose appropriate values for $\alpha$, $\beta$, and $\gamma$ prior to the arrival of any request $r_j$ and VNF instance $k$, $1 \leq k \leq K$ as follows.

$$
2n+2 \leq \alpha \leq \min_{1 \leq k \leq K} \{2^{\frac{\mu^{(k)}}{\rho_j}}\} \tag{16}
$$

$$
2n+2 \leq \beta \leq \min_{1 \leq k \leq K} \min_{v \in V} \{2^{\frac{C_v}{C(f^{(k)})}}\} \tag{17}
$$

$$
2n+2 \leq \gamma \leq \min_{e \in E} \{2^{\frac{B_e}{\rho_j \cdot b_e}}\} \tag{18}
$$

**Lemma 4** *Let $\mathcal{T}(j)$ be the set of requests that are rejected by* Algorithm Online_Multi *but admitted by the optimal offline algorithm $OPT$ prior to the arrival of request $r_j$. Then, for any request $r_{j'} \in \mathcal{T}(j)$, we have*

$$\sum_{v \in V} \sum_{l=1}^{L_{j'}} \sum_{i \in F_v^{(\lambda(j',l))}} \omega_{v,i}^{(\lambda(j',l))}(j') + \sum_{v \in V} \omega_v(j') + \sum_{e \in E} \omega_e(j') > \min\{\sigma_1, \sigma_2, \sigma_3\} = n.$$

*Proof* Consider a request $r_{j'}$ that is admitted by the optimal offline algorithm $OPT$ yet rejected by Algorithm Online_Multi. A request $r'_j$ will be rejected by Algorithm Online_Multi by one of the four cases: (1) at least one VNF instance does not have sufficient processing capacity to admit request $r_{j'}$; (2) there is no sufficient computation resource in edge servers to create new VNF instances for request $r_{j'}$ as required; (3) there is no sufficient bandwidth in $G$ for routing its data traffic; or (4) the sum of normalized usage costs is too high, in other words, the admission control policy is not met.

Case (1)   At least one VNF instance $i'$ of type $k'$ in edge server $v'$ does not have sufficient processing capacity to process data traffic of request $r_{j'}$, i.e., $\mu_{v',i'}^{(k')}(j') < \rho_{j'}$. We then have

$$\sum_{v \in V} \sum_{l=1}^{L_{j'}} \sum_{i \in F_v^{(\lambda(j',l))}} \omega_{v,i}^{(\lambda(j',l))}(j') \geq \omega_{v',i'}^{(k')}(j') \qquad (19)$$

$$= \alpha^{1-\frac{\mu_{v',i'}^{(k')}(j')}{\mu^{(k')}}} - 1 > \alpha^{1-\frac{\rho_{j'}}{\mu^{(k')}}} - 1, \quad \text{since } \mu_{v',i'}^{(k')}(j') < \rho_{j'}$$

$$\geq \alpha^{1-\frac{1}{\log \alpha}} - 1 = \frac{\alpha}{2} - 1 \geq n, \quad \text{by Ineq. (16)}$$

Case (2)   At least one edge server $v' \in V$ does not have sufficient capacity to create a new instance for a VNF of type $k'$ in $SFC_{j'}$ as required, i.e., $C_{v'}(j') < C(f^{(k')})$. Similarly, we have $\sum_{v \in V} \omega_v(j') \geq \omega_{v'}(j') \geq \frac{\beta}{2} - 1 \geq n$.

Case (3)   If request $r_{j'}$ is rejected, then there is an edge $e' \in E$ that does not have sufficient residual bandwidth to accommodate the request. This implies that $B_{e'}(j') < \rho_{j'} \cdot b_e$. Therefore, the normalized cost sum of $E$ is greater than $\sigma_3$, i.e., $\sum_{e \in E} \omega_e(j') \geq \omega_{e'}(j') \geq \frac{\gamma}{2} - 1 \geq n$.

Case (4)   Although there are sufficient resources to admit request $r_{j'}$, $r_{j'}$ is rejected by Algorithm Online_Multi due to not meeting the admission control policy. That is

$$\sum_{v \in V} \sum_{l=1}^{L_{j'}} \sum_{i \in F_v^{(\lambda(j',l))}} \omega_{v,i}^{(\lambda(j',l))}(j') + \sum_{v \in V} \omega_v(j') + \sum_{e \in E} \omega_e(j') > \min\{\sigma_1, \sigma_2, \sigma_3\} = n.$$

$$(20)$$

Lemma 4 thus follows.

□

We finally analyze the competitive ratio of Algorithm Online_Multi.

**Theorem 3** *Given an MEC network $G = (V, E)$ with a set $V$ of APs in which each $v \in V$ is attached an edge server with computing capacity $C_v$, each link $e \in E$ has bandwidth capacity $B_e$, there is an online algorithm, Algorithm Online_Multi, with competitive ratio of $O(\log n)$ for the online throughput maximization problem, and the algorithm takes $O((L_j \cdot |V|)^{\frac{1}{\epsilon}} |D_j|^{\frac{2}{\epsilon}})$ time to admit each request $r_j$ where $n = |V|$, $L_j = |SFC_j|$, and $\epsilon$ is a constant with $0 < \epsilon \le 1$.*

**Proof** Denote by $D_{max}$ and $\rho_{max}$ the maximum cardinality of destination set $D_{j'}$ and the maximum packet rate of request $r_{j'}$ among all requests respectively, prior to the arrival of request $r_j$, i.e., $D_{max} = \max_{1 \le j' \le j}\{D_{j'}\}$, and $\rho_{max} = \max_{1 \le j' \le j}\{\rho_{j'}\}$. We first analyze the competitive ratio of the proposed online algorithm. We here abuse the notation $OPT$ to denote the optimal offline algorithm $OPT$ and the number of requests admitted by it. Let $\mathcal{A}(j)$ be the set of admitted requests when request $r_j$ arrives, we have

$$\frac{n}{D_{max}^\epsilon}(OPT - |\mathcal{A}(j)|) \le \frac{n}{D_{max}^\epsilon} \sum_{r_{j'} \in \mathcal{T}(j)} 1 \le \sum_{r_{j'} \in \mathcal{T}(j)} n \qquad (21)$$

$$\le \sum_{r_{j'} \in \mathcal{T}(j)} \sum_{v \in V} \sum_{l=1}^{L_{j'}} \sum_{i \in F_v^{(\lambda(j',l))}} \omega_{v,i}^{(\lambda(j',l))}(j') + \sum_{r_{j'} \in \mathcal{T}(j)} \sum_{v \in V} \omega_v(j') + \sum_{r_{j'} \in \mathcal{T}(j)} \sum_{e \in E} \omega_e(j')$$

$$\le \sum_{r_{j'} \in \mathcal{T}(j)} \sum_{v \in V} \sum_{l=1}^{L_j} \sum_{i \in F_v^{(\lambda(j,l))}} \omega_{v,i}^{(\lambda(j,l))}(j) + \sum_{r_{j'} \in \mathcal{T}(j)} \sum_{v \in V} \omega_v(j) + \sum_{r_{j'} \in \mathcal{T}(j)} \sum_{e \in E} \omega_e(j), \qquad (22)$$

$$= \sum_{r_{j'} \in \mathcal{T}(j)} \sum_{v \in V} \sum_{l=1}^{L_j} \sum_{i \in F_v^{(\lambda(j,l))}} \frac{W_{v,i}^{(\lambda(j,l))}(j)}{\mu^{(\lambda(j,l))}} + \sum_{r_{j'} \in \mathcal{T}(j)} \sum_{v \in V} \frac{W_v(j)}{C_v} + \sum_{r_{j'} \in \mathcal{T}(j)} \sum_{e \in E} \frac{W_e(j)}{B_e}$$

$$= \sum_{v \in V} \sum_{l=1}^{L_j} \sum_{i \in F_v^{(\lambda(j,l))}} W_{v,i}^{(\lambda(j,l))}(j) \sum_{r_{j'} \in \mathcal{T}(j)} \frac{1}{\mu^{(\lambda(j,l))}} +$$

$$\sum_{v \in V} W_v(j) \sum_{r_{j'} \in \mathcal{T}(j)} \frac{1}{C_v} + \sum_{e \in E} W_e(j) \sum_{r_{j'} \in \mathcal{T}(j)} \frac{1}{B_e} \qquad (23)$$

$$\leq \sum_{v \in V} \sum_{l=1}^{L_j} \sum_{i \in F_v^{(\lambda(j,l))}} W_{v,i}^{(\lambda(j,l))}(j) + \sum_{v \in V} W_v(j) + \sum_{e \in E} W_e(j) \tag{24}$$

$$\leq 2n\mathbb{B}(j) \log \alpha + 2n L_{max} C(f_{max}) \log \beta \cdot |\mathcal{A}(j)| + 2n\mathbb{B}(j) \log \gamma$$

$$\leq 2n|\mathcal{A}(j)|\big(\rho_{max} \log \alpha + L_{max} \cdot C(f_{max}) \log \beta + \rho_{max} \cdot b_e \log \gamma\big).$$

Ineq. (21) holds since $D_{max} \geq 1$, and $0 < \epsilon \leq 1$, thus $D_{max}^\epsilon \geq 1$. Ineq. (22) holds since the resource utilization ratio does not decrease and thus the usage cost of each VNF instance, each edge server, and each link does not decrease with more request admissions. Ineq. (23) holds because $\sum_{i=1}^m \sum_{j=1}^n A_i \cdot B_j \leq \sum_{i=1}^m A_i \cdot \sum_{j=1}^n B_j$, for all $A_i \geq 0$ and $B_j \geq 0$. Ineq. (24) holds because all algorithms, including the optimal offline algorithm $OPT$, the accumulated usage of resources in any VNF instance, edge server and link is no greater than its capacity.

Recall that $\mathcal{A}(j)$ is the set of requests admitted by Algorithm Online_Multi, and $\mathcal{T}(j)$ is the set of requests rejected by Algorithm Online_Multi but accepted by the optimal offline algorithm $OPT$. We have $\frac{OPT - |\mathcal{A}(j)|}{|\mathcal{A}(j)|} \leq 2D_{max}^\epsilon(\rho_{max} \log \alpha + L_{max} \cdot C(f_{max}) \log \beta + \rho_{max} \cdot b_e \log \gamma)$. Thus, we have $\frac{OPT}{|\mathcal{A}(j)|} \leq 2D_{max}^\epsilon(\rho_{max} \log \alpha + L_{max} \cdot C(f_{max}) \log \beta + \rho_{max} \cdot b_e \log \gamma) + 1 = O(\log n)$ when $\alpha = \beta = \gamma = O(n)$.

$\square$

# 8 Performance Evaluation

In this section, we evaluate the performance of the proposed algorithms for the admissions of NFV-enabled multicasting requests through experimental simulations.

## 8.1 Experiment Settings

We consider an MEC network $G = (V, E)$ consisting of from 10 to 250 APs (edge servers). All network topologies are generated by the tool GT-ITM [8]. The computing capacity of each edge server is set in the range from 2000 to 5000 MHz [13], while the bandwidth capacity of each link varies from 2000 to 20,000 Mbps [15]. The number of different types of network functions $K$ is set at 30. The computing resource demand of each network function is set from 300 to 600 MHz randomly, and their processing rate is also randomly drawn from 50 to 100 data packets per millisecond [22]. Recall that the admission cost of an NFV-enabled multicast request consists of three components: the VNF instance processing cost, the VNF instance instantiation cost, and the bandwidth usage cost, where the instantiation cost of a VNF instance in an edge server is randomly drawn in the interval [0.50, 2.0], while the processing cost of per packet by a VNF instance is a random value drawn from [0.01, 0.1] [27]. The routing cost per data packet along a link is a value drawn randomly from the interval [0.01, 0.1]. To generate request $r_j$, one AP node in $V$ is randomly selected as its source $s_j$, and a set of AP nodes in $V$ are randomly chosen as its destination set $D_j$. The data packet rate is drawn from 2 to 10 packets per millisecond [17], where each data packet is of size 64 KB. The length of its service function chain is set from 5 to 20, and each network function is randomly drawn from the $K$ types. The value in each figure is the mean of the results out of 30 MEC instances of the same size. The running time of an algorithm is obtained on a machine with 4.0 GHz Intel i7 Quad-core CPU and 32 GB RAM. Unless otherwise specified, these parameters will be adopted in the default setting.

In the following, we first evaluate the performance of the proposed approximation algorithms for the minimum cost problem against four baseline heuristics Alg_One_Server, CostMinGreedy, ExistingGreedy, and NewGreedy. Algorithm Alg_One_Server [35] only uses a single edge server to implement service chain $SC_k$ of each NFV-enabled multicast request $r_k$. Algorithm CostMinGreedy considers network functions in the service function chain one by one, it always chooses the edge server with the minimum admission cost (including the processing cost, instance instantiation cost, and routing cost) for the next network function. Algorithm ExistingGreedy considers network functions one by one and tries to admit the request by existing VNF instances with the minimum admission cost as long as there is a VNF instance with sufficient residual processing capacity, while algorithm NewGreedy always aims to instantiate a new VNF instance for the request providing sufficient computation resource in an edge server. We finally evaluate the performance of the proposed online algorithm against a benchmark OnlineLinear for the online throughput maximization problem, where for each arrived request, algorithm OnlineLinear first excludes those VNF instances, edge servers and links that do not have sufficient residual resources to accommodate the admission of the request from the consideration, it then assigns a cost to each VNF instance, each edge server, and each link, and constructs an auxiliary directed acyclic graph for the request. It finally finds a multicast tree rooted at the source node and spanning all destination nodes for the request.

## 8.2 Performance Evaluation of Approximation Algorithm for the Cost Minimization Problem with Consolidated VNFs

We first evaluate the performance of algorithm Appro_Multi against that of algorithm Alg_One_Server by varying the network size from 50 to 250 and the ratio of the maximum number $D_{max}$ of destinations of each request to the network size $|V|$ from 0.05 to 0.2. The operational cost and running time curves delivered by algorithms Appro_Multi and Alg_One_Server are drawn in Fig. 6, where the

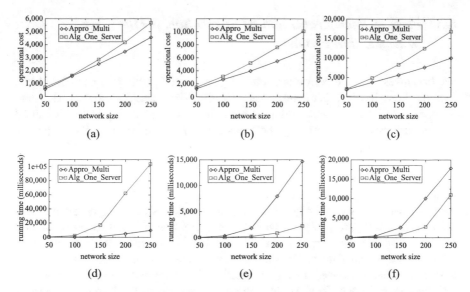

**Fig. 6** The performance of algorithms Appro_Multi and Alg_One_Server with different ratios of $D_{max}/|V|$. (**a**) Operational costs with $D_{max}/|V| = 0.05$. (**b**) Operational costs with $D_{max}/|V| = 0.1$. (**c**) Operational costs with $D_{max}/|V| = 0.2$. (**d**) Running times with $D_{max}/|V| = 0.05$. (**e**) Running times with $D_{max}/|V| = 0.1$. (**f**) Running times with $D_{max}/|V| = 0.2$

operational costs and running times are the average of admitting 1000 NFV-enabled multicast requests. Specifically, we can see from Fig. 6a that the operational cost by algorithm Appro_Multi is around 80% of that of algorithm Alg_One_Server. The reason is that algorithm Appro_Multi may use multiple edge servers that are close to the destinations of the request to implement the service chain of the request, which can significantly reduce the cost of bandwidth resource usage. Furthermore, it can be seen from the figure that the performance gap between the two algorithms becomes larger and larger, with the increase on the network size. The rationale behind is that algorithm Appro_Multi has more chances to select a set of edge servers that are closer to the destinations of each request, considering that more edge servers in larger networks are to be chosen. The similar performance behavior can be observed from Fig. 6b and c. Furthermore, it can be seen from Fig. 6d–f that approximation algorithm Appro_Multi takes a slightly more time than that of algorithm Alg_One_Server, as different combinations of edge servers in $V_S$ are to be considered.

## 8.3 Performance Evaluation of Algorithms for the Cost Minimization Problem

We first investigate the performance of Algorithm Appro_Multi_CH against that of three baseline heuristics CostMinGreedy, ExistingGreedy, and NewGreedy, for the cost minimization problem of a single NFV-enabled request admission, by varying the network size from 10 to 250. Figure 7 illustrates the admission cost and running time of the four mentioned algorithms. From Fig. 7a, we can see that Algorithm Appro_Multi_CH achieves a much lower admission cost than those three benchmarks. Specifically, Algorithm Appro_Multi_CH is only 43.1%, 24.0%, and 14.4% of the admission costs of algorithms NewGreedy, ExistingGreedy, and CostMinGreedy, respectively, when the network size is 250. The reason behind is that Algorithm Appro_Multi_CH jointly considers the placement of VNF instances and data traffic routing for a request admission, it also makes a smart decision between using an existing VNF instance or creating a new VNF instance. Figure 7b plots the running time curves of the four comparison algorithms. It can be seen that algorithm NewGreedy achieves the least running time, as it gives priority to create new VNF instances in edge servers, while Algorithm Appro_Multi_CH takes the most running time due to the fact that it strives for finding a multicast tree with the least cost while passing through VNFs in its service function chain at the same time.

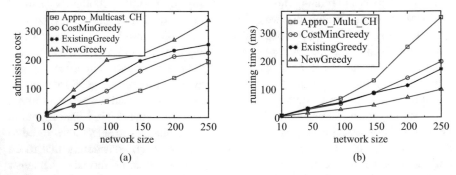

**Fig. 7** Performance of Algorithm Appro_Multi_CH, CostMinGreedy, Existing Greedy, and NewGreedy, by varying the network size. (**a**) The admission cost. (**b**) The running time

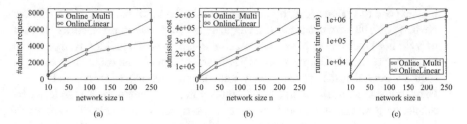

**Fig. 8** Performance of Algorithm Online_Multi and OnlineLinear by varying the network size from 10 to 250. (**a**) The network throughput. (**b**) The admission cost. (**c**) The running time

## 8.4 Performance Evaluation of the Online Algorithm for the Throughput Maximization Problem

We now evaluate the performance of Algorithm Online_Multi for the online throughput maximization problem, by varying the network size from 10 to 250 for a sequence of 10,000 requests. Figure 8 plots the performance curves of different algorithms, from which we can see that Algorithm Online_Multi outperforms the baseline algorithm OnlineLinear in all cases, and Algorithm Online_Multi can admit 38.6% more requests than that by algorithm OnlineLinear when the network size is 200. Figure 8c shows the running time of the two comparison algorithms.

## 9 Conclusion and Future Research Directions

In this chapter, we studied NFV-enabled multicast request admissions in an MEC network by formulating three novel optimization problems. We first proposed approximation algorithms with provable approximation ratios for the cost minimization problem of a single NFV-enabled multicast request admission. We then investigated the online throughput maximization problem where NFV-enabled multicast requests arrive one by one without the knowledge of future request arrivals, by devising an online algorithm with a provable competitive ratio for it. We finally evaluated the performance of the proposed algorithms through experimental simulations. Simulation results demonstrate that the proposed algorithms are very promising, and exhibits better performance compared with their counterparts.

Apart from the above studies on secure multicasting in MEC networks, there are several potential directions in MEC networks built upon this research, which are listed as follows.

- **Security-aware multicasting:** Besides of the placement and chaining of virtualized security functions for multicasting, the virtualized security functions may expose to various security attacks, such as denial of services (DoS). In the chaining of VNFs for each multicast request, how to consider the security statuses of VNF instances and edge servers is a challenging issue. Specifically, a potential research direction is to jointly place VNFs and find multicast routes for multicast requests, such that the security, delay, and chaining requirements are met.

- **Privacy-aware multicasting:** Most security network functions are stateful, by maintaining states of connection, session or transaction levels. The states of security network functions are very sensitive and may cause security breaches if the state data is exposed to attackers. To guarantee the correct operation of security network functions in the process of multicasting data traffic, a fundamental problem is how to place VNF instances to edge servers with privacy guarantees of their status data is another future research direction.
- **Security-aware VNF migrations:** VNF instances in an MEC environment may be exposed to various unpredictable attacks. Once such attacks happen, resilient migrations of the VNF instances of each multicast request is an important problem. Specific challenges include: (1) how to find secure edge servers for each multicast request, considering the uncertain arrivals of attacks, and (2) how to identify low-cost back-up paths for each multicast request to enable fast flow migrations when security breach happen.

# References

1. Abbas, N., Zhang, Y., Taherkordi, A., Skeie, T.: Mobile edge computing: a survey. IEEE Internet Things J. **5**(1), 450–465 (2018)
2. Alhussein, O., Do, P.T., Li, J., Ye, Q., Shi, W., Zhuang, W., Shen, X., Li, X., Rao, J.: Joint VNF placement and multicast traffic routing in 5G core networks. In: 2018 IEEE Global Communications Conference (GLOBECOM). IEEE, Piscataway (2018)
3. Ceselli, A., Premoli, M., Secci, S.: Mobile edge cloud network design optimization. IEEE/ACM Trans. Netw. **25**(3), 1818–1831 (2017)
4. Chang, C.S., Zajic, T.: Effective bandwidths of departure processes from queues with time varying capacities. In: Proceedings of INFOCOM'95 . IEEE, Piscataway (1995)
5. Charikar, M., Chekuri, C., Cheung, T.-Y., Dai, Z., Goel, A., Guha, S., Li, M.: Approximation algorithms for directed Steiner problems. J. Algor. **33**(1), 73–91 (1998)
6. Chen, M., Hao, Y.: Task offloading for mobile edge computing in software defined ultra-dense network. IEEE J. Sel. Areas Commun. **36**(3), 587–597 (2018)
7. Feng, H., Llorca, J., Tulino, A.M., Raz, D., Molish, A.F.: Approximation algorithms for the NFV service distribution problem. In: IEEE INFOCOM 2017-IEEE Conference on Computer Communications. IEEE, Piscataway (2017)
8. GT-ITM (2019). http://www.cc.gatech.edu/projects/gtitm/
9. He, T., Khamfroush, H., Wang, S., La Porta, T., Stein, S.: It's hard to share: joint service placement and request scheduling in edge clouds with sharable and non-sharable resources. In: 2018 IEEE 38th International Conference on Distributed Computing Systems (ICDCS). IEEE, Piscataway (2018)
10. Jia, M., Liang, W., Xu, Z., Huang, M.: Cloudlet load balancing in wireless metropolitan area networks. In: IEEE INFOCOM 2016-The 35th Annual IEEE International Conference on Computer Communications. IEEE, Piscataway (2016)

11. Jia, M., Cao, J., Liang, W.: Optimal cloudlet placement and user to cloudlet allocation in wireless metropolitan area networks. IEEE Trans. Cloud Comput. **5**(4), 725–737 (2017)
12. Jia, M., Liang, W., Huang, M., Xu, Z., Ma, Y.: Throughput maximization of NFV-enabled unicasting in software-defined networks. In: GLOBECOM 2017-2017 IEEE Global Communications Conference. IEEE, Piscataway (2017)
13. Jia, M., Liang, W., Xu, Z.: QoS-aware task offloading in distributed cloudlets with virtual network function services. In: Proceedings of the 20th ACM International Conference on Modelling, Analysis and Simulation of Wireless and Mobile Systems . ACM, New York (2017)
14. Jia, M., Liang, W., Huang, M., Xu, Z., Ma, Y.: Routing cost minimization and throughput maximization of NFV-enabled unicasting in software-defined networks. IEEE Trans. Netw. Serv. Manag. **15**(2), 732–745 (2018)
15. Knight, S.. Knight, S., Nguyen, H.X., Falkner, N., Bowden, R., Roughan, M.: The internet topology zoo. IEEE J. Sel. Areas Commun. **29**, 1765–1775 (2011)
16. Kou, L., Markowsy, G., Berman, L.: A faster algorithm for Steiner trees. Acta Inform. **15**, 141–145 (1981)
17. Li, Y., Phan, L.T.X., Loo, B.T.: Network functions virtualization with soft real-time guarantees. In: IEEE INFOCOM 2016-The 35th Annual IEEE International Conference on Computer Communications. IEEE, Piscataway (2016)
18. Ma, Y., Liang, W., Xu, Z.: Online revenue maximization in NFV-enabled SDNs. In: 2018 IEEE International Conference on Communications (ICC). IEEE, Piscataway (2018)
19. Ma, Y., Liang, W., Wu, J.: Online NFV-enabled multicasting in mobile edge cloud networks. In: 2019 IEEE 39th International Conference on Distributed Computing Systems (ICDCS). IEEE, Piscataway (2019)
20. Ma, Y., Liang, W., Xu, Z., Guo, S.: Profit maximization for admitting requests with network function services in distributed clouds. IEEE Trans. Parall. Distrib. Syst. **30**(5), 1143–1157 (2019)
21. Mao, Y., You, C., Zhang, J., Huang, K., Letaief, K.: A survey on mobile edge computing: the communication perspective. IEEE Commun. Surv. Tutor. **19**, 2322–2358 (2017)
22. Martins, J., Martins, J., Ahmed, M., Raiciu, C., Olteanu, V., Honda, M., Bifulco, R., Huici, F.: ClickOS and the art of network function virtualization. In: 11th USENIX Symposium on Networked Systems Design and Implementation (NSDI 14) (2014)
23. Pahlavan, K., Krishnamurthy, P.: Principles of Wireless Access and Localization. Wiley, Hoboken (2013)
24. Qin, Y., Xia, Q., Xu, Z., Zhou, P., Galis, A., Rana, O.F., Ren, J., Wu, G.: Enabling multicast slices in edge networks. IEEE Internet Things J. (2020). https://doi.org/10.1109/JIOT.2020. 2991107
25. Rossem, S.V., Tavernier, W., Sonkoly, B., Colle, D., Czentye, J., Pickavet, M., Demeester, P.: Deploying elastic routing capability in an SDN/NFV-enabled environment. In: 2015 IEEE Conference on Network Function Virtualization and Software Defined Network (NFV-SDN), pp. 22–24 (2015)
26. Xia, Q., Liang, W., Xu, W.: Throughput maximization for online request admissions in mobile cloudlets. In: Proceedings of the 38th Annual IEEE Conference on Local Computer Networks (LCN'13). IEEE, Piscataway (2013)
27. Xu, Z., Liang, W., Galis, A., Ma, Y.: Throughput maximization and resource optimization in NFV-enabled networks. In: 2017 IEEE International Conference on Communications (ICC). IEEE, Piscataway (2017)
28. Xu, Z., Liang, W., Huang, M., Jia, M., Guo, S., Galis, A.: Approximation and online algorithms for NFV-enabled multicasting in SDNs. In: Proceeding of the 37th International Conference on Distributed Computing Systems (ICDCS'17). IEEE, Piscataway (2017)
29. Xu, Z., Liang, W., Huang, M., Jia, M., Guo, S., Galis, A.: Efficient NFV-enabled multicasting in SDNs. IEEE Trans. Commun. **67**(3), 2052–2070 (2019)
30. Xu, Z., Liang, W., Jia, M., Huang, M., Mao, G.: Task offloading with network function services in a mobile edge-cloud network. IEEE Trans. Mobile Comput. **18**(11), 2672–2685 (2019)

31. Xu, Z., Zhang, Y., Liang, W., Xia, Q., Rana, O., Galis, A., Wu, G., Zhou, P.: NFV-enabled multicasting in mobile edge clouds with resource sharing. In: Proceedings of the 48th International Conference on Parallel Processing. ACM, Berlin (2019)
32. Xu, Z., Gong, W., Xia, Q., Liang, W., Rana, O.F., Wu, G.: NFV-enabled IoT service provisioning in mobile edge clouds. IEEE Trans. Mobile Comput. (2020). https://doi.org/10.1109/TMC.2020.2972530
33. Xu, Z., Zhang, Z., Liang, W., Xia, Q., Rana, O.F., Wu, G.: QoS-Aware VNF placement and service chaining for IoT applications in multi-tier mobile edge networks. ACM Trans. Sensor Netw. 13(3), Article 23:1–23:27 (2020)
34. Yi, S., Li, C., Li, Q.: A survey of fog computing: concepts, applications and issues Proceedings of the Workshop on Mobile Big Data'15, pp. 37–42. ACM, New York (2015)
35. Zhang, S.Q., Zhang, Q., Bannazadeh, H., Garcia, A.L.: Routing algorithms for network function virtualization enabled multicast topology on SDN. IEEE Trans. Netw. Serv. Manag. 12(4), 580–594 (2015)

# Blockchain-Based Security Services for Fog Computing

Arvind W. Kiwelekar, Pramod Patil, Laxman D. Netak, and Sanjay U. Waikar

## 1 Introduction

Emerging technologies are impacting our lives in two different ways. First, these technologies are improving our *standard of living*. For example, Artificial and Machine Learning are the technologies behind personalized health care, intelligent transport services, free and open education to all. Second, they are also improving the *quality of service* we expect from service providers. Technologies such as the internet and mobile communication are providing the quality of services which was unimaginable a few years back. For example, these technologies enable 24 × 7 banking services, global-market for selling local products, and opportunities to monetize excess personal resources through aggregated services like Airbnb.

In this chapter, we review the impact of two such emerging technologies called Blockchain Technology and Fog Computing. Both technologies improve the standard of living and the quality of services offered to us through the internet.

Diverse domains such as Crowd Surveillance and Public Safety, Geospatial Data Analysis, Intelligent Transport Service, Smart Grid and Smart Healthcare have started adopting Fog Computing in recent times. Adoption of Fog Computing mainly aims to reduce the response time required for accessing critical services like energy, healthcare and transportation.

Deep penetration of information and communication technologies in our social life is also raising concerns about the security and privacy of the personal data collected through them. In recent times, use of Blockchain technology has increased for protecting personal data so that trustworthy system can be built.

A. W. Kiwelekar (✉) · P. Patil · L. D. Netak · S. U. Waikar
Department of Computer Engineering, Dr. Babasaheb Ambedkar Technological University, Lonere, India
e-mail: awk@dbatu.ac.in; ldnetak@dbatu.ac.in

© Springer Nature Switzerland AG 2021
W. Chang, J. Wu (eds.), *Fog/Edge Computing For Security, Privacy, and Applications*, Advances in Information Security 83,
https://doi.org/10.1007/978-3-030-57328-7_11

271

The chapter contributes by presenting an evaluation of Blockchain Technology in the context of Fog Computing. We first identify the security requirements for various application domains of Fog Computing. Then we present a detailed analysis of the strengths and weaknesses of Blockchain Technology to meet these security requirements.

Rest of the chapter is organized as below: (1) A brief overview of fog computing (2) Fog computing use cases and their Security requirements (3) Generic Security requirements for Fog Applications (4) A Blockchain Primer. (5) Blockchain-based Security Solutions (6) Conclusion.

## 2 Fog Computing: Introduction

The Fog Computing (FC) [36] is emerging as a complementary computing paradigm for Cloud Computing (CC) to meet the computing, storage, and network requirements of *resource-constrained* computing nodes. Smartphones, tablets, the Internet of Things (IoT), wireless sensors and actuators are some of the examples of *resource-constrained* computing devices. Such kinds of devices have limited computing power, small memory, and access to the network through wireless media. Despite their limited capacity, such types of devices are transforming the nature of computing from an enterprise phenomenon into a pervasive phenomenon.

In this section, we describe limitations of CC to meet the requirements of resource-constrained devices followed by a description of distinct characteristics of the Fog in comparison with the Cloud.

### 2.1 Limitations of Cloud Computing

The Cloud Computing (CC) is now an established alternative to meet the computing, storage and networking requirements of enterprises in the presence of the reliable Internet. The cloud provides computing resources and services to remote machines on a *pay-per-use* billing model. Additionally, the CC environment offers flexible deployment models such as Platform as a Service (PaaS, e.g., Google's Cloud Services), Infrastructure as a Service (IaaS, e.g., Amazon's Elastic Computing Cloud), and Software as a Service (SaaS, e.g., Salesforce's Cloud Services). This flexibility makes CC a cost-effective solution to host resources and services for enterprise computing needs [13].

The CC paradigm has been found useful especially for enterprise resource planning [33], customer relations management, e-business owing to its characteristics such as high scalability, ease-of system administration, and support for rich programming models.

However, the CC environment falls short to meet various requirements of *resource-constrained devices* which include IoT, wearable devices, wireless sensors and actuators. Some of these requirements identified in the Reference [31] are described below.

(i) *Latency:* Video streaming, gaming, smart factories, and connected vehicles are some of the application scenarios which use devices like IoT and wireless sensors. The latency requirements of such applications fall in the range of microseconds to tens of milliseconds. The average latency experienced by resource-constrained devices when they are connected to the cloud falls in the range of hundreds of millisecond. This high latency is undesirable in such application scenarios.

(ii) *Bandwidth:* The resource-constrained devices typically access the network through a wireless medium. At the same time, applications enabled by these devices such as smart factories produce data at the rate of thousands of gigabyte per second. The cloud computing environments fall short to meet such high bandwidth requirements.

(iii) *Privacy and Security:* In some of the application contexts such as health monitoring and control, devices need to transmit private and personal information for remote processing. The resource-constrained devices lack the computing power to execute complex encryption algorithms needed to secure data when it is transmitted over the public Internet as in the case of cloud computing. Hence, securing such information becomes a challenge when resource-constrained devices are connected to the cloud.

(iv) *Context Awareness:* In application scenarios such as connected vehicles, Intelligent Transport Systems(ITS) need to transfer context information. For example, information about traffic conditions, weather information, location and information. When resource-constrained devices are connected to a distant cloud data centre, transmitting such local information has little temporal and spatial relevance.

## 2.2 Distinct Features of Fog Computing

From the functional point of view, Cloud Computing and Fog Computing are similar phenomena which provide computing, networking and storage resources to remote machines. Both environments include resource-rich devices such as high-end servers accessed through either public or private networks. Although the business model for the FC is currently evolving, similar to CC, the business model of the FC in future may be centred around pay-per-use billing mechanisms and hosting of resources by a third party.

In terms of software engineering terminology, Fog computing and Cloud computing differ regarding non-functional requirements. It includes Performance requirements, Reliability requirements, Deployment models and Security requirement. These requirements are also known as operational requirements. Table 1 shows a comparison of Edge, Fog and Cloud Requirement [40].

**Table 1** A comparison of edge, fog and cloud computing [40]

| Feature | Edge computing | Fog computing | Cloud computing |
|---|---|---|---|
| Latency | Low | Medium | High |
| Bandwidth | Low | Medium | High |
| Compute capacity | Low | Medium | High |
| Reliable compute | Low | Medium | High |
| Reliable connectivity | Low | Medium | High |
| Data longevity | Low | Medium | High |

Hence, to handle these non-functional requirements emanating from the requests of edge devices, a new computing paradigm has emerged in recent times called Fog Computing. The Fog Computing which has introduced a new application management layer in the middle between cloud and edge devices referred to as a Fog layer. The Fog layer extends the cloud management services and brings them nearer to the network.

Fog and Cloud mainly differ in terms of latency. The latency to transfer data from a Fog to edge devices is lower than when data transfer occurs from an edge device to a Cloud. This lower latency is because of edge devices are a one-hop topological distance from fog servers. Also, the network bandwidth between edge devices and the Fog is much higher through a wireless link than between edge devices and the Cloud.

Additionally, the Fog stores the data transferred from edge devices for a shorter period; the Fog periodically pushes the data to the Cloud for archival purposes.

Mobility is another distinct non-functional parameter in which Cloud and Fog Computing differ. The servers and computing nodes hosting cloud management services are centralized one. When they are geographically distributed, often the computing nodes reside in an office premise and not mobile. Unlike this configuration, a Fog may host computing nodes and services in mobile vehicles. Also, the number of requests that a Fog may have to handle from mobile clients are enormous.

Additionally, it is also essential to know the differences between Fog Computing and Edge Computing. Although, the differences between fog and edge computing are blurred one, we discuss here some of them. An Edge Computing node supports the computing requirement of edge devices which include wireless sensors and actuators. Edge computing nodes are directly interfaced with edge devices. An edge computing node communicates with edge devices through conventional communication mode such as pooling and interrupts in contrast to client-server communication used in Fog Computing. The edge computing node supports hardware-enabled security, unlike application-level security provided in Fog Computing. Further, the Edge Computing nodes typically use flash storage devices, unlike spinning storage disks used in Fog Computing.

# 3  Fog Computing Use Cases and Their Security Requirements

Many application domains such as listed in Table 2 have started adopting Fogs over Clouds to meet their computation, storage and networking requirements. For these application domains, Fog platforms meet their requirements of low-latency, high bandwidth and context-awareness. At the same time, these application domains have stringent security requirements. A brief description of the security requirements specific to these domains follows.

1. **Urban Surveillance and Public Safety** Low-cost surveillance technologies such as CCTVs and sensors enable to collect and monitor data about people living in urban areas. For example, law enforcement agencies can track the movement of suspicious people in designated sensitive areas to prevent any public damage and crimes. The collected data is location-specific and relevant to take timely decisions. Hence the fog computing paradigm is an appropriate alternative for storage and analysis purposes.

    Though the data is collected to provide public-safety, it is susceptible for misuse either by the fog service providers or edge operators who transmit the surveillance data. One of the frequently cited threats includes a Fog node operator may share the collected information about the movement of a person to a third party without informing the concerned person. Another example of threat includes denial of service attacks through flooding the network by malicious edge operators. At the same time, such systems are giving rise to a panoptic system which continuously monitors citizens.

2. **Smart Power Grid** In the energy sector, the increased thrust upon the adoption of renewable energy sources (e.g., solar, wind) has changed the relationship between energy generators and consumers. The conventional energy systems are

**Table 2** Security requirements for Fog computing use-cases

| Sr. | Application domains | Security requirements |
|-----|--------------------|-----------------------|
| 1 | Urban surveillance and public safety | Privacy and autonomy [12], Panoptic systems [39, 41] |
| 2 | Smart power grid | Denial of service attacks [7], integrity attacks [25], Malware attacks, power thefts, billing manipulations [4] |
| 3 | Geospatial data analysis (UAV) | Secured communication [15, 17], man in the middle attack, privacy [32] |
| 4 | Intelligent transportation systems (ITS) and connected vehicles | Authentication, availability, non-repudiation, integrity [37], denial of service, sybil, black-hole attack [34] |
| 5 | Smart healthcare | Data confidentiality, data authentication, data integrity, availability for wireless body network [35] |
| 6 | Industry 4.0 | Enterprise cyber-espionage, denial of service attacks, and phishing attacks [30] |

mostly fuel or coal-based, centralized, and information flows from the generator to the consumer. The modern energy sector is increasingly using renewable energy sources, a large number of energy distributors are dispersed along a wide geographical area, and the information flows in both the direction. The network of energy generators, distributors and consumers called smart grid [22] is formed through the use of information and communication technologies, sensors, and actuators to effectively operate the energy grid.

To effectively operationalize smart grids, Fog computing has emerged as a preferred distributed paradigm in comparison with cloud computing in recent times [28]. The guaranteed response time, a large number of decentralized grid operators and stringent privacy requirements from the consumer point of view are some of the factors behind the preference of fog computing over cloud computing.

The security requirements in Smart grid arise from the domain-specific concerns such as assuring the integrity of the data communicated between grid operators and consumers [7]. This data includes valuable information such as billing information, and, energy usage patterns of consumers. Further, a malicious smart meter can overload the network to disrupt and deny services to authorized customers from accessing the services provided by a Fog service provider.

3. **Geospatial Data Analysis**: Low-cost technologies such as Unmanned Aerial Vehicles (UAV), Radio Frequency Identifiers (RFID) and GPS enabled devices are producing a large amount of geospatial data [18]. Geographic Information Systems (GIS) manage and analyze such geospatial data to support urban planning, agriculture and environment monitoring.

   The requirements for reduced storage space, reduced transmission power, reduced latency and increased throughput are driving software engineers to adopt the Fog computing paradigm to build GIS applications [5].

   The geospatial data need to be protected from different types of security attacks to ensure regional security and privacy of persons who share the data. The commonly employed techniques are trust management in GIS service provider, data integrity checks, and authentication of GIS users [6].

4. **Intelligent Transportation Systems (ITS) and Connected Vehicles** The Intelligent Transport System (ITS) refers to the use of Information and Communication Technologies (ICT) for improving the efficiency and effectiveness of transport services. Some of the technologies that form the backbone of ITS are Wireless Sensors and actuators, Cloud Computing, and GPS controlled vehicles [29]. The Connected Vehicle (CV) is another related concept that is enabling the evolution of the next generation of ITS and Internet of Vehicles(IoV). The connected vehicle refers to using wireless technologies for communicating with other vehicles and the infrastructure offering transport services[23].

   The ITS and Connected Vehicle have started utilizing the advantages of Fog Computing such as scalability, low latency, and context awareness to improve the Quality of Services. The use of Fog Computing for ITS reduces the average trip time, $CO_2$ emissions and fuel consumption [8].

Jin Cui et al. identify and catalogue various kinds of security attacks for which autonomous vehicles and ITS need to protect. These include authentication, availability, data integrity, confidentiality and privacy[9].

5. **Smart Healthcare** To make healthcare more personalized and precise, medical systems have started adopting technologies such as wearable health monitoring devices, IoT, big data analysis, and Artificial Intelligence. Such health care systems, referred to as smart healthcare systems, have to address computational and security challenges.

   In the context of smart healthcare, the Fog-based platforms tackle the computational challenges by bringing resources closer to the patients, reducing response time and by providing energy-efficient data processing[1].

   Preserving the privacy of the patient's data and making health care services available round the clock are some of the security challenges that need to be addressed effectively [35].

6. **Industry 4.0** The combination of ICT, IoT and intelligent systems have revolutionized manufacturing and production systems in recent times. This industrial revolution is named as Industry 4.0 [19]. Industry 4.0 has brought a transformation into the nature of manufacturing units from the automated one to an autonomous one.

   The Fog Computing is a technology that is leading this 4th industrial revolution because of its inherent strengths such as low latency rate [27], low power consumption and proximity to wireless sensors and actuators which monitor and control various production processes.

   Some of the common security attacks observed in smart manufacturing systems are: (1) the leakage of critical production information, and (2) withholding access to a manufacturing unit. These security threats intend either to disrupt the production process or the production schedule [30].

# 4  Generic Security Requirements for Fog Applications

The previous section briefly surveys security requirements for various use cases of Fog Computing. Some of the security requirements are common across more than one application domains. For example, protecting end users from the *denial of service attacks* is a requirement of ITS, Industry 4.0, and other domains. This section identifies and explains such generic requirements common across various Fog applications.

1. **Authentication** Authentication is the primary service in distributed and networked environment. The purpose of authentication is to verify and validate the identity of end users. An end user may be a person or a device or an application who would like to access a service. The task of authenticating is a primitive operation because it ensures that only legitimate users can enter the network.

Some of the mechanisms that are commonly used for authenticating users in cloud computing are: passwords, hard/soft tokens, device identification, biometric identification or a combination of these techniques [43].

While devising effective authentication services for Fog Computing constraints such as resource limitations of edge devices, high mobility of fog nodes and edge devices, network heterogeneity and availability of wired wireless communication need to be considered [43].

2. **Secured Communication** Assuming a fool-proof underlying secure communication channel leads to many security attacks such as eavesdropping, spoofing, and information leakage at application level. Hence, Cloud as well as Fog applications need to protect the integrity of application data by providing a secured communication channel on top of underlying un-secured medium.

   Two types of communications are observed in Fog networks. First, a communication between edge devices and fog nodes. This communication can be secured through symmetric key cryptography. Maintaining an public key infrastructure and reducing message overhead are some of the challenges that need to be addressed considering resource constraints of Fog networks.

   Second, providing end-to-end security in the presence of multiple hops in a fog network and mobility of fog nodes are some of the challenges that need to considered while securing communication among fog nodes.

3. **Availability** One of the critical requirements that is common across the domains is that the services offered as Fog services need to be made available round the clock. Malicious users adopt techniques such as flooding the network with illegal packets or re-routing network traffic to a wrong destinations for denying requested services to legitimate users. Promptly detecting and protecting against such threats can save lives in Health and IIS domains.

4. **Privacy** Most of the Fog applications track personal information to provide personalized services. Few examples follow. First, systems like ITS and urban surveillance monitor mobility patterns of citizens which have personal value. Second, in case of smart grids, energy usage patterns are tracked and monitored by grid operators. Third, smart healthcare systems store personal and medical history of patients. Privacy is at stake when service providers use such critical personal information for monetary gains or for competitive advantage without the consent of service users. Designing fog layer which protects unintended usage of such personal information is a challenge.

5. **Trust Management** Trust in network-centric systems is a bidirectional phenomenon. Service providers need to earn the trust of service users by providing timely and secure responses. Also, service users need to demonstrate to service providers that they are the legitimate and non-malicious users. Such bidirectional trust is built through a series of interactions among service providers and users.

   Quantifying reputations of service providers, opinions of service users and service level agreements are some of the techniques used in case of cloud-based service providers.

Dynamic nature of fog nodes i.e. a fog node leaves and joins network dynamically and mobility of edge devices are some of the factors which need to be considered while implementing a trust management system at Fog layer.

The emerging blockchain technology has potential to address these security concerns in the context of Fog Computing. Before discussing blockchain-based solutions, we are briefly reviewing the essential elements of blockchain technology follows.

## 5 A Blockchain Primer

The necessity of blockchain technology can be understood by evaluating potential and pitfalls of the Internet as a platform for business.

The Internet has introduced an information-centric model of business, and it has revolutionized the way people transact online. For example, the emergence of e-commerce sites (e.g., Amazon) has been attributed to the growth and widespread presence Internet.

The Internet has bridged the information gap that exists between a service provider and service consumer by creating a third-party for information exchange called intermediaries or agents or service providers. These agents which are e-commerce sites, hold the information about who sells what, i.e. seller's information and who wants what, i.e. buyers profile and their needs thus bringing together consumers of services or goods with that of producers.

Some of the advantages of doing business online include the process of business transactions is simplified, and the time required for businesses is reduced.

Despite the various benefits of the Internet, it has always remained an unreliable platform to share valuable personal information because of its mediator-centric model for information exchange. A server or mediator may be a payment gateway or an e-commerce site. The information shared with such sites is always susceptible to breach of security and privacy attacks.

The emerging blockchain technology removes these pitfalls by laying a trust layer on top of the existing Internet technology. It replaces the mediator-centric model of information exchange with the peer-to-peer model or decentralized model. It transforms the Internet into a trustworthy platform for doing business when transacting parties do not trust each other. It eliminates the role of mediator responsible for authenticating the identities of transacting parties. Initially emerged as a platform to exchange digital currency over the Internet, now the blockchain technology is gradually emerging as a general-purpose platform for sharing and protecting information.

The four fundamental concepts common across the blockchain implementation are [10]: (1) Distributed Ledger, (2) Cryptography, (3) Consensus Protocols, and (4) Smart Contracts

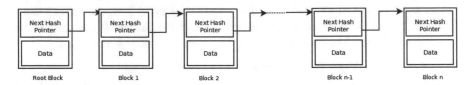

**Fig. 1** Blockchain

## 5.1 Distributed Ledger

In a conventional sense, ledgers are the registers or logbooks employed for account-keeping or book-keeping operations. Similarly, in the context of a blockchain-based information system, ledgers are the databases storing up-to-date information about business transactions. These are distributed among all the nodes participating in the network. So multiple copies of a ledger exist in a business network. When a node in a network updates its local copy, all other nodes synchronize their copy with the updated one. Hence, each copy is consistent with each other.

These ledgers are used to store information about valuable assets. In the Bitcoin implementation, the first blockchain-based system, ledgers are used to store digital currencies. It may be used to store information about other valuable assets such as land records, diamonds, student's academic credentials and others.

In a blockchain-based information system, records in a distributed ledgers are arranged in a chain-format, as shown in Fig. 1 for storage purpose. Here, multiple transactions related to an asset are grouped in a block. The $(n + 1)$th block in the chain links to the $n$th block and the $n$th block links to the $(n - 1)$th block and so on. Due to this peculiar storage arrangement, the distributed ledgers are also known as Blockchain. The blockchain data structure permits only append of new records. Updating and deletion of records are not permissible.

The most critical design feature of blockchain-based information system is the use of hash pointers instead of physical memory based pointers to link blocks in a chain. A hash pointer is a message digest calculated from the information content of a block. Whenever a node attempts to tamper the information content, a small change in the information leads to a ripple effect of changes in hash-pointers making it impossible to change the information once it has been recorded in the blockchain.

Facilitating mediator-less business transactions and supporting immutability of stored information are the two significant quality attributes associated with blockchain-based information systems. These quality attributes are derived from replicating ledgers on all the nodes in a network and linking blocks in a chain through hash pointers.

## 5.2 Cryptography

Blockchain technology makes heavy use of cryptographic functions to assure trust among the users transacting over a blockchain-based business network. A typical business network includes many un-trustworthy elements. These cryptographic functions address various purposes. Some of them are:

(1) *Authenticating the identity of agents involved in a business transaction*: Blockchain-based systems use a kind of asymmetric key cryptography. These protocols use two different keys called public and private keys. The public keys are open and used as addresses for performing business transactions while private keys are secret and used for validating the transactions. SHA-256 (e.g., Bitcoin) and ECDSA (e.g., Hyperledger) are some of the cryptographic protocols used for this purpose. Cryptographic functions such as digital signature are also used to authenticate a particular transaction.

(2) *Ensuring Privacy*: The blockchain technology adopts various mechanisms to preserve the privacy of a transaction. Below we discuss these mechanisms and their intentions behind the design.

1. *Decentralised Privacy.* The blockchain technology adopts decentralization as one of the guiding design principles. It eliminates the role of mediator to store transaction information at a central place. The transaction information is distributed throughout a business network. Thus the threat of a mediator sharing the transaction information with a third party is eliminated.

2. *Use of Asymmetric Cryptography.* The blockchain technology uses asymmetric key cryptography to protect the identity of transaction owners and to authenticate a transaction. Transactions are delinked from the real-world identity of transaction owners. The transaction owners are identified through using public keys which an owner can generate multiple times. In this way, transactions are pseudo-anonymous. The private keys are used to authenticate a transaction.

3. *Additional Mechanism for Anonymity*: In the majority of blockchains implementations, transaction owners are identified through pseudo-anonymous identity. To provide full anonymity, additional mechanisms such are mixing transaction information, and a cryptographic technique called Zero-Knowledge proof can be used. In zero-knowledge proof, is a verification technique which assures the validity of information without disclosing additional information.

## 5.3 Consensus Protocols

In decentralized systems, agreeing upon the global state of the transaction is a challenge. In a centralized system, this is not an issue because only one copy of transaction history is present at the central authority (e.g., Banks main Server machine). Blockchain being a decentralized system, holds multiple replicas of

**Table 3** Comparison of consensus algorithms [42]

| Sr. No. | Consensus algorithm | Tolerated power of adversary | Throughput |
|---------|---------------------|------------------------------|------------|
| *Public blockchain* | | | |
| 1 | PoW | 50% | Low |
| 2 | PoS | 50% | Good |
| *Private blockchain* | | | |
| 1 | Paxos/Raft | 50% | Good |
| *Consortium blockchain* | | | |
| 1 | PBFT | 33% | Low |

transactions at several nodes. Agreeing upon the unique state of the transaction is an issue which is solved by executing a consensus process involving all the nodes in the system. This process is typically carried out in three stages. In the first phase, a node is elected/selected as a leader node to decide upon a unique state. In the second stage, transactions are validated. In the third stage, transactions are committed. A variety of consensus algorithms exists in blockchain-based system. These are often compared based upon how scalable the algorithm is and several malicious nodes it tolerates. The Proof-of-Work (PoW) algorithm used in Bitcoin is one example of the consensus protocol. It selects the leader node responsible for deciding upon a global state by solving a cryptographic puzzle. It takes about 10 min for solving the puzzle requiring extensive computational work and much electric energy. It can work in the presence of 50% of malicious nodes in the network.

The Proof-of-Stake (PoS) is another consensus protocol in which a leader is selected with the highest stakes in the network. It has been found as scalable as compared to PoW, and it also works in the presence of 50% of malicious nodes in the network.

The Practical Byzantine Fault Tolerant (PBFT) is the third example of consensus protocol which has been found scalable and works in the presence of 33% (1/3) malicious nodes in the network.

Table 3 compares various consensus protocols used in private, public, and consortium blockchain.

## 5.4 Smart-Contracts

Smart-contracts are the most significant element in the blockchain-based system because it provides configuring the behaviour of such systems. Blockchain programmers can customize the working of blockchain systems by writing programs called *Smart-Contract*. The smart contracts are scripts which are executed when a specific event occurs in a system. For example, in the context of Bitcoin, a coin may be released when more than one signatures are validated, or when miners solve a cryptographic puzzle.

These scripts can be written in a native language provided by blockchain systems or general-purpose programmable language. For example, Bitcoin provides a simple and less expressive native language to write a smart contract while Ethereum provides a Turing complete native language called Solidity to write smart contracts. In Hyperledger, blockchain programmers can write a smart contract in a general-purpose language such as Java/Go.

# 6 Blockchain Based Security Solutions

This section describes blockchain-based approaches used to provide the solutions for the generic security requirements identified in Sect. 4 in context of fog or cloud computing.

## 6.1 *Blockchain Based Authentication*

In a networked system such as cloud and fog environment, two modes of authentications exist. These are *centralized authentication*, and *decentralized authentication*. For example, OAuth 2.0 is a centralized authentication protocol. In such protocols, a centralized authentication server verifies the credentials submitted by a client, and it authorizes to access the third party the requested services when it successfully validates the client. Majority of cloud service providers adopt this mode of authentication. Authentication services from Google, Facebook, and Twitter act as authentication servers with the login id and password on these platforms play the role of the client's credentials. Such kind of centralized authentication servers suffer from a single point of failure, and it also invades the privacy of clients [3].

Decentralized authentication protocols overcome the limitations of a centralized scheme. Pretty Good Privacy (PGP) and Web of Trust (WoT) are some of the examples of decentralized protocols. Blockchain technology is a platform supporting decentralized application development. Hence, it facilitates the development of decentralized authentication services. This section reviews some of the techniques that use blockchain technology for authentication purpose.

Fog systems or IoT use blockchain technology to implement in many ways. In the first kind of implementation, Fog nodes authenticate a client or edge device through a *smart-contract* running on the fog nodes. The smart-contract stores a mapping of edge devices and authorized users along with their credentials. Upon the receipt of an authentication request, the smart contract running on any of the Fog nodes can validate the submitted credentials [3].

In the second kind of blockchain-based authentication protocol, the system makes use of *distributed ledgers* for storing credential information and authorized device mapping. Typically the credential information includes asymmetric key cryptography or digital signatures. Any fog node running blockchain instance known as miners can authenticate a request to access the desired service [26].

In the third kind blockchain-based variant, edge devices are grouped into a cluster called *bubbles of trust*. The edge devices can send/receive messages within the *bubbles- of- trust*. A master node administers each bubble-of-trust. A request for send or receive is a transaction to be recorded in the blockchain. The master node validates a send/receive request similar to the case of a certification authority [14].

The blockchain-based authentication mechanisms have been evaluated for various kinds of security threats, and they are found robust for denial of service attacks. Also, these protocols have been scalable as compared to centralized ones[3, 14, 26].

## 6.2 Blockchain Based Secured Communication

The Fog/Cloud systems which adopt blockchain technology to implement authentication services also use the same for secured communication. As discussed in Sect. 5.2, the blockchain technology uses cryptography algorithms to communicate between nodes and to store data in distributed ledgers.

As seen earlier in Sect. 4, two kinds of communication need to secure: (1) from an edge device to a fog node and, (2) between one fog node to another fog node. Typically blockchain is implemented as a fog service running on fog nodes.

The communication between edge devices and fog nodes (i.e. blockchain service) is secured by assigning a public address. In the case of Ethereum, an edge device is identified through a 20-byte address. This address can be leveraged to establish an SSL session between an edge device and a fog node [3]. By default, all the communications between fog-nodes use asymmetric-key cryptography.

The blockchain system adopting secured communication have been found resilient to attacks such as man-in-the middle and replay attacks. Thus ensuring data confidentiality, data integrity and communication integrity.

## 6.3 Blockchain Based Availability

Distributed ledgers and smart contracts are the two storage and computational elements in a blockchain-based system. Multiple copies of these elements exist throughout the blockchain network. Consensus protocols maintain a consistent global state of storage and computational elements. Because of these inherent design properties, blockchain-based fog services are resilient to a single-point failure. Hence, they are fault-tolerant, thus reducing down-time.

Denial of service attacks is another means to disrupt the functioning of fog services. A blockchain-based system can adopt hierarchical mechanisms to defend itself from such an attack. One such mechanism is implemented in [11]. At the device level, blockchain miners protect the edge devices against deploying malware on edge devices by malicious users. Because all miners authorise and validate an access to edge devices. At the network level, it is the responsibility of blockchain to validate each communication emanating from edge devices and among the fog nodes. Further, as explained in [14], a blockchain-based system can dynamically form *bubbles-of-trust* to limit send/receive operations with a group of trusted edge devices or to isolate a malicious node/device.

## 6.4   Blockchain Based Privacy

Cloud/Fog service developers who adopt *client-server* model for interaction have a limited set of primitives (e.g., storing personal information in the encrypted format) at their disposal to protect the privacy of the personal information shared by their users. Unlike this, blockchain, is a peer-to-peer system, provides a range of mechanisms to protect the privacy of personal data. Below, we explain some of these primitives.

1. *Pseudo-anonymity* Blockchain-based system facilitates de-linking of user's real-life identity from its system identity. A user can use as many public keys as s/he wishes to perform an interaction. Also, s/he can use a hash of some of its real-time identity for performing an interaction. The approach, explained in [2], adopts this technique to protect health records of patients.
2. *Data Ownership* In blockchain-based system, it is possible to own and control access to the personal data by the concerned user[44]. Unlike centralized systems, data is owned and controlled by service providers.
3. *Fine-grained Authorization* Data access can be authorized at multiple levels (e.g. file, record, field) by data owners. Also, one-time data access in contrast to perpetual data access is possible to grant [44].
4. *Encrypted Storage* Data is always stored in an encrypted format. Data owner's public and private keys are required to decrypt the data.
5. *Data Transparency* Data owner is aware of what kind of data about him is collected, and it's intended use.
6. *Incentives for maintaining Privacy* The Reference [20] explains an application of blockchain which forwards safety-critical information (e.g., news of an accident) in a transport system without disclosing the identity of the forwarder. System rewards such *good* behaviour through incentives in the form of a coin which adds to their reputation.
7. *Data Provenance* It refers to maintaining metadata about the creation and each access operation performed on the data. Such kind of metadata is useful for accountability and forensics purposes, which also increases data privacy. The Reference [21] describes an application blockchain technology for data provenance.

## 6.5    Blockchain Based Trust Management

Computing trust is a challenging task. Blockchain technology provides various mechanisms to handle it. This section reviews some of the blockchain-based approaches to computing trust in decentralized systems.

First computational challenges arise from its subjectivity. Trust is subjective. To handle the subjectivity, trust is either computed for an entity or for the delivered data or in a combined way. For example in case of a vehicular ad-hoc network, the trustworthiness of a vehicle needs to be defined or trustworthiness of a received message such as a notification about road accident, or trustworthiness of both message as well as who sent it. In Reference [24], the blockchain-based anonymous reputation system is explained which computes trustworthiness of a sender and the received message. Historical interactions and indirect opinions of other participating nodes are used to calculate the trustworthiness of a message and a sender.

The second computational challenge arise from the fact that trust changes over the period of time. To address this challenge a blockchain-based solution is developed in Reference [26]. The approach calculates the trustworthiness of a node, in the context of wireless sensor networks. The reputation of a node is calculated based on how it responds to an event. A reputation factor is associated to every event. To make it relevant with respect to time, reputation factor is a continuously decreasing function. The immutability feature of the blockchain plays a role to assign a reputation factor to nodes based on its historic interaction.

The third computational challenge is to develop a trust model which is generic in the sense that computational process is applicable to multiple domains. This challenge is addressed in Reference [16] which provides a blockchain-based solution by identifying diverse attributes for calculating trust. These attributes includes: reputation, context, environment, goals, expectations, social relationships, willingness and timeliness of evaluation. The approach further demonstrates the applicability of the model in the domain of Social Internet of Vehicle (SIoV). It further states that the emerging technologies such as blockchain and fog computing are appropriate for providing scalable solution for managing trust in the dynamic environment such as (SIoV).

## 7    A Performance Analysis of Blockchain and Fog Computing Integration

The blockchain computing, particularly public blockchains such as Bitcoin, is known for its high energy consumption and low throughput. In this context, the use of blockchain technology in a resource-constrained environment such as Fog and Edge Computing is questionable. In this section, we discuss the performance analysis of implementing blockchain as a fog service.

As seen from Table 2, the security requirements for the majority of the Fog Computing use cases are of the type to authenticate users, to ensure the privacy of data, to check the integrity of data, and to provide secured communication.

For the well-known use cases of Blockchain technology such as in the financial sector, asset management and supply-chain management, the additional security requirements are to maintain consistency of data in a decentralized network, the provenance of data and seamless execution of business processes. To realize all such security requirements, the elements of blockchain technology, such as consensus protocol and smart contract, in addition to distributed ledger and cryptography, are necessary to implement. In such contexts, the computation demand and energy consumption are typically high.

But, a lightweight blockchain implementation that includes minimal elements of blockchain technology such as distributed ledger and cryptography can meet the majority of the security requirements of the Fog Computing use cases.

Further, such a lightweight implementation supports different configurable deployment options. For example, a blockchain service can be deployed with (e.g., Cloud+Fog deployment) or without Cloud (e.g., Fog only deployment model). Such a lightweight implementation additionally shall realize the tasks of encrypting and decrypting data in the hardware with a secured wireless protocol (e.g., Zigbee) to achieve secured communication.

Performance of one such lightweight implementation has been reported in [38]. It demonstrates the use of Blockchain technology in Fog computing context for the smart-healthcare use case. It observes that the energy consumption and latency requirement is acceptable for the health care use case even when blockchain service is implemented in the Cloud+Fog integration environment.

However, the performance of Blockchain and Fog Computing integration with various tuning parameters needs to be evaluated in other application domains of IoT.

# 8 Conclusion

Identifying security requirements for an emerging computing platform is a challenging task. In this chapter, we address this challenge in the context of Fog Computing. The emerging paradigm of Fog computing assures to deliver reduced latency time, better throughput and increased scalability to many applications designed around resource-constrained edge devices.

Due to this assured performance, Fog Computing is increasingly preferred over Cloud Computing platform in various safety-critical application domains. Few examples of such application domains include Urban Surveillance and Public Safety, Smart Grid, Geospatial Data Analysis, Intelligent Transport Systems, Smart Health care, and Industry 4.0.

All these domains have stringent security requirements. Hence a trustworthy platform is required to process information in these domains. Despite the blockchain technology's numerous drawbacks such as high energy consumption, an evolving ecosystem of developers, and legal constraints in the deployment of blockchain-based solutions; software developers prefer to adopt the blockchain technology as a robust platform to meet many security requirements.

The chapter describes blockchain-based solutions for authentication, secured communication, availability privacy, trust management in the context of fog computing. It assumes that the blockchain as a service is available either at the layer of Cloud or Fog. Such deployments of blockchain-based solutions have been found scalable and robust to many known security attacks.

# References

1. Ahmad, M., Amin, M.B., Hussain, S., Kang, B.H., Cheong, T., Lee, S.: Health fog: a novel framework for health and wellness applications. J. Supercomput. **72**(10), 3677–3695 (2016)
2. Al Omar, A., Rahman, M.S., Basu, A., Kiyomoto, S.: MediBchain: A blockchain based privacy preserving platform for healthcare data. In: International Conference on Security, Privacy and Anonymity in Computation, Communication and Storage, pp. 534–543. Springer, Berlin (2017)
3. Almadhoun, R., Kadadha, M., Alhemeiri, M., Alshehhi, M., Salah, K.: A user authentication scheme of iot devices using blockchain-enabled fog nodes. In: 2018 IEEE/ACS 15th International Conference on Computer Systems and Applications (AICCSA), pp. 1–8. IEEE, Piscataway (2018)
4. Aloul, F., Al-Ali, A.R., Al-Dalky, R., Al-Mardini, M., El-Hajj, W.: Smart grid security: Threats, vulnerabilities and solutions. Int. J. Smart Grid Clean Energy **1**(1), 1–6 (2012)
5. Barik, R.K., Dubey, H., Samaddar, A.B., Gupta, R.D., Ray, P.K.: FogGIS: Fog computing for geospatial big data analytics. In: 2016 IEEE Uttar Pradesh Section International Conference on Electrical, Computer and Electronics Engineering (UPCON), pp. 613–618. IEEE, Piscataway (2016)
6. Bertino, E., Thuraisingham, B., Gertz, M., Damiani, M.L.: Security and privacy for geospatial data: concepts and research directions. In: Proceedings of the SIGSPATIAL ACM GIS 2008 International Workshop on Security and Privacy in GIS and LBS, pp. 6–19 (2008)
7. Bou-Harb, E., Fachkha, C., Pourzandi, M., Debbabi, M., Assi, C.: Communication security for smart grid distribution networks. IEEE Commun. Mag. **51**(1), 42–49 (2013)
8. Brennand, C.A.R.L., da Cunha, F.D., Maia, G., Cerqueira, E., Loureiro, A.A.F., Villas, L.A.: Fox: A traffic management system of computer-based vehicles fog. In: 2016 IEEE Symposium on Computers and Communication (ISCC), pp. 982–987. IEEE, Piscataway (2016)
9. Cui, J., Liew, L.S., Sabaliauskaite, G., Zhou, F.: A review on safety failures, security attacks, and available countermeasures for autonomous vehicles. Ad Hoc Netw. **90**, 101823 (2019)
10. Dinh, T.T.A., Liu, R., Zhang, M., Chen, G., Ooi, B.C., Wang, J.: Untangling blockchain: a data processing view of blockchain systems. IEEE Trans. Knowl. Data Eng. **30**(7), 1366–1385 (2018)
11. Dorri, A., Kanhere, S.S., Jurdak, R., Gauravaram, P.: Blockchain for iot security and privacy: The case study of a smart home. In: 2017 IEEE International Conference on Pervasive Computing and Communications Workshops (PerCom Workshops), pp. 618–623. IEEE, Piscataway (2017)
12. Elmaghraby, A.S., Losavio, M.M.: Cyber security challenges in smart cities: Safety, security and privacy. J. Adv. Res. **5**(4), 491–497 (2014)
13. Goutas, L., Sutanto, J., Aldarbesti, H.: The building blocks of a cloud strategy: evidence from three SaaS providers. Commun. ACM **59**(1), 9097 (2015)

14. Hammi, M.T., Hammi, B., Bellot, P., Serhrouchni, A.: Bubbles of trust: a decentralized blockchain-based authentication system for IoT. Comput. Security **78**, 126–142 (2018)
15. He, D., Chan, S., Guizani, M.: Communication security of unmanned aerial vehicles. IEEE Wirel. Commun. **24**(4), 134–139 (2016)
16. Iqbal, R., Butt, T.A., Afzaal, M., Salah, K.: Trust management in social internet of vehicles: factors, challenges, blockchain, and fog solutions. Int. J. Distrib. Sensor Netw. **15**(1), 1550147719825820 (2019)
17. Javaid, A.Y., Sun, W., Devabhaktuni, V.K., Alam, M.: Cyber security threat analysis and modeling of an unmanned aerial vehicle system. In: 2012 IEEE Conference on Technologies for Homeland Security (HST), pp. 585–590. IEEE, Piscataway (2012)
18. Lee, J.-G., Kang, M.: Geospatial big data: challenges and opportunities. Big Data Res. **2**(2), 74–81 (2015)
19. Lee, J., Bagheri, B., Kao, H.-A.: A cyber-physical systems architecture for industry 4.0-based manufacturing systems. Manufact. Lett. **3**, 18–23 (2015)
20. Li, L., Liu, J., Cheng, L., Qiu, S., Wang, W., Zhang, X., Zhang, Z.: Creditcoin: a privacy-preserving blockchain-based incentive announcement network for communications of smart vehicles. IEEE Trans. Intell. Transp. Syst. **19**(7), 2204–2220 (2018)
21. Liang, X., Shetty, S., Tosh, D., Kamhoua, C., Kwiat, K., Njilla, L.: Provchain: A blockchain-based data provenance architecture in cloud environment with enhanced privacy and availability. In: 2017 17th IEEE/ACM International Symposium on Cluster, Cloud and Grid Computing (CCGRID), pp. 468–477. IEEE, Piscataway (2017)
22. Liserre, M., Sauter, T., Hung, J.Y.: Future energy systems: integrating renewable energy sources into the smart power grid through industrial electronics. IEEE Ind. Electr. Mag. **4**(1), 18–37 (2010)
23. Lu, N., Cheng, N., Zhang, N., Shen, X., Mark, J.W.: Connected vehicles: solutions and challenges. IEEE Internet Things J. **1**(4), 289–299 (2014)
24. Lu, Z., Wang, Q., Qu, G., Liu, Z.: Bars: A blockchain-based anonymous reputation system for trust management in vanets. In: 2018 17th IEEE International Conference on Trust, Security and Privacy in Computing and Communications/12th IEEE International Conference on Big Data Science and Engineering (TrustCom/BigDataSE), pp. 98–103. IEEE, Piscataway (2018)
25. Metke, A.R., Ekl, R.L.: Security technology for smart grid networks. IEEE Trans. Smart Grid **1**(1), 99–107 (2010)
26. Moinet, A., Darties, B., Baril, J.-L.: Blockchain based trust & authentication for decentralized sensor networks (2017). Preprint. arXiv:1706.01730
27. O'donovan, P., Gallagher, C., Bruton, K., O'Sullivan, D.T.J.: A fog computing industrial cyber-physical system for embedded low-latency machine learning industry 4.0 applications. Manufact. Lett. **15**, 139–142 (2018)
28. Okay, F.Y., Ozdemir, S.: A fog computing based smart grid model. In: 2016 International Symposium on Networks, Computers and Communications (ISNCC), pp. 1–6. IEEE, Piscataway (2016)
29. Perallos, A., Hernandez-Jayo, U., Onieva, E., Zuazola, I.J.G.: Intelligent Transport Systems: Technologies and Applications. Wiley, Hoboken (2015)
30. Pereira, T., Barreto, L., Amaral, A.: Network and information security challenges within industry 4.0 paradigm. Procedia Manufact. **13**, 1253–1260 (2017)
31. Puliafito, C., Mingozzi, E., Longo, F., Puliafito, A., Rana, O.: Fog computing for the internet of things: a survey. ACM Trans. Internet Technol. **19**(2), 1–41 (2019)
32. Rodday, N.M., Schmidt, R. de O., Pras, A.: Exploring security vulnerabilities of unmanned aerial vehicles. In: NOMS 2016–2016 IEEE/IFIP Network Operations and Management Symposium, pp. 993–994. IEEE, Piscataway (2016)
33. Saini, S.L., Saini, D.K., Yousif, J.H., Khandage, S.V.: Cloud computing and enterprise resource planning systems. In: Proceedings of the World Congress on Engineering, vol. 1, pp. 681–684 (2011)
34. Sakiz, F., Sen, S.: A survey of attacks and detection mechanisms on intelligent transportation systems: VANETs and IoV. Ad Hoc Netw. **61**, 33–50 (2017)

35. Saleem, S., Ullah, S., Yoo, H.S.: On the security issues in wireless body area networks. Int. J. Digital Content Technol. Appl. **3**(3), 178–184 (2009)
36. Satyanarayanan, M., Bahl, P., Caceres, R., Davies, N.: The case for VM-based cloudlets in mobile computing. IEEE Pervas. Comput. **8**(4), 14–23 (2009)
37. Siegel, J.E., Erb, D.C., Sarma, S.E.: A survey of the connected vehicle landscape— architectures, enabling technologies, applications, and development areas. IEEE Trans. Intell. Transp. Syst. **19**(8), 2391–2406 (2017)
38. Tuli, S., Mahmud, R., Tuli, S., Buyya, R.: FogBus: a blockchain-based lightweight framework for edge and fog computing. J. Syst. Softw. **154**, 22–36 (2019)
39. Van Brakel, R., De Hert, P.: Policing, surveillance and law in a pre-crime society: understanding the consequences of technology based strategies. Technol.-led Policing **20**, 165–92 (2011)
40. Varshney, P., Simmhan, Y.: Demystifying fog computing: Characterizing architectures, applications and abstractions. In: 2017 IEEE 1st International Conference on Fog and Edge Computing (ICFEC), pp. 115–124. IEEE, Piscataway (2017)
41. Xu, R., Nikouei, S.Y., Chen, Y., Blasch, E., Aved, A.: BlendMAS: A blockchain-enabled decentralized microservices architecture for smart public safety. In: 2019 IEEE International Conference on Blockchain (Blockchain), pp. 564–571. IEEE, Piscataway (2019)
42. Zhang, R., Xue, R., Liu, L.: Security and privacy on blockchain. ACM Comput. Surv. **52**(3), 1–34 (2019)
43. Ziyad, S., Rehman, S.: Critical review of authentication mechanisms in cloud computing. Int. J. Comput. Sci. Issues **11**(3), 145 (2014)
44. Zyskind, G., Nathan, O., et al.: Decentralizing privacy: Using blockchain to protect personal data. In: 2015 IEEE Security and Privacy Workshops, pp. 180–184. IEEE, Piscataway (2015)

# Part V
# Applications of Fog/Edge Computing

# Industrial Internet of Things (IIoT) Applications of Edge and Fog Computing: A Review and Future Directions

G. S. S. Chalapathi, Vinay Chamola, Aabhaas Vaish, and Rajkumar Buyya

## 1 Introduction

**The Internet of Things (IoT)** [1] refers to a system of smart devices that are connected through the Internet. The basic structure of IoT systems involves the use of a large number of smart devices that can acquire, process, transmit, and receive data between one another. IoT devices thereby enable us to reliably monitor and precisely control any environment, control system, or device through this system of interconnected smart devices. With forecasts predicting an estimated 28.5 billion network-connected devices to become active by 2022 [2], the IoT technology is poised to make a total economic impact between $3.9 trillion and $11.1 trillion per year in 2025 [3]. While most of the IoT systems developed until now have been consumer-centric, the disruptive nature of this technology has enabled the adoption of this technology in a gamut of industrial settings thus leading to the development of **Industrial Internet of Things (IIoT) technology** [4]. IIoT technology, in essence, refers to a system of interconnected smart devices in an industrial

G. S. S. Chalapathi (✉)
Department of Electrical and Electronics Engineering, BITS Pilani, Pilani, Rajasthan, India

Cloud Computing and Distributed Systems (CLOUDS) Lab, School of Computing and Information Systems, The University of Melbourne, Parkville, VIC, Australia
e-mail: gssc@pilani.bits-pilani.ac.in

V. Chamola · A. Vaish
Department of Electrical and Electronics Engineering, BITS Pilani, Pilani, Rajasthan, India
e-mail: vinay.chamola@pilani.bits-pilani.ac.in; f2016370@pilani.bits-pilani.ac.in

R. Buyya
Cloud Computing and Distributed Systems (CLOUDS) Lab, School of Computing and Information Systems, The University of Melbourne, Parkville, VIC, Australia
e-mail: rbuyya@unimelb.edu.au

© Springer Nature Switzerland AG 2021
W. Chang, J. Wu (eds.), *Fog/Edge Computing For Security, Privacy, and Applications*, Advances in Information Security 83,
https://doi.org/10.1007/978-3-030-57328-7_12

setting. IIoT connects industrial resources including sensors, actuators, controllers, machines with each other as well as intelligent control systems. These intelligent control systems analyse the acquired data and optimize the ongoing industrial processes to improve execution speed, reduce involved costs, and dynamically control the industrial environment [4].

One of the most important reasons behind the meteoric rise of IIoT systems in various industries is that IIoT systems can lead to a significant improvement in efficiency, throughput, and response time of operations inside these industries [5]. IIoT has already revolutionized companies in many major industries across the globe, including the mining industry where IIoT systems have led to the installation of wireless access points in mining tunnels, and RFID tracking technology has helped companies in tracking vehicles leading to an increase in production levels by 400% [6]. Proposed IIoT systems in agricultural settings can help farmers in nutrient monitoring as well as automated irrigation to improve crop yield [7]. The medical industry can also benefit from the capabilities of Industrial IoT systems where emergency services can access data from patients, ambulances, and doctors to help all stakeholders in making informed decisions and improve resource utilization [8]. Pilot projects in China have successfully implemented an NB-IoT (Narrow Band IoT) system for smart electrical meters which allows real-time collection of power consumption data thereby enabling the energy grid officials to improve the electricity supply strategy in any area [9]. Similarly, NB-IoT smart parking systems have been deployed in cities to help drivers easily find parking spaces. Further, integration of this parking system with payment solutions has led to automated transaction authorization for parking payment which has subsequently improved utilization of parking bays [10]. The railway industry can also leverage the power of IIoT solutions to improve the functioning of surveillance systems, signaling systems, predictive maintenance, and passenger or freight information systems to improve services and safety [11]. Supply Chain Management (SCM) can also benefit by adopting IIoT based systems which will directly enhance tracking and traceability while also aiding in the optimization of shipment routes based on rapidly changing customer requirements [12]. While the IIoT shows immense potential as a transformative technology, it is important to know the critical requirements that must be validated and verified in the design of IIoT systems to maximize the efficiency and performance of these systems [13, 14]. These requirements arise from the challenges often faced by Cyber-Physical Systems (CPS). The requirements of IIoT systems include Scalability, Fault Tolerance or Reliability, Data Security, Service Security, Functional Security, and Data Production and Consumption Proximity (Fig. 1).

With the rise in computational power being offered by computing systems in general in recent years, the focus of most industries has shifted towards garnering practical and useful patterns from their data which has been aided by the rapid development in statistical analysis and learning-based algorithms. Today, industries that are making use of IIoT solutions want to utilize the massive amount of data being generated to collect useful insights which can help in the reduction of unplanned downtimes, improve the efficiency of production, lower energy

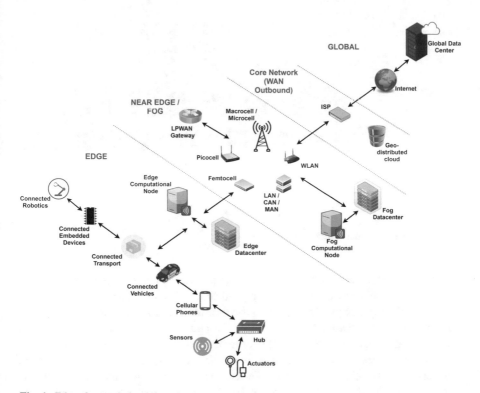

**Fig. 1** Edge, fog, and cloud tiers

consumption, etc. However, to process such massive amounts of data, IIoT systems generally require cloud computing services which often experience large round-trip delays and poor Quality of Service (QoS) as a large amount of data needs to be transferred to centralized data-centers for computation [15]. Since most sensors and data acquisition devices in IIoT systems operate at the periphery of the network, more data tends to be produced near the periphery of the network, which implies that processing the data at the edge of the network would be more efficient [16]. Therefore, efforts in shifting the computational power towards the periphery of the network have given rise to the edge and fog computing paradigms.

**Edge Computing** refers to the computing paradigm in which computations are performed at the edge of the network instead of the core of the network. In this scenario, the "edge" refers to any resource located on any network path between data acquisition devices (situated near the periphery of the network) and the cloud datacentre (situated at the core of the network) [16]. The basis of the edge computing paradigm is that the computations should be done on the "edge" which is in the proximity of the data sources and this avoids the latency associated with data transfer to the network's core.

The **Fog Computing** paradigm is similar to edge computing in that it also has a decentralized architecture for computation but with the fundamental difference

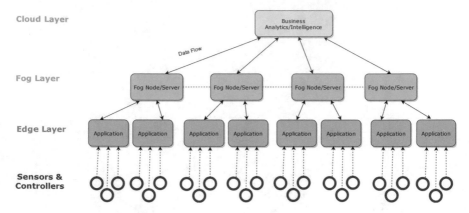

**Fig. 2** Industrial IoT data processing layer stack

being that Fog Computing can expand to the core of the network as well [17]. This means that resources located at both edge and core can be used for computations and consequently, fog computing can aid in the development of multi-tier solutions that can offload service demand to the core of the network as the load [17]. However, in most fog computing systems, the computational power is concentrated with the LAN resources which are closer to the data sources and further away from the network core, thus reducing the latency associated with data transfer to the core as seen in edge computing as well. Therefore, the fundamental difference between the edge and fog computing paradigms is basically in the location where the computational power and intelligence are stored. In the case of edge computing, this computational power is concentrated at the edge of the network usually in powerful embedded devices like wireless access points or bridges whereas, in the case of fog computing, the computational power is usually in the LAN resources. The rest of this chapter is organized as follows: Sect. 2 discusses the background of edge and fog computing systems and how these paradigms address the requirements of modern IIoT systems. Section 3 describes various applications of edge computing in industrial settings. Section 4 elaborates on fog computing applications. In Sect. 5 we present several outstanding issues and challenges with these computing paradigms that can be interpreted as future directions for research in this domain. Finally, in Sect. 6 we conclude with the salient points of this chapter (Fig. 2).

## 2 Relevant Computing Paradigms and Requirements

IoT is seen as a major technological turn-around in various applications. However, due to the high volume of data which is generated by several IoT devices, it is extremely difficult to forward all this data to a central cloud server for processing as it lays heavy stress on the network. Also, it increases the latency involved in processing the data on the cloud server and receiving the results or carrying out a response on the IoT devices. Edge computing paradigm is a computing technology that enables data to be processed almost exclusively on the "edge" of the network, which refers to locations between the end devices (like sensors, controllers, and actuators) and the centralized cloud servers. The rationale behind the development of this technology is that computations performed closer to the end devices will lead to lower latency in the system. This is because the system does not need to transfer data between edge devices and central cloud servers as the computations have been offset to closer locations on the edge. Therefore, in edge computing systems, edge devices can not only request content and services from the cloud servers but can also perform computational offloading, caching, storage, and processing, thereby making the edge devices both data producers and consumers [16].

The fog computing paradigm can be understood as an extension of the traditional cloud computing model wherein additional computational, data handling, and networking resources (nodes) are placed at locations on the network which are close to the end devices [18]. The consequence of this extension is that processes involving data management, data processing, networking, and storage can occur not only on the centralized cloud servers but also on the connections between end devices and the cloud servers [19]. Fog computing, therefore, can be extremely useful for low latency applications as well as applications that generate an enormous amount of data that cannot be practically transferred to cloud servers in real-time due to bandwidth constraints [20].

As discussed in the previous section, there are many requirements that cyber-physical systems need to maintain to become a viable supplement for real-world operations and applications. These include the following:

1. **Scalability** which ensures that the increased data transfer between nodes does not degrade latency or response time.
2. **Fault tolerance and reliability** which guarantees that the system functions normally under variable external factors like under high load conditions.
3. **Data security** which ensures that the system is resistant to external attacks attempting to steal confidential information stored in the system or network.
4. **Service security** to make the system resistant to external attacks which are attempting at disrupting the service provided by the system to the industry such as through Denial-of-service (DoS) attacks or Blackhole attacks.
5. **Functional security** so that physical accidents such as fires, explosions, leaks do not occur at any time especially in industries handling potentially hazardous substances such as nuclear plants, chemical plants, and oil rigs
6. **Data production and computation proximity** which ensures that the devices collecting the data and the systems processing the data are close to each other over the network to reduce latency.

To realize the benefits offered by the edge and fog computing paradigms, IIoT systems must be designed as per the network structures of these paradigms since these paradigms adhere to all the requirements of cyber-physical systems:

1. Edge and fog computing-based systems are **scalable** since increased data transfer between nodes can be addressed by the introduction of additional edge devices to compensate for the added computational load without degrading the network's latency since these devices function in proximity to end devices, and hence, do not increase the data transfer delays over the network.

2. Edge and fog computing systems are **reliable and fault tolerant** especially when compared with cloud-based systems since faults in the centralized cloud servers would result in a total loss of service but the decentralized nature of Edge and Fog Computing systems ensures that even if some of the computational nodes fail, the remaining healthy nodes can still maintain partial service. Furthermore, if the computational load of the failed nodes can be offset to the remaining healthy nodes, then the system can still run full service while corrective action is undertaken.

3. Edge and fog computing systems maintain **data security** within the system due to data decentralization which means that if an adversary wants to breach the system, it would need to breach each one of a large number of decentralized computing nodes to collect the entire system's data.

4. Edge and fog computing systems maintain **service security** by using advanced defense mechanisms such as per-packet-based detection, data perturbation, and isolation networks for the identification of and defense against attacks [21].

5. Edge and fog computing systems ensure **functional security** since these systems as they can be used to create extremely stable and robust multi-loop control systems for functionally sensitive industrial operations such as temperature control [22].

6. Edge and fog computing systems were developed with the rationale that **data consumption** (processing, storing, caching, etc.) and **production** are always in **proximity** which is ensured by the fundamental structure of these systems where computational nodes are located on the edges of the network, which are close to the end devices at the periphery of the network.

The distributed nature of edge and fog computing systems leads to several advantages in terms of reduced communication times and improved reliability, which makes these systems especially useful in a variety of industrial settings that require reliable, latency-sensitive networks for process automation. By realizing the inherent advantages of these paradigms, a large number of industries have started to utilize these paradigms in their system designs and we shall look at several such use cases in the following sections of this chapter.

# 3 Industrial Applications of Edge Computing

## 3.1 Manufacturing Industry

To understand the applications of edge computing in manufacturing, we will be considering the system architecture for a manufacturing-based setup as presented in Fig. 3. After describing this architecture, a case study is presented which is based on the implementation of an active maintenance system on a prototype platform. Finally, this subsection concludes with a summary of the tests and results from this case study, as presented in [23].

### 3.1.1 System Architecture

As depicted in Fig. 3 the architecture has been divided into four domains as follows:

a. **The application domain** is responsible for providing a comprehensive oversight over the entire manufacturing system to aid in the active administration of the system. This oversight includes services such as monitoring of data flow and network health, as well as the capacity for control of the system. The application domain, therefore, allows the system to provide flexible, generalized, and inter-operable intelligent applications while also aiding in the maintenance of service security.

b. **The data domain** is responsible for providing services such as data cleaning, feature extraction, and intelligent inference, which enables the system to optimize

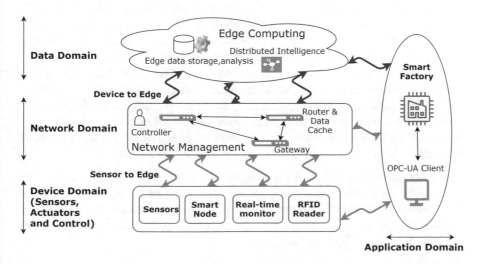

**Fig. 3** Architecture of IoT and edge computing-based manufacturing

system operations to improve the system's throughput and efficiency. Another important feature of this domain is that it allows end nodes to quickly access data, due to the proximity of the edge computing node and the end devices, which aids in generating real-time responses for specific events. Therefore, this is a critical part of dynamically controlled manufacturing systems.

c. **The network domain**, in essence, is responsible for connecting the end devices with the data platform and this domain utilizes the Software-Defined Networking (SDN) architecture [24] to manage operations involved in the control plane and network transmission. A Time-Sensitive Network (TSN) protocol is also employed within this domain to handle time-sensitive information and is used extensively in processing the information related to the network in sequence. This domain also offers universal standards for sustaining and supervising the time-sensitive nodes, making it a critical part of the overall system architecture.

d. **The device domain** refers to the devices located or embedded within the field apparatus like machine tools, controllers, sensors, actuators, and robots. This domain must be able to sustain an infrastructure for flexible communication models to maintain a variety of communication protocols by maintaining nodes that change the system's execution strategies dynamically based on the inputs obtained from the sensors. We normally observe that on the edge nodes, the information model is built with popular protocols such as OPC UA [25] and Data Distributed Service (DDS) [26]. Finally, the unified semantics of information communication is realized within this domain of the system architecture, and it is also responsible for maintaining data privacy and security.

### 3.1.2 Active Maintenance Case Study

With the proliferation of cyber-physical systems, a wide variety of industrial projects are being migrated to edge computing-based frameworks because of the promise of improved efficiency, ease of maintenance, and real-time adaptability offered by this computing paradigm. We shall be reviewing a case study on a customized production line for candy packaging, as entailed in [27]. In this study, a private cloud was used to provide service to customer orders. To make stable and high-speed communications possible, an ad-hoc network was built connecting the edge nodes. Furthermore, to achieve a proper exchange of information, a standardized version of the DDS protocol and ethernet were integrated before the deployment of the system. The functioning of the system can be summarized as:

i. Candy packaging tasks were associated with each robot and these tasks were also linked to the cloud. After getting their assigned tasks, the robots were required to pick up the particular candy assigned to them and keep the candy into the relevant open packaging. In this operation, backbone network nodes were represented by the robots.

ii. System was also capable of shifting nodes to different positions on the production line in case of any failures. Therefore, a system with multiple agents was established to improve the self-governing functionality in this scenario.

iii. The agents of the system, physically represented by robots, were independent and self-directed which means that their objective and behaviour was not constrained by other agents of the system.

iv. This system of multiple agents was deployed to complete tasks efficiently by assigning different agents with various tasks and procedures.

v. CNP (Contract Net Protocol) was used to assign different tasks to different agents by using techniques such as winning modes, bidding and open tendering.

vi. By the means of contests and discussions the agents can bargain and resolve their conflicts and so this self-organized system can efficiently complete the assigned tasks.

The implementation of this scenario was made possible with various setups, which include the following:

i. With the help of the Hadoop architecture, a distributed data processing system was built wherein at the local database level, real-time mining and analysis was performed with the help of Hadoop MapReduce and Hadoop Distributed File System (HDFS).

ii. Information such as machine status and logs constituted the sensory data which was used to create a reasoning-based model which was loaded onto a Raspberry Pi system.

iii. On the Raspberry Pi, an OPC UA server was made functional to perform pre-processing tasks on the transmission data that was acquired from different sensory devices. This data was raw in nature and hence, had to be transmitted safely and reliably which was made possible by the use of OPC UA server.

iv. To integrate the data received from multiple sources, a semantic model was also built which reformed the data to maintain consistency, accuracy, and merit of the information. This semantic model used data fusion to provide generate features as inputs from the acquired data. Finally, this data was used as input to the reasoning-based model.

### 3.1.3 Tests Performed

To estimate the difference in performance obtained by using an edge computing-based system instead of a centralized cloud computing system, a cloud-based system was also set up in this system [27]. This system had a centralized control server that managed the different agents of the system. To test the time of operation on the systems, both were tasked with completing the same orders under similar conditions of distribution of candy types. The number of candies to be packed was varied and the average time for robot operation completion was recorded for both systems. The results are summarized in the following two points:

i. With an increase in the number of orders, we observe that the self-organized version built on edge nodes is far more efficient and agile than the centralized system when the number of orders rises above 2000, as the operation completion time for the self-organized system becomes consistently lower that of the centralized system.
ii. With a stable production line, the speed of the backbone network in the centralized version was observed to be around 16 Mb/s. However, after the deployment of the self-organized system, the backbone network speed dropped to around 5–6 Mb/s which represents a 65% drop in speed.

The results of this study [27] suggest, that a decentralized and self-organizing system can become extremely useful in mass-production scenarios due to the reduced operation completion time. While the study shows that a decentralized system leads to a reduction in transmission speeds within the backbone network, the system can still function efficiently as the reduced operation completion time outweighs the drop in the backbone network speed thereby increasing the effective system throughput.

## 3.2 Supply Chain Management

Supply Chain Management (SCM) can be understood as a set of activities that are used to control, plan, and monitor the flow of products from their production to their distribution in the most efficient manner. While modern industries have already adopted cloud-based technologies to support their supply chains, an increasing number of these chains have begun to generate massive amounts of data from a diverse set of sensors and end devices located at different points along the supply chain. In such situations, it becomes impractical to store and process data in remote servers due to several reasons such as network bandwidth restrictions, large latency, and need for better fault tolerance. These restrictions, coupled with the proliferation of Radio Frequency Identification (RFID) technology, have given rise to edge computing-based solutions for the supply chain management.

Using the case study of a blackberry (fruit) supply chain as proposed in [28], we shall attempt to explain how industries can augment their supply chain management systems to leverage the power of edge computing. The proposed system has a three-layer architecture which is explained below:

1. **Layer 0**: This layer includes the data producing end-devices (primarily RFID embedded sensors) responsible for generating relevant data such as the Electronic Product Code (EPC), temperature, internal pressure, humidity, air quality, and other important parameters.
2. **Layer 1**: This layer is primarily responsible for monitoring and control purposes which entails the generation of actuator commands, execution of the control logic, and generation of relevant alarms. With the use of active and smart edge

nodes along with onboard decision support units, this layer aid administrators in improved quality monitoring as well as in the execution of real-time corrective actions.

3. **Layer 2**: This layer consists of the traditional, centralized servers which can be used for long-term pattern recognition and analysis of offloaded sensor data, giving valuable insights that can be useful while optimizing production and distribution pipelines.

As illustrated in the case study, the introduction of edge computing-based technology can enable efficient monitoring and actuation in all three stages of the supply chain:

- **In the field**: Edge nodes deployed at farms can aid in the real-time monitoring of blackberries. Through sensor information, the edge nodes can predict and notify farmers when the blackberries are ready for harvesting, thus improving shelf-life for the berries while also ensuring that all berries are harvested at the correct time.
- **In transit**: Edge processing nodes and sensors installed in transport vehicles can monitor various environmental parameters of berries such as temperature, relative humidity, and light. While these systems can continuously provide updates to the system managers, they can also execute instant corrective actuation methods in response to variations in environmental parameters such as controlling the air conditioning of the vehicle, adjustment of air filters, and notifying the driver about a possible opening of the vehicle doors.
- **At the packing location**: The data from the fog nodes can be used to determine the priority of cooling of incoming crates or pallets of berries which can enhance the freshness of the products while also minimizing any wastage resulting from spoilt berries.

This case study illustrates how an edge computing-based system can drastically improve the quality of monitoring for supply chains while also offering low-latency actuation techniques for system managers. Furthermore, due to the proximity of computational resources and end-devices, the amount of data transferred to the cloud servers is reduced drastically, thereby reducing the strain on the network. This leads to improved efficiency of these supply chains and while also resulting in reduced delays associated with the networks supporting these supply chains.

## 3.3  Food Industry

Modern food manufacturing industries have started to rely heavily on automated food production systems in factories to improve the quality and speed of production of consumable items. However, unlike other industries, the food industry constantly deals with perishable items—whether it is milk or sugar as raw materials or chocolates as finished products. Therefore, the food industry must invest in resources and systems that help in product traceability in all stages of production, processing, and distribution. These resources not only aid in the optimization of the manufacturing and distribution pipeline but also enable the industry to perform product recalls (such as in the case of some contamination) with minimal losses. In this regard, edge computing solutions have emerged as viable frameworks due to their distributed nature and the introduction of these systems can be extremely beneficial for the food manufacturing industry.

In the system proposed in [29], food manufacturing industries can rely on QR codes, barcodes, RFID tags, or transponders implanted onto objects such as primary and secondary packaging, pallets, trucks or containers, throughout the supply chain to aid in their identification and tracking along the production and supply pipeline. Edge-computing enabled sensors can be used in the process of product identification at different points along the production and supply pipelines to ensure that the flow of products is maintained. Within this system, the edge devices can rely on ad-hoc networks to communicate with each other to determine bottle-necks along the production and supply pipelines and automatically optimize these pipelines. The centralized cloud database can also be linked with this ad-hoc network and can maintain a global database of the products for administrative supervision. Therefore, with the use of such an edge-computing powered system, the industry can rely on a latency-sensitive system that functions with reduced response times, unlike a traditional cloud computing-based system.

## 3.4 Distributed Synchronization Services

One of the biggest use cases of cloud computing-based storage is distributed data storage, commonly referred to as cloud storage services wherein files can be accessed from anywhere on the planet by connecting a system with cloud storage servers which periodically synchronize data on different devices to enable access of files. However, even for small applications like office suite software, cloud storage services can often lead to unnecessary bandwidth consumption while also compromising latency. The EdgeCourier [30] is a file storage solution that can overcome the problems of traditional cloud computing-based distributed storage options by making use of the edge-hosted personal services (EPS) technique in conjunction with the *ec-sync* incremental synchronization approach. The essence of EPS is to make use of computational resources on the edge nodes (like access points or base stations) to provide localized services for mobile wireless users connected to these edge nodes. The *ec-sync* synchronization approach requires two participants: the *sync-sender* and *sync-receiver*, both of which are instrumental in the synchronization process which is explained as follows:

- The *sync-sender* detects if there is any document that requires synchronization with the receiver and is responsible for capturing the changes made within the document, by going through every sub-document within the document.
- To capture sub-document changes, the *sync-sender* compares two files: the edited document and the last-synced version of the same file.

- Thereafter, the *sync-sender* places the detected changes into a file known as *edit-patch*, which is transmitted to the *sync-receiver*.
- Upon receiving the *edit-patch* file, the *sync-receiver* applies the edit-patch differences to the relevant sub-documents from the last-synced version of the same file to obtain the edited document.
- This edited document is then also shared with the cloud storage services to transmit it to various EPS instances or nodes across the network for global synchronization.

Furthermore, an important advantage of having different EPS instances is that they can be managed by a centralized management service (on a cloud service), which can migrate data to and from the edge nodes if needed. This, therefore, leads to better oversight and increased fault tolerance as data can be migrated to different resources for analysis or in response to outages experienced at edge nodes. The overview of the EdgeCourier system can be seen in Fig. 4. Laboratory-based studies on the Edge Courier system [30] showed that with the rise in the size of documents that need to be synchronized, the time spent on network transmission becomes notably lower for the EdgeCourier system as seen with a document size of 1 MB which takes 0.6 s lesser on the EdgeCourier system than on the direct sync system. Such distributed synchronization systems can be particularly useful in the software development industry for real-time code synchronization in large team projects. Similarly, the banking industry can also derive some critical applications from these systems such as in the real-time synchronization of transactions and other banking data. These examples clearly show that edge computing powered data synchronization systems find a lot of applications in modern industries that require reliable network services. As we have seen, these systems lead to reduced data transmission over the network, resulting in reduced latency and lesser strain on the network's bandwidth capabilities, hence leading to dependable network services.

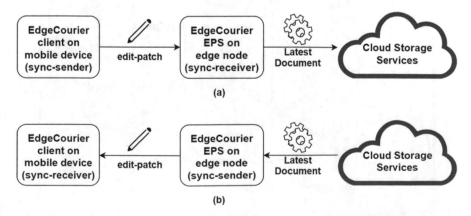

**Fig. 4** System overview for EdgeCourier. (**a**) Upstream document synchronization. (**b**) Downstream document synchronization

## 3.5   Healthcare

With the recent advancements made in the domain of medical IoT devices, the
healthcare industry has started to adopt IoT solutions that provide vital medical
services such as the monitoring of Electrocardiogram (ECG) data and processing
of Magnetic Resonance Imaging (MRI) data. However, most of the traditional IoT
based solutions for healthcare rely heavily on cloud-based processing as well as
storage which has started to create problems for these solutions as the massive
amount of data being generated is straining the communication network's capacity.
This often leads to unpredictable delays in communication while also promoting
increased latency in the network which can significantly impact healthcare operations within the hospital or clinic especially in time-sensitive situations that require
urgent reactions such as in heart attacks or strokes. Therefore, modern medical IoT
systems require a flexible multi-level network architecture that can cohesively work
with heterogeneous sensors and process the relevant data with minimal latency to
produce relevant results and responses. These requirements have led to the adoption
of the edge computing paradigm in medical IoT systems due to the benefits it can
provide in terms of reduced latency and improved reliability, both of which are
critical for these systems. In this subsection, we will be reviewing the BodyEdge
architecture [31] as shown in Fig. 5, which is structured and inspired by the edge
computing paradigm to achieve the following goals:

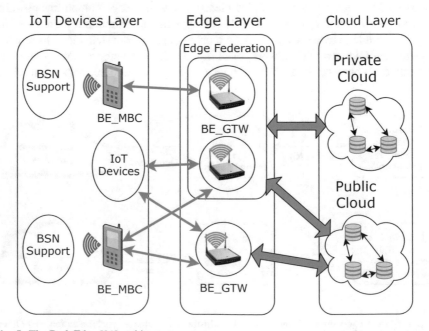

**Fig. 5** The BodyEdge [31] architecture

- Reduced communication delay and latency.
- Wide support for scalability and responsiveness.
- Limited cost in terms of bandwidth for data transmission (i.e. only limited statistics data needs to be transmitted to the cloud).
- Improved Privacy (since the edge network may be interpreted as a private cloud).

This architecture consists of two complementary parts. The first is a mobile client called BodyEdge Mobile BodyClient (BE-MBC) which is primarily responsible as a relay node for communication between the sensors and the edge client using multi-radio communication technology. The second is a performing gateway known as the BodyEdge Gateway (BE-GTW), which is placed at the edge of the network and is primarily responsible for acquiring device data and locally processing it to produce valuable insights and patterns that can be relayed back to the end devices or sensors. In addition to this, the gateway also ensures communication with the cloud to allow users to maintain oversight over this system.

To validate the BodyEdge architecture, it was physically implemented in [31] and compared with a cloud-based architecture for the task of stress detection using cardiac sensors. Within the implementation, the BE-MBC module was installed on a smartwatch which was paired with a chest band to acquire ECG signals. The BE-GTW was installed on an independent hardware platform (Raspberry Pi3) as well as on an Azure cloud virtual machine to perform the comparative study. Finally, the edge-based system with the BE-GTW installed on the Raspberry Pi3 was tested on 100 athletes to determine stress levels using the Heart Rate Variability (HRV) technique [32] and the average round trip delay time (RTT) for this case was 152 ms. The same experiment was then conducted with the cloud-based system which yielded an average round trip delay time (RTT) of 338 ms. This result, therefore, corroborates our assumptions about the performance benefits offered by edge-computing-based systems in terms of reduced latency and indicates that medical IoT systems should indeed adopt edge computing-based network architectures.

## 3.6 Agriculture

Modern agriculture has extensively embraced automation and modern technology to improve and optimize existing agricultural processes due to the improved connectivity between agricultural resources. As technology is becoming increasingly interconnected, edge computing-based infrastructures have started to dominate most network-based applications, and to tackle the growing amount of data being generated by end devices, the agricultural industry has also started edge computing-based architectures to create latency-sensitive applications for agricultural processes. The concept of Precision Agriculture (PA) has seen a significant rise in popularity due to the improvement in sensor technologies, and several systems based on edge computing have been proposed, like the precision agriculture platform [33]. These systems make use of intelligent algorithms in conjunction with smart sensors and actuators in the field to providing real-time monitoring services that enable control services to maintain optimal environments for crop growth. In the system proposed in [33], the architecture is divided into 3 tiers namely: crop (Cyber-Physical System or CPS) tier, edge computing tier, and the cloud tier. The architecture has been illustrated in Fig 6. The crop (CPS) tier is majorly comprised of sensors that aid in real-time monitoring of various environmental parameters such as temperature, humidity, pH, $CO_2$ levels, solar radiation, and other important factors. In addition to sensors, this tier also supports various actuation devices such as soil nutrition pumps, valves, irrigation devices, ventilation devices, and light-control devices. Within this architecture, operations at this tier require low latency and high reliability in communication so that emergency services can be enacted without human intervention, which is made possible through the edge computing-based computational nodes situated closer to the data sources. In continuation, edge nodes within the edge computing tier are responsible for executing commands through actuation devices based on inputs received from sensor networks in the crop tier. Therefore, this layer is responsible for the control of irrigation, climate control, nutrition control, and other auxiliary tasks like alarm and energy management. Finally, the cloud tier is responsible for long-term data analytics and system management services. The physical implementation of this system showed savings of more than 30% in terms of water consumption along with savings of nearly 80% in terms of some soil nutrients when compared with a regular open crop. In addition to environmental monitoring, edge computing powered systems can also be employed for video analytics through UAVs that can help farmers in optimized weeding and harvesting. This clearly illustrates the impact of automation on the agricultural industry and shows how edge computing-based architectures can replace cloud computing frameworks especially in applications that require low-latency and high reliability.

# 4 Industrial Applications of Fog Computing

## 4.1 Smart Grids

Conventional energy grid systems have been powering industries and countries for the past 100 years, and with the tremendous rise in demand for electrical power, the domain of IoT has emerged to be the pioneering technology that is leading developments in the smart grid systems. Traditional grid operations relied on simple analog meters to record units of power flowing per month to each household or industry, but with the evolution of intelligent and autonomous systems, modern smart grids offer solutions that allow comprehensive oversight over energy distribution which is beneficial to both consumers and producers. For power producers, these smart grid solutions allow accurate monitoring of energy demands and supplies which

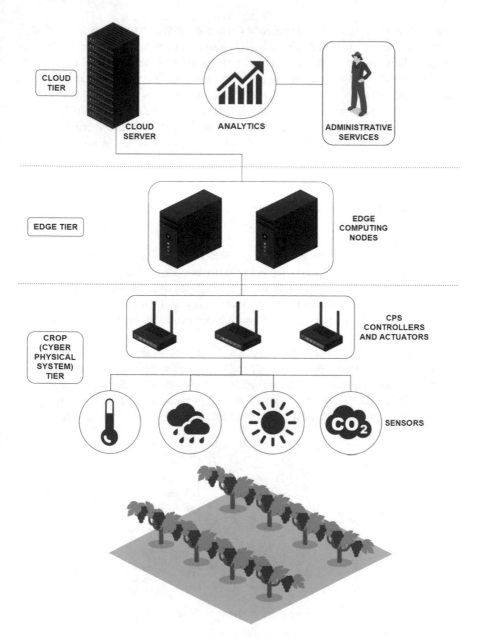

**Fig. 6** Architecture overview for agricultural monitoring system

allows them to effectively control pricing as well as load balancing to sustain the healthy functioning of the grid. On a similar note, consumers can monitor their energy consumption in real-time for each device which allows them to effectively and reliably manage their energy spending. The framework of such a smart grid, therefore, involves a heavy dependence on the collection and aggregation of real-time data from every device within each household or industry that is powered by the grid. This will inevitably lead to the generation of a large amount of data that needs to be efficiently managed and analyzed while maintaining the security of the data. To manage such massive amounts of data, it is easy to perceive that the cloud computing paradigm cannot be a viable network architecture for these IoT powered smart grid solutions since the sheer volume of the data would not adhere to any conventional network's transmission capacity. To reduce the strain on the network capacity, fog computing-based grid systems can become a viable option since the fog computing architecture allows computational offloading from the centralized cloud servers to fog nodes that are situated closer to the end devices. This distributed nature allows the network to function with low latency and improved reliability while also maintaining data security, and these are exactly the properties that a modern smart grid system requires.

The basic architecture of smart grid systems is generally composed of advanced metering infrastructure (AMI) along with area networks, data centers, and integrated substation centers. Within this architecture, AMI ensures two-way communication is maintained between the end devices and the fog nodes which leads to a secure, reliable, and cost-effective service. The model proposed in [34] is a three-tier architecture as shown in Fig. 7.

The first tier is comprised of the smart meters which are responsible for collecting data regarding energy consumption as well as for inter-tier and intra-tier communication. The second tier comprises the resource-rich fog nodes which are responsible for delivering the majority of computational services to the network. Finally, the third tier comprises the traditional cloud servers which are usually

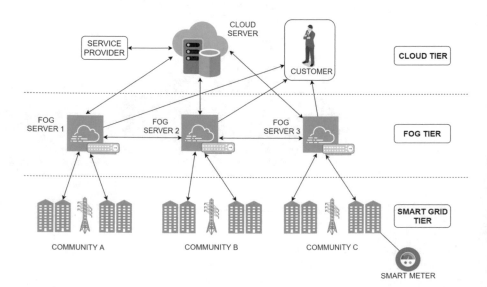

**Fig. 7** Structure of fog computing enabled smart grid

responsible for oversight and maintenance of the entire grid. This structure allows inter-tier communication within the first and second tiers which enables different geographical sub-grids to communicate with one another.

Through the following points, we can appreciate the benefits offered by fog computing architecture:

- The smart metering technology enables the energy producers to monitor power loads in real-time which helps them in drafting an effective load-balancing methodology, with extremely low latencies and transmission delays.
- The smart meters allow consumers to monitor the energy consumed by each device in real-time and this can aid them in controlling device usage dynamically to minimize their energy costs.
- While the smart meters maintain a local database of the profile of energy consumed by each device, they aggregate this data for the complete household or industry and forward this encrypted aggregate to the fog servers. These fog servers can then store this data securely within storage systems that are localized in that geographical area, and because the encryption key is only known to the fog node and the respective smart meter, the system maintains privacy even if the data is accessed by someone through the cloud server.
- Finally, the varied geographical location of fog computing nodes can be beneficial to the grid in an interesting way: specifically for the case of electric vehicles which can be charged at any location inside the grid while the grid maintains the correct billing information. For instance, if an electric vehicle is charged in any neighbourhood, the smart meter deployed in that neighbourhood can identify the owner of the car using a unique identifier and transfer the billing information via the fog node tier to the owner's smart meter, thereby ensuring consistency in billing within the smart grid.

## 4.2   Satellite Communication

With the recent advances made in satellite technology, the communication industry has started to rely heavily on satellites to provide access to people situated in remote locations. Satellite-Terrestrial Networks (STN) are communication networks that have emerged as one of the most promising low-cost technology which can lead to ubiquitous access to internet connectivity across the globe. A majority of STN setups make use of Low Earth Orbit (LEO) satellites to provide connectivity to sparsely distributed users by interconnecting small terrestrial terminal stations which are placed in remote locations to ensure maximum area coverage, as shown in Fig. 8. But, with the evolution of smartphones and tablets, the amount of data that needs to be transferred across the network has increased drastically, particularly because of an increase in the number of applications such as speech recognition and gaming that make use of cloud services to process user-generated data. This puts a strain on the network's data transfer capacity, and so we must look towards

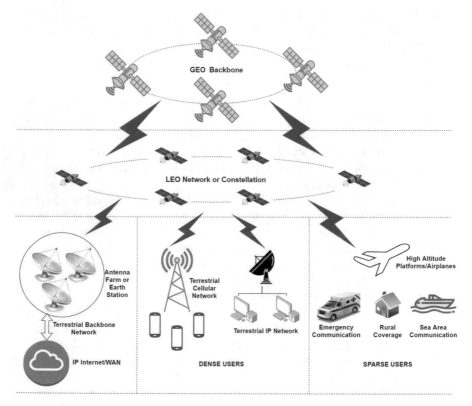

**Fig. 8** Traditional satellite terrestrial network

computational offloading to help alleviate this problem. In this situation, satellite mobile edge computing (SMEC) [35] can be a possible solution that can offload computation as well as storage to local servers, thereby leading to an improved QoS, increased reliability, and reduced latency. This technology, although dubbed as edge computing, is better classified as a fog computing-based technology as computational resources are essentially an extension of the cloud servers. Therefore, the introduction of fog computing resources near the end-devices can lead to content caching and other storage facilities which effectively reduces that traffic in the overall STN. In terms of computational offloading, the fog sites can be located at 3 different locations, and these are:

- **Proximal Terrestrial Offloading (PTO)**: In this situation, satellite mobile fog computing servers are deployed at terrestrial stations, as shown in Fig. 9b. The advantage of this system is that the communication latency is significantly reduced because backhaul transmission through the satellite is avoided. While such a system would be extremely useful for terrestrial terminal stations that cater to dense user areas, it would not be practical for terrestrial terminal stations

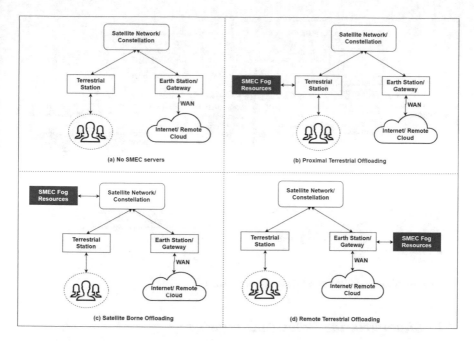

**Fig. 9** SMEC with offloading at different fog sites

that are placed in spare user areas especially because these stations do not hold extra computational facilities and are remote.

- **Satellite Borne Offloading (SBO)**: In this situation, the satellite mobile fog computing servers are deployed in LEO satellites, as shown in Fig. 9c. With this network extension, both sparse and dense users will benefit from reduced latencies while the traffic in the terrestrial backbone network will also reduce significantly. However, the latency in this situation would be higher than that of PTO and it would significantly increase the power consumption of satellites which will be performing the offloaded computations which will not be practical for satellites with limited power sources.
- **Remote Terrestrial Offloading (RTO)**: In this situation, the satellite mobile fog computing servers are deployed to the terrestrial backbone network, as shown in Fig. 9d. In this situation, the delays in transmission over the WAN IP that connects with the Remote Cloud servers can be avoided and this translates to a reduced latency when compared to the situation with no edge computing offloading. The latency in this network scheme is higher than PTO and SBO, but it is the most practical scheme to implement and maintain.

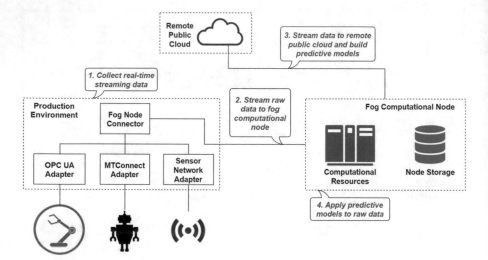

**Fig. 10** Architecture for the process monitoring system

## 4.3 Manufacturing Process Monitoring

With rapid globalization, industries across the globe have started to adopt modern process control systems which rely heavily on sensor networks that efficiently monitor production lines and processes while collecting valuable data which can be used to identify faults before they occur while also aiding in optimization efforts to improve the throughput and performance of the industry. In this regard, we shall be looking at a fog computing-based framework for process monitoring in different production environments. The proposed system architecture in [36] is shown in Fig. 10 and is described sequentially below:

- Step 1: Collect machine data from the production environment that streams real-time data from various sensor networks and communication adapters that function on protocols such as Simple Object Access Protocol (SOAP), MTConnect, and Open Platform Communications Unified Architecture (OPC UA).
- Step 2: Stream the raw data to a private computational fog node that is responsible for real-time monitoring and providing time-sensitive control signals to the production environment. This allows the system to function with low response times, improves reliability, and reduces the strain on the network's capacity as data is processed in a fog computing node that is situated close to the production environment.
- Step 3: Also, various samples from this data can be sent to high-performance cloud data centers which can be used to build models for predictive maintenance and process optimization. Since these samples are small in size and sporadically transferred to the cloud, the strain on the network's capacity is minimal while the

models built with the sampled data can be extremely beneficial for the industry in terms of improved throughput and reduced unplanned downtimes.

- Step 4: Apply these predictive models to raw data and obtain tangible insights into the production environment's real-time health and performance.

# 5 Future Research Directions

The edge and fog computing paradigms are considered as powerful extensions to the cloud computing paradigm, however, they face some common challenges [16] that are yet to be addressed. In this section, we describe some of the major issues and challenges faced by these paradigms as well as some potential research directions for these paradigms.

## 5.1 Programmability and Task Partitioning

In the traditional cloud computing-based architectures, users generally program their back-end applications on an abstract platform, without worrying about the exact configuration of the cloud server. The benefit of this abstraction is programmability since the user is not aware of the exact configuration of the platform which means that the cloud service providers can easily compile the application and run it on a single runtime of the cloud server which can have a variable configuration, unknown to the user. However, with the rise of the edge and fog computing paradigms, back-end processing is distributed across an array of distributed computational nodes—all of which can have slightly different run-times. This creates interesting and challenging problems for system designers, who need to design optimized methods for distributing computation as well as storage across nodes while making sure that synchronization processes do not impact the network's transmission capacities and ensure low latency in intra-network transmissions.

An important issue that arises with the evolution of distributed computing paradigms like edge and fog computing, is the issue of task partitioning. Within these paradigms, it is imperative that the system design takes into account the optimization of task partitioning and process scheduling, to facilitate concurrent execution across distributed nodes. An optimized task partitioning scheme allows the system to autonomously locate edge or fog nodes in real-time, and intelligently allocate tasks to these nodes, while taking into account various factors such as the computational power associated with the nodes as well as the associated overheads involved in exchanging data between these nodes.

In consideration of these issues, system designers should also think about control - whether the system should allow users to implicitly or explicitly control computational resources. In case of implicit control, which can be seen in the case of Amazon's Lambda@Edge [37], where the users need not worry about server administration, as the web services are responsible for running and scaling the application at resources available closest to the end-users. This leads to reduced complexity of programming for the users, while giving system administrators greater control over the network. In contrast, explicit control of the network gives users greater flexibility in terms of resource allocation, which can often lead to improved efficiency and increased reliability. This explicit control, however, comes at the cost of increased complexity in terms of programmability and goes to illustrate how designers need to make trade-offs while planning the layout of edge and fog computing-based systems.

## 5.2 Security and Privacy

With an increased interest in the edge and fog computing paradigms, people have started to appreciate the capabilities of these paradigms which enable the extension of storage, networking, and processing resources of cloud computing servers toward the edge of the network. However, with this rise in flexibility and distribution leads to many security and privacy concerns [38] that must be addressed by system designers. After analyzing several different aspects of network security, we can summarize the major security and privacy concerns as follows:

1. **Trust and Authentication**: Edge and Fog Computing-based networks are expected to provide secure and reliable services to all users and this leads to an important requirement in that all devices on the network should be able to trust one another. Therefore, trust plays a two-way role within edge and fog computing-based networks. This implies that fog or edge nodes that offer services to the network must be in a position to validate whether the resources requesting these services are indeed genuine. Similarly, edge or fog nodes that are transmitting data to or requesting services from network resources should also be able to verify whether these resources are genuine or not. These concerns have given rise to various authentication mechanisms that can be used to authenticate network resources before transmissions and requests. Systems can employ mechanisms such as permissioned blockchain networks like TrustChain [39] for authentication, cryptographic authentication schemes like SAKA-FC [40], and hardware-based authentication schemes like Physically Unclonable Functions (PUF) [41], to authenticate network resources.
2. **Integrity**: Edge and Fog Computing systems should always ensure that data transmission within the network should be done securely so that transmitted data is not altered or modified by attackers. The most prominent method to ensure the integrity of data in networks is through cryptographic signature verification systems like the GNU Privacy Guard (GPG) system [42] which is used to digitally sign transmitted data. The received data is then verified at the receiving station to establish the integrity of the data, which is extremely important in edge and fog computing-based systems as they rely heavily on intra-network data transfers due to their distributed topology.

3. **Availability**: The availability of information refers to the ability of the system to ensure that authorized parties can access relevant information whenever needed. The biggest concern concerning the availability of information is Denial of Service (DoS) attacks that hamper or eliminate accessibility to information. Edge and Fog Computing-based systems are generally well equipped to handle DoS attacks since these systems have distributed computational resources, however, Distributed Denial of Service (DDoS) attacks can still impact these systems and to protect networks or applications against DoS attacks, designers often make use of smart DNS resolution services, Web Application Firewalls (WAF), and other intelligent traffic management techniques to ensure service security.

4. **Confidentiality**: The confidentiality of information represents the ability of the system to protect information from being disclosed to unauthorized parties. This implies that edge and fog computing paradigms should ensure that information is stored securely to prevent data leaks, which is especially likely due to the distributed architecture of these paradigms. Edge and Fog Computing-based architectures often use homomorphic encryption schemes as well as crypto-graphic hashing techniques to store confidential data at different distributed locations within the network [50]. Due to the use of these techniques, even if attackers can gain access to secure databases, they will not be able to understand the data as it will be in an encrypted format.

5. **Data Ownership**: This issue extends from the fact that unlike cloud computing-based systems, edge and fog computing-based systems store data in distributed locations across the network which means that the system can store data locally at the computational nodes, thereby providing complete access and ownership to the end-users. However, these paradigms often involve the transmission of data between nodes especially when processing or computations have been offloaded to different nodes on the network, and this creates a problem in the data ownership. Thus system designers should take this behaviour into account while drafting the privacy policy of the network. This also involves thinking about legal jurisdictions, such as when data crosses international borders, it may be subject to different regulations. This means that data transfer methods should consider the compatibility of data with two different data regulation policies for the source and destination.

## 5.3 Blockchain for IIoT

In recognition of the prevalent security concerns within the domain of edge and fog computing, researchers have started looking into Blockchain as a potential solution for these security concerns [43]. Following this technology's conceptualization in 2008, industries across the world have extensively adopted it for several applications such as in the authentication of financial transactions, in the preservation of digital contracts, and in the identification of agents in distributed systems, making it an extremely viable solution for the security concerns of edge and fog computing.

The essence of the blockchain technology lies in the decentralized transaction ledger that maintains the record of all exchanges and transactions. The most essential components of the distributed ledger are the blocks that comprise the blockchain, where each block acts as a record of the transactions or exchanges which are being monitored by the blockchain. To preserve the authenticity of transactions and exchanges, blockchains rely on two main aspects of the blockchain framework. First, each block within the ledger possesses a cryptographic hash that uniquely identifies that block. This hash acts as a fingerprint that is referenced by the next block after being added to the ledger. This reference-linking structure allows the distributed ledger to easily identify distortion within the blockchain, since changing the transaction within any block in the blockchain leads to change in the block's unique cryptographic hash, thereby leading to a break in the chain as the references for each of the following blocks in the ledger need to be changed to remake the blockchain. Second, each transaction is verified by different agents that maintain independent copies of the ledger, which makes tampering even more difficult as any unwarranted changes in the ledger by one agent can be easily identified by other agents through a block-by-block comparison. These two aspects allow the blockchain to effectively control distortion to maintain a transparent and verifiable transaction history.

The tamper-proof nature of blockchain can be extremely useful in addressing the trust and authentication problems of edge and fog computing. Specifically, these paradigms can adopt blockchain-based authentication systems such as BSeIn [44] that allows fine-grained access control, user anonymity, and mutual authentication while allowing networks of these paradigms to scale as usual. Similarly, the tamper-proof nature of the blockchain ledger allows the system to detect adversarial end devices which may be attempting to manipulate raw sensor data, allowing the system to take immediate countermeasures. Furthermore, the distributed nature of the edge and fog computing paradigms match the decentralized nature of the blockchain technology, making such systems resistant to node failures as the loss of any one node does not compromise the health of the complete system as data remains distributed across the network.

Although the blockchain technology is relatively new, several interesting systems have been proposed in the past, which have integrated the blockchain technology with IIoT architectures, to improve the security and reliability of such systems. In the BPIIoT platform [45], systems can make use of a peer-to-peer network as well as smart contracts to allow end devices to share information across the network after successfully verifying the authenticity of their counterparts within the network. This enables machine-to-machine (M2M) communication that is safe, transparent, and verifiable thereby improving the security of systems that utilize this technology. Similarly, to design edge and fog computing systems that are suited for distributed data storage, the blockchain technology can be extremely useful in verifying device identity, while advanced encryption schemes can be used to encrypt transactions and store them within the blockchain in a chronological fashion. In such a situation, if an adversarial end device attempts to alter the blockchain, it will be identified immediately by the other agents in the system while the chronological arrangement of transactions can be used to determine the exact time of the attack by the adversary - leading to easier data recovery. Similarly, edge and fog computing-based IIoT systems that handle a large variety of goods, identities, and credentials, can use blockchains to store information relating to these domains. Furthermore, for physical assets such as end devices, the cryptographic hashes of the device firmware can be stored in a separate private blockchain to ensure the authenticity of the hardware within the system.

While the integration of the blockchain technology with edge and fog computing-based IIoT systems is a viable future direction, we must understand at the existing limitations of the blockchain technology and work to improve the effectiveness of this technology. These limitations include:

1. **Poor performance on scaling**: In contrast with traditional centralized databases, blockchains tend to slow down with scale due to the complexity of the consensus mechanisms leading to lower transactional throughput and increased latency.
2. **Energy inefficient algorithms**: The complexity of proof-of-work (PoW) increases as the number of transactions conducted increases which makes these algorithms extremely energy-hungry leading to several issues on battery-powered IoT devices which are power-constrained and cannot afford to make such computations.
3. **Lack of flexible test platforms**: It is imperative to have flexible test platforms in place that allow people to experiment with different configurations of blockchains in various IIoT applications to test the stability, performance, scalability, and security of these applications.

## 5.4  Virtualization

Due to the resource-constrained nature of fog and edge devices, most applications utilizing these distributed computing paradigms need to run multiple operating environments on a single edge or fog device. Typically, each edge or fog device needs to run two different environments for different users, leading to two important requirements in this respect. First, we require separation of services, wherein different tasks and user environments must be maintained separately on every node within the system. Second, we want to ensure application fairness, wherein resource allocation and distribution of computational power should be monitored by resource management algorithms that enforce fairness in allocation procedures.

In this regard, virtualization technologies for task partitioning can be viable options for the encapsulation of services from various users and applications into separate Virtual Machines (VM). Through the use of VMs, virtualization technologies can run different operating environments on each node. While extensively developed virtualization technologies for cloud computing exist [46], their support for edge and fog based systems remains limited. Therefore, within the field of virtualization for edge and fog computing paradigms, there are many standing issues and challenges, including:

1. **Service Encapsulation**: In traditional cloud-oriented virtualization, techniques for service encapsulation to VMs tend to be resource-intensive which makes them infeasible for virtualization at edge and fog layers. An interesting approach can be through the deployment of services in containers [47–49]. In this approach, we use containers as operating system-level virtualization objects to execute different services on resource-constrained computational nodes, allowing service encapsulation and improved task allocation procedures.

2. **Resource or Container Allocation**: The allocation of resources in the form of containers running on different nodes requires sophisticated allocation algorithms to ensure virtualization in edge and fog computing systems. These virtualization algorithms should be able to perform resource (computational capacity) estimation and overhead estimation in real-time to identify the optimal strategy for task partitioning. Furthermore, these algorithms should not be resource-intensive and must be able to aggregate the results obtained from distributed computations to accomplish the overall task, thereby leading to efficient virtualization.

3. **Container Migration**: As seen in the earlier sections, the distributed and decentralized nature of the edge and fog computing paradigms enables us to design fault-tolerant and reliable systems. To maintain this fault-tolerance, virtualization techniques must be responsive enough to ensure the migration of services to other containers (nodes) in response to the failure of nodes that are currently executing tasks or applications. Therefore, these virtualization techniques should be able to identify, allocate, and migrate tasks quickly, so that the system experiences minimal downtime while maintaining a reliable service. A robust container migration policy will be similarly beneficial for systems under dynamic workloads and will improve the efficiency of these systems.

Given the success of virtualization techniques in the scenario of traditional cloud computing services, the development of robust virtualization techniques for edge and fog computing-based IIoT systems can become an influential force in the future development of these paradigms.

## 5.5 Resource Allocation

We already know that edge and fog nodes are constrained in terms of computational resources, memory, network elements, and energy, and therefore, efficient management of resources is imperative for the success of these paradigms and to achieve efficient resource management, resource allocation mechanisms should work in conjunction with virtualization techniques [20]. To facilitate the process of task partitioning through virtualization, resource estimation mechanisms, running as software middleware, should be able to accurately estimate the various resources available with the different edge or fog nodes in the system and must be able to do so with low computational overhead. Similarly, resource allocation mechanisms should work to ensure the highest QoS for the end-user, and at the same time, it should ensure fairness of allocation for distinct services, wherein tasks with higher priority (like real-time content streaming) should receive larger bandwidth when compared with lower priority tasks. Finally, resource allocation should also take care of dynamic network situations where IoT devices enter and exit the network at will, making technologies such as software-defined networking (SDN) important factors in system design. SDN can aid the system in the management of dynamic network resources to ensure continued connectivity between them while also improving oversight and control. By working towards the improvement of resource allocation mechanisms, system designers can establish better virtualization techniques which, in turn, will lead to the superior edge and fog computing-based IIoT systems.

## 5.6  System Metrics

While there exists a large variety of advantages that arise due to the architecture of the edge and fog computing paradigms, there are some associated metrics that also need to be considered while designing these paradigms. Importantly, system designers often deal with the design of policies which govern task partitioning and work offloading from one computational node to others, and in such situations, they should give importance to the following metrics:

1. **Energy**: Edge and fog nodes often consist of embedded devices such as wireless access points, routers, or switches, which often have power sources in the form of batteries. Due to the limited capacity of the batteries, system designers should always consider if it would be energy efficient to offload some task to a particular node, while also taking into account the computational power associated with that node and the expected amount of computation that is required for the task being offloaded. An important environmental benefit in this regard is that the energy requirement of fog and edge nodes is smaller than that of cloud servers. This means that the edge and fog nodes can use renewable energy sources for their power requirements, leading to an overall reduction in $CO_2$ emissions, which shows that the edge and fog computing paradigms are also much more eco-friendly when compared to cloud computing.
2. **Cost**: While migrating applications to edge and fog computing-based architectures often leads to reduced latency, improved reliability, and increased fault tolerance, it still comes at the expense of increased cost. With thousands of embedded computational nodes in modern edge and fog computing-based systems, the cost is generally much higher than traditional cloud services, which means that systems within the edge and fog computing paradigms should be cost-efficient, to justify their development in response to improved user experience.

3. **Bandwidth**: The edge and fog computing paradigms need to be designed while considering bandwidth especially for low-cost systems which generally have low bandwidths within the network. In the edge and fog computing paradigms, we see a lower amount of data transmission whenever a larger amount of data is processed closer to the edge since no data needs to reach the remote cloud server. However, the distributed nature of the system can often increase the number of transmissions within the system, especially in co-operative systems that rely heavily on inter-node communication. Therefore, system designers can appreciate these major factors that influence bandwidth consumption and can organize their system accordingly.

# 6  Summary and Conclusions

With the recent advances within the domain of IIoT, people have started to observe strong trends that indicate rapid growth in the number of smart devices connected to IIoT networks and this growth cannot be supported by traditional cloud computing platforms. In response to this, edge and fog computing systems have emerged as important frameworks that have the potential to support the growing demands of automation in different industrial settings. As these paradigms are inherently distributed in nature, their resources are distributed along the edges of the network. This in turn leads to reduced latency and improved reliability of services associated with edge and fog computing-based systems. Through this chapter, we have described the fundamentals of the edge and fog computing paradigms while comprehensively exploring the benefits offered by these systems over the traditional cloud-based platforms. Furthermore, the chapter discusses several industrial applications for both edge and fog computing through an in-depth analysis of proposed system architectures for the different industrial use cases. With several supporting case studies and experiments explained in the chapter, we practically demonstrate the superiority of these computing paradigms and build a strong case for the adoption of these paradigms in modern industrial systems. Finally, we present the major issues and challenges faced by these paradigms, along with some plausible solutions which serve as future research directions.

# References

1. Gubbi, J., Buyya, R., Marusic, S., Palaniswami, S.: Internet of Things (IoT): A vision, architectural elements, and future directions. Future Gener. Comput. Syst. **29**(7), 1645–1660 (2003)
2. Cisco Systems Inc.: Cisco Visual Networking Index: Forecast and Trends, pp. 2017–2022 (2019). https://www.cisco.com/c/en/us/solutions/collateral/service-provider/visual-networking-index-vni/white-paper-c11-741490.pdf Accessed: 19 June 2020

3. Manyika, J., Michael Chui, M.: Open interactive popup McKinsey Global Institute. The Internet of Things: Mapping the value beyond the Hype (2015). https://www.mckinsey.com/mgi/overview/in-the-news/by-2025-internet-of-things-applications-could-have-11-trillion-impact Accessed: 19 June 2020
4. Sisinni, E., Saifullah, A., Han, S., Jennehag, U., Gidlund, M.: Industrial internet of things: challenges, opportunities, and directions. IEEE Trans. Ind. Inform. **14**(11), 4724–4734 (2018)
5. Lu, Y.: Industry 4.0: A survey on technologies, applications and open research issues. J. Ind. Inf. Int. **6**, 1–10 (2017)
6. Cisco Systems, Inc.: Mining firm quadruples production, with internet of every-thing (2014). https://www.cisco.com/assets/global/BE/tomorrow-starts-here/pdf/c36-730784-01_dundee_precious_metals_cs_v3a_en_be.pdf Accessed: 28 November 2019
7. Tzounis, A., Katsoulas, N., Bartzanas, T., Kittas, C.: Internet of things in agriculture, recent advances and future challenges. Biosyst. Eng. **164**, 31–48 (2017)
8. Xu, B., Xu, L.D., Cai, H., Xie, C., Hu, J., Bu, F.: Ubiquitous data accessing method in IoT-based information system for emergency medical services. IEEE Trans. Ind. Inform. **10**(2), 1578–1586 (2014)
9. GSMA: China Mobile Electric Smart Metering – Internet of Things Case Study (2018). https://www.gsma.com/iot/wp-content/uploads/2018/03/iot_china_mobile_metering_04_18.pdf Accessed: 19 June 2020
10. GSMA: China Mobile Smart Parking – Internet of Things Case Study (2018). https://www.gsma.com/iot/wp-content/uploads/2018/03/iot_china_mobile_parking_04_18.pdf Accessed: 19 June 2020
11. Fraga-Lamas, P., Fernández-Caramés, T.M., Castedo, L.: Towards the internet of smart trains: A review on industrial IoT-connected railways. Sensors **17**(6), 1457 (2017)
12. Shah, S., Ververi, A.: Evaluation of Internet of Things (IoT) and its Impacts on Global Supply Chains In: Proceedings of the 2018 IEEE International Conference on Technology Management, Operations and Decisions (ICTMOD), Marrakech, Morocco, pp. 160–165 (2018)
13. Antão, L., Pinto, R., Reis, J., Gonçalves, G.: Requirements for testing and validating the industrial internet of thing. In: Proceedings of the IEEE International Conference on Software Testing, Verification and Validation Workshops (ICSTW), Vasteras, pp. 110–115 (2018)
14. Breivold, H.P., Sandström, K.: Internet of things for industrial automation – challenges and technical solutions. In: Proceedings of the IEEE International Conference on Data Science and Data Intensive Systems, Sydney, pp. 532–539 (2015)
15. Chamola, V., Tham, C., Chalapathi, G.S.S.: Latency aware mobile task assignment and load balancing for edge cloudlets In: Proceedings of the IEEE International Conference on Pervasive Computing and Communications Workshops, Kona, HI, pp. 587–592 (2017)
16. Shi, W., Cao, J., Zhang, Q., Li, Y., Xu, L.: Edge computing: vision and challenges. IEEE Internet Things J. **3**(5), 637–646 (2016)
17. Mahmud, R., Kotagiri, R., Buyya, R.: Fog Computing: A Taxonomy, Survey and Future Directions. In: Di Martino, B., Li, K.C., Yang, L., Esposito, A. (eds.) Internet of Everything (Algorithms, Methodologies, Technologies and Perspectives), pp. 103–130. Springer, Singapore (2018)
18. Vaquero, L.: Finding your way in the fog: Towards a comprehensive definition of fog computing. ACM SIGCOMM Comput. Commun. Rev. **44**(5), 27–32 (2014)
19. Yousefpour, A., Fung, C., Nguyen, T., Kadiyala, K., Jalali, F., Niakanlahiji, A., Kong, J., Jue, J.P.: All one needs to know about fog computing and related edge computing paradigms: A complete survey. J. Syst. Arch. **98**, 289–330 (2019)
20. Ahmed, A., Arkian, H., Battulga, D., Fahs, A., Farhadi, M., Giouroukis, D., Gougeon, A., Gutierrez, F., Pierre, G., Souza, Jr. P., Ayalew Tamiru, M., Wu, L.: Fog Computing Applications: Taxonomy and Requirements (2019). https://arxiv.org/pdf/1907.11621.pdf. Accessed: 19 June 2020
21. Xiao, Y., Jia, Y., Liu, C., Cheng, X., Yu, J., Lv, W.: Edge computing security: state of the art and challenges. Proc. IEEE **107**(8), 1608–1631 (2019)

22. Lyu, L., Chen, C., Zhu, S., Cheng, N., Yang, B., Guan, X.: Control performance aware cooperative transmission in multiloop wireless control systems for industrial IoT applications. IEEE Internet Things J. **5**(5), 3954–3966 (2018)
23. Chen, B., Wan, J., Celesti, A., Li, D., Abbas, H., Zhang, Q.: Edge computing in IoT-based manufacturing. IEEE Commun. Mag. **56**, 103–109 (2018)
24. Wan, J., Tang, S., Shu, Z., Li, D., Wang, S., Imran, M., Vasilakos, A.V.: Software-defined industrial internet of things in the context of industry 4.0. IEEE Sens. J. **16**(20), 7373–7380 (2016)
25. Gîrbea, A., Nechifor, S., Sisak, F., Perniu, L.: Design and implementation of an OLE for process control unified architecture aggregating server for a group of flexible manufacturing systems. Softw. Lett. **5**(4), 406–414 (2011)
26. Kang, W., Kapitanova, K., Son, S.H.: RDDS: A real-time data distribution service for cyber-physical systems. IEEE Trans. Ind. Inform. **8**(2), 393–405 (2012)
27. Wang, S., Wan, J., Zhang, D., Li, D., Zhang, C.: Towards smart factory for industry 4.0: A self-organized multi-agent system with big data based feedback and coordination. Comput. Netw. **101**, 158–168 (2016)
28. Musa, Z., Vidyasankar, K.: A fog computing framework for Blackberry supply chain management. Procedia Comput. Sci. **113**, 178–185 (2017)
29. Industrial Internet Consortium White Paper: Introduction to Edge Computing in IIoT (2018). https://www.iiconsortium.org/pdf/Introduction_to_Edge_Computing_in_IIoT_2018-06-18.pdf
30. Hao, P., Bai, Y., Zhang, X., Zhang, Y.: Edgecourier: an edge-hosted personal service for low-bandwidth document synchronization in mobile cloud storage services In: Proceedings of the Second ACM/IEEE Symposium on Edge Computing (SEC '17), San Jose/Silicon Valley, CA, pp. 1–14 (2017)
31. Pace, P., Aloi, G., Gravina, R., Caliciuri, G., Fortino, G., Liotta, A.: An edge-based architecture to support efficient applications for healthcare industry 4.0. IEEE Trans. Ind. Inform. **15**(1), 481–489 (2019)
32. Bernardi, L., Wdowczyk-Szulc, J., Valenti, C., Castoldi, S., Passino, C., Spadacini, G., Sleightp, P.: Effects of controlled breathing, mental activity and mental stress with or without verbalization on heart rate variability. J. Am. College Cardiol. **35**(6), 1462–1469 (2000)
33. Zamora-Izquierdo, M.A., Santa, J., Juan, A., Martínez, J.A., Martínez, V., Skarmeta, A.F.: Smart farming IoT platform based on edge and cloud computing. Biosyst. Eng. **177**, 4–17 (2019)
34. Okay, F.Y., Ozdemir, S.: A fog computing-based smart grid model In: Proceedings of International Symposium on Networks, Computers and Communications (ISNCC), Yasmine Hammamet, pp. 1–6 (2016)
35. Zhang, Z., Zhang, W., Tseng, F.: Satellite mobile edge computing: Improving QoS of high-speed satellite-terrestrial networks using edge computing techniques. IEEE Netw. **33**(1), 70–76 (2019)
36. Wu, D., Liu, S., Zhang, L., Terpenny, J., Gao, R.X., Kurfess, T., Guzzo, J.A.: A fog computing-based framework for process monitoring and prognosis in cyber-manufacturing. J. Manuf. Syst. **43**, 25–34 (2017)
37. Amazon Web Services Lambda@Edge: https://aws.amazon.com/lambda/edge/. Accessed: 28 November 2019
38. Mukherjee, M., Matam, R., Shu, L.: Security and privacy in fog computing: challenges. IEEE Access **5**, 19293–19304 (2107)
39. Jayasinghe, U., Lee, G.M., MacDermott, Á., Rhee, W.S.: TrustChain: A privacy preserving blockchain with edge computing. Wirel. Commun. Mob. Comput. (2019). https://doi.org/10.1155/2019/2014697
40. Wazid, M., Das, A.K., Kumar, N., Vasilakos, A.V.: Design of secure key management and user authentication scheme for fog computing services. Future Gener. Comput. Syst. **19**, 475–492 (2019)

41. Huang, B., Cheng, X., Cao, Y., Zhang, L.: Lightweight hardware-based secure authentication scheme for fog computing. In: Proceedings of the IEEE/ACM Symposium on Edge Computing (SEC), Seattle, WA, USA, pp. 433–439 (2018)
42. GNU Privacy Guard: https://www.gnupg.org/. Accessed: 28 November 2019
43. Tuli, S., Redowan Mahmud, R., Tuli, S., Buyya, R.: FogBus: A blockchain-based lightweight framework for edge and fog computing. J. Syst. Softw. **154**, 22–36 (2019)
44. Lin, C., He, D., Huang, X., Choo, K.R., Vasilakos, A.V.: BSeIn: A blockchain-based secure mutual authentication with fine-grained access control system for industry 4.0. J. Netw. Comput. Appl. **116**, 42–52 (2018)
45. Bai, L., Hu, M., Liu, M., Wang, J.: BPIIoT: A light-weighted blockchain-based platform for industrial IoT. IEEE Access **7**, 58381–58393 (2019)
46. Nurmi, D., Wolski, R., Grzegorczyk, C., et al.: The eucalyptus open-source cloud-computing system. In: Proceedings of the 9th IEEE/ACM International Symposium on Cluster Computing and the Grid, Shanghai, pp. 124–131 (2009)
47. Kaur, K., Dhand, T., Kumar, N., Zeadally, S.: Container-as-a-service at the edge: Trade-off between energy efficiency and service availability at fog nano data centers. IEEE Wirel. Commun. **24**(3), 48–56 (2017)
48. Yin, L., Luo, J., Luo, H.: Tasks scheduling and resource allocation in fog computing-based on containers for smart manufacturing. IEEE Trans. Ind. Inform. **14**(10), 4712–4721 (2018)
49. Santoro, D., Zozin, D., Pizzolli, D., De Pellegrini, F., Cretti, S.: Foggy: A platform for workload orchestration in a fog computing environment. In: Proceedings of the IEEE International Conference on Cloud Computing Technology and Science (CloudCom), pp. 231–234 (2017)
50. Roman, R., Lopez, J., Mambo, M.: Mobile edge computing, fog et al.: A survey and analysis of security threats and challenges. Future Gener. Comput. Syst. **78**, 680–698 (2018)

# Leveraging Edge Computing for Mobile Augmented Reality

Sarah M. Lehman and Chiu C. Tan

## 1 Introduction

Augmented reality (AR) is the insertion of virtual content into a view of the real world based on environmental context. The virtual content may take the form of characters, textures, or labels, and is dynamically updated based on the user's location, viewing direction, behavior, and other contextual data. The virtual content utilized in AR systems is traditionally visual in nature, but may be supplemented with auditory and haptic feedback as well.

AR systems differ from virtual reality (VR) and mixed reality (MR) systems in several key ways. Virtual reality systems seek to replace the physical world completely with virtual content, while augmented reality inserts virtual content into the real world. The range from augmented to mixed reality is a sliding scale; an application which simply places a virtual character on a tabletop would be considered augmented reality, whereas an application which identifies a plethora of grocery and kitchen items and displays possible recipes could be considered mixed reality. While the difference between AR and MR systems is one of degrees, for the purposes of this chapter, we will focus on augmented reality systems.

Augmented reality has made great impacts in recent years in both research and commercial domains, such as education [3, 10, 17], tourism [11, 13, 25, 56], entertainment [46, 58], healthcare [7, 20, 24, 27, 41], and manufacturing [19, 38, 57]. Contemporary platforms for AR systems are as varied as the systems themselves; smartphones and tablets are popular tools for AR, but alternative platforms such as

S. M. Lehman (✉) · C. C. Tan
Temple University, Philadelphia, PA, USA
e-mail: smlehman@temple.edu; cctan@temple.edu

© Springer Nature Switzerland AG 2021
W. Chang, J. Wu (eds.), *Fog/Edge Computing For Security, Privacy, and Applications*, Advances in Information Security 83,
https://doi.org/10.1007/978-3-030-57328-7_13

head-mounted displays (HMDs) like the Microsoft HoloLens,[1] and Google Glass,[2] and augmented windshields for smart vehicles[3] are also gaining popularity. This wider availability of low-cost AR-enabled hardware has encouraged market growth for AR systems with global profits of 6.1 billion USD in 2016. As the cost of hardware decreases and the availability of development tools increases, this profit is predicted to grow to an estimated 18.8 billion USD in 2020.[4]

As AR systems increase in prevalence and popularity, the security risks of such systems must also be considered. There are several points of an AR system that are vulnerable to different kinds of attacks. Attackers could target user authentication or input data collection, compromising how the system extracts and prepares data for processing. They could also target the actual processing of the data itself by constructing system inputs in such a way that forces the application to produce a result that is both unexpected to the developer and beneficial to the attacker. Finally, the attacker could target the output of the application, either to interfere with outputs from other applications or to compromise the user's interactions with their environment. These attacks can be mitigated when the AR system is moved to the edge, thanks to expanded processing resources and data storage capabilities on edge servers. However, when moving operations to the edge, an AR system must first and foremost be able to guarantee low end-to-end latency in order to preserve a high quality user experience. Understanding how user experience can be measured, which system metrics impact it, and to what degree, is key to designing an effective AR system. Ensuring that a system meets these requirements while also providing a high level of security is even more important.

The rest of the chapter is organized as follows. First, we provide an background of augmented reality research and development, introducing the AR processing pipeline. Next, we discuss the security implications for AR systems at large, including examples of specific vulnerabilities within each phase of the AR processing pipeline. Then, we give a brief overview of research efforts focused on moving AR systems to the network edge as well as the metrics for evaluating them, followed by an exploration of the security issues for AR at the edge. We wrap up with a discussion of open problems in this research space, and present our conclusions.

# 2  Background of AR Systems

The fundamental requirement for an augmented reality system is the ability to sense and respond to changes in the user's environment and behavior, typically via the

---

[1] https://www.microsoft.com/en-us/hololens.

[2] https://www.google.com/glass/start/.

[3] https://www.motorauthority.com/news/1120806_hyundai-turns-the-windshield-into-an-augmented-reality-nav-system.

[4] https://www.statista.com/statistics/591181/global-augmented-virtual-reality-market-size/.

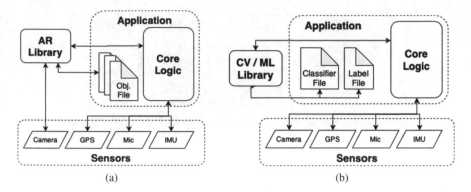

**Fig. 1** App architectures when using (**a**) AR libraries or (**b**) computer vision/machine learning libraries

delivery of visual or auditory cues to the user. Because of this, AR systems rely on live data feeds from various sensors, such as the camera, microphone, GPS, and more. The exact implementation for processing these data feeds depends on the type of underlying library being utilized. For systems built with specialized AR libraries like ARCore[5] and Vuforia[6] (Fig. 1a), the library acts as an intermediary between the core application logic and the target device sensors (such as the camera). The developer is responsible for initializing the library instance with object files representing the targets to be recognized, and then subscribing to event listeners that will be raised when a target is located. This ease of integration comes with a price, however, as developers are limited to only those functions which the library publisher explicitly supports. On the other hand, developers using more general-purpose libraries such as OpenCV[7] and TensorFlow[8] (Fig. 1b) can exert more fine-grained control over the functionality of their systems, executing any operations for which they are able to collect data and train models. However, the downside is that these operations are remarkably resource-intensive, making them difficult to integrate efficiently into resource-constrained systems.

The pipeline of operations for an AR system is split in three major phases, as demonstrated by Fig. 2. The first phase is the **input phase**, during which the AR device is responsible for collecting data from the on-board sensors, including visual data from the camera, audio from the microphone, and environmental data such as location, ambient light levels, and more. The second phase is the **transformation phase**, during which the data that has been collected is utilized in a computer vision or machine learning operation, such as object recognition. This phase includes preprocessing the data to get it ready for the operation, performing the operation, and

---

[5] https://developers.google.com/ar/.

[6] https://www.vuforia.com.

[7] https://opencv.org.

[8] https://www.tensorflow.org.

**Fig. 2**  AR processing pipeline

then aggregating the results as appropriate. The final phase is the **output phase**, during which the transformation results are used to generate and display virtual content, such as labels, sounds, or haptic feedback.

There are many types of computer vision and machine learning operations that can be utilized during the transformation phase of an AR system. **Classification** refers to the act of determining the subject of a given image. The drawing of **bounding boxes** refers to, not only identifying objects of interest within an image, but denoting the space they embody by drawing a box around it. **Pose estimation** looks at the posture of an entity within an image, such as a human subject, and recreates the posture of the underlying skeleton. **Face detection** identifies features of human faces within a given image in order to judge location, expression, gaze direction, and more. **Target tracking** refers to the monitoring of an actor's position as they move through a scene, including behind other objects or actors in that scene. All of these operations and more can be utilized by AR systems to identify users' context and place responsive content into the environment.

# 3   Security Risks of AR

Due to the richness of the data collected by AR applications, they are subject to a number of security and privacy vulnerabilities that non-AR applications do not face. These vulnerabilities stem from the always-on nature of the microphone and camera sensors in combination with atypical input methods, such as voice commands or physical gestures, and environmental or context data, such as GPS location and accelerometer readings.

Each phase of the AR pipeline is subject to different types of attack. Attacks during the input phase can target user authentication processes or collection of data from system sensors or remote servers. Attacks during the transformation phase can focus on how this collected data is pre-processed, manipulate or compromise the actual computer vision operation being performed, or contaminate the returned result. Finally, attacks during the output phase can target the content being generated or the placement and styling of that content in the system display. We will look at examples of different kinds of attacks in each of these phases, and wrap up with a quick discussion of other concerns in AR security beyond the processing pipeline.

## 3.1   Input Phase Attacks

### 3.1.1   User Authentication

AR systems rely on gestures and voice commands to maintain an immersive experience in a system without traditional input operations, such as typing on a keyboard or clicking with a mouse. Systems utilizing these non-traditional input methods rely on user authentication to identify valid inputs and filter out environmental noise, ensuring that subsequent system operations are germane to the current user. Proper user authentication can prevent both the creation of incorrect or unnecessary virtual content, as well as incorrect blocking of desired content.

For voice commands, a third-party attacker can control a victim's system through a number of methods. For instance, inaudible or unintelligible commands can be played through speakers; alternatively, a specific user's voice can be recorded or synthesized from recordings and then played back. The typical defense against attacks like this is called liveness detection, in which the presence of a valid human user is determined before executing a command. One particular solution presented in [52] suggested the use of contact microphones in AR headsets to compare waves travelling through the speaker's head with those received through the air at the headset microphone.

Physical gestures are also vulnerable in AR systems. Because gestures are readily visible by observers, simply making a given gesture is insufficient for user authentication. The exact manner in which the gesture is completed must be analyzed to distinguish between users. An example of this is in [53], in which the IMU sensors of an AR headset are used to monitor and analyze a user's walking patterns. The user's gait is then utilized as a unique identifier for authentication. Gestures can also be used when pairing devices among multiple users [18, 54], though extra steps must be taken to prevent man-in-the-middle attacks during the pairing process.

However, voice- and gesture-based authentication are only really viable with wearable sensors. For more pervasive AR environments, such as augmented wind-screens in smart cars, the display is part of the environment and accessible to all persons in the immediate area. Therefore, performing authentication becomes a more delicate process than attempting to validate a single user in isolation.

### 3.1.2   Data Collection

After the AR system has authenticated the user, it can start collecting data. However, once data is collected by an application, any control that the user has over that data is gone. Therefore, limiting a malicious application's ability to collect data in the first place becomes a major concern. For example, a retail application which places virtual furniture in the user's home would have a justifiable need to recognize planes, but might also recognize text in the background without indicating as such to the user. The primary method of defending against this is to rely on the underlying operating system or commercial libraries to enforce data access limitations. For example, instead of allowing applications to subscribe directly to device sensors such as the camera and perform operations on the raw frame, the operating system could expose access to abstracted objects such as skeletons, faces, and hands [22]. In this way, applications have access to the objects and associated meta data that they truly need without accessing raw frames from the camera feed. Similarly, wrappers could be implemented for popular computer vision libraries such as OpenCV to enforce a customizable level of sanitation upon frames being processed by a given application [23].

There are also situations in which data access is based, not on the operation being performed, but on the user's environment and the objects therein. Data collection in this situation can be managed by incorporating access control policies directly into the user's environment. One potential approach is to modify a device's camera subsystem directly to blank out sections of a 2D plane encompassed by a particular boundary, or to build a trusted application into the OS to support the identification of 3D objects to avoid [47, 48]. Another approach would be to instrument individual spaces or objects within the environment to broadcast their own privacy preferences, either with physical markers or wireless signals [50]. Once a privacy policy has been received and the source has been verified, whether as a broadcast or from one of the device's own subsystems, it is the operating system's responsibility to enforce the necessary response logic, and sanitize the raw sensor data according to the received policy.

Unfortunately, collecting data from system sensors and nearby devices is not the source of information that must be secured. AR systems are also subject to side-channel attacks, in which externally observable data such as power consumption or network traffic patterns are used to infer private information about the user. For example, it is possible to infer a user's location based on patterns in network traffic when downloading AR data [39]. Additionally, end users typically have a poor understanding of what data is being collected and how it is being processed [12, 14], making them more likely to agree to privacy-invading permissions or to overlook suspicious activity from malicious applications.

However, the primary drawback to these solutions is that they require users to provide explicit privacy preferences, or for devices and environments to establish and broadcast individualized security policies. Additionally, application developers are expected to monitor their own resource consumption or network traffic patterns to combat side-channel attacks. There is no coordination between users, devices, or environments, and as such, privacy and security decisions must be made on a case-by-case basis.

## 3.2 Transformation Phase Attacks

Attacks during the transformation phase are largely focused on compromising the internal computer vision logic of the system, usually by modifying inputs either to produce an unexpected output or to prevent recognition entirely. Computer vision modules have long been known to suffer under less than ideal environmental conditions such as low light, inclement weather, or oblique viewing angles. However, research has also shown that image classification and other computer vision operations are also vulnerable to small changes in input images [15, 16, 32, 40]. These minor changes or "perturbations", so small they're practically imperceptible to humans, are interpreted by the system in such a way that the underlying vision module can return unexpected results, either preventing the desired result or producing a result more profitable to the attacker.

Additionally, system inputs can be manipulated to prevent recognition altogether. For example, certain patches can be attached to people's clothing in order to prevent surveillance systems from recognizing them as human [55]. Alternatively, clear bubble-like masks can be worn,[9] distorting the facial features enough to fool facial recognition software but still allow another person to infer facial expression. "Phantom" glasses also exist,[10] which reflect visible and infrared light back at cameras to block out the area of the face.

Each of these attacks requires the adversary to have some degree of knowledge of the underlying computer vision model in order to exploit it. Therefore, traditional solutions typically involve making some change to the underlying model that negates the attacker's knowledge of that model. However, updating computer vision components on distributed devices such as those found in AR systems can be expensive and unreliable, especially when the devices are resource-constrained.

## 3.3 Output Phase Attacks

Finally, attacks during the output phase manipulate the type and styling of virtual content being served to the user. Such attacks can be characterized in terms of *who* can display content, *what* content can be generated, *when* content will be displayed, and *where* content will be placed [28]. By compromising system output, attackers can distract their victims from the real-world, make them ill from motion sickness, or interfere with content from other trusted applications. As AR systems using head-mounted displays or smart windscreens (such as vehicles or helmets) become more

---

[9]https://www.businessinsider.com/clothes-accessories-that-outsmart-facial-recognition-tech-2019-10.

[10]https://mashable.com/review/review-reflectacles-phantom-anti-facial-recognition-technology-glasses-frames/.

|        |        |        |
|:------:|:------:|:------:|
|  (a)   |  (b)   |  (c)   |

**Fig. 3** Examples of output attacks for AR systems. (**a**) Original. (**b**) Blocking other Apps. (**c**) Blocking real world

popular, dealing with output phase attacks also becomes more important in order to ensure user safety.

A malicious application can compromise the "who" and "what" aspects via a number of methods. Attacks made during the transformation phase can modify the result such that unintended content (or no content at all) is displayed. A malicious application running on a multi-program system can display its own content on top of the content from another application, effectively rendering that content obsolete (Fig. 3b). If the application developers themselves are the attackers, then the "when" and "where" aspects of system output can also be leveraged to attack the user. Timing of content can be managed so that it is distracting or irrelevant to the user's current context. Content can also placed poorly, either so close to the user's viewpoint so as to take up most or all of the view space, or locked in place relative to the user's head so that she cannot turn away from it, or placed over top more relevant real-world content such as a staircase or stop sign (Fig. 3c).

Some researchers have proposed OS-level solutions to manage output security in AR applications. In [29, 30], the authors designed an AR-specific middle layer between installed applications and the underlying system sensors and drivers. This middle layer, called Arya, would be responsible for intercepting raw sensor data, translating it to high level recognizer objects, and applying security policies to any resulting application output. Supported policy logic includes detection of and subsequent remediating logic for a variety of security and safety-violating situations, such as content that moves too quickly, takes up too much of the user's view, blocks road signs or pedestrians while driving, and more. Application output found to be in violation of a policy can either be modified to satisfy the policy, or blocked altogether.

However, these defenses suffer from a number of limitations. While the concept of an AR-specific operating system is not new, it has yet to gain sufficient traction for commercial and research platforms. Thus, any device running solutions such as these will be doing so on top of more commercially available operating systems, and in doing so, incurring non-trivial amounts of processing overhead. Additionally, output policies must also be explicitly written and applied on a frame-by-frame basis for pairwise combinations of apps and users. Some work has been done to pave the way for securely sharing output in multi-user systems [31, 51], but the focus of such work has been on pairs of interacting users rather than larger groups.

The limitations described above can be resolved by moving portions of the AR system to the edge. In the following sections, we will describe the nature of AR systems at the edge, the impact that moving to the edge has on AR system requirements, and finally, how the edge can help resolve these security concerns.

## 4  AR Security at the Edge

The architecture of an edge-enabled AR system consists of several actors, as displayed in Fig. 4. The first is the **AR device**, which may be a smartphone, head-mounted display (HMD), or other "smart" display, such as a vehicular windshield or helmet faceplate. The AR device connects to the network through a **wireless access point** or base station, which will then forward the data from the device to a **server**, either there on the edge or deeper within the network.

Figure 5 shows the updated AR pipeline when offloading operations to the edge, the most commonly offloaded operations being those in the transformation phase. First, the input phase now includes any necessary pre-processing of the collected data, such as downsampling an image or video clip. The AR device then forwards the data to its associated wireless access point, which in turn forwards the data to the edge server. The transformation phase then begins, wherein the server performs the necessary computer vision or machine learning operations on the data, such as object classification or pose estimation. The results of the operation are then sent back to the wireless access point, which forwards them to the AR device. Finally, in the output phase, the AR device processes and responds to the results of the offloaded operation, usually by providing a visual cue to the user, such as displaying a particular label.

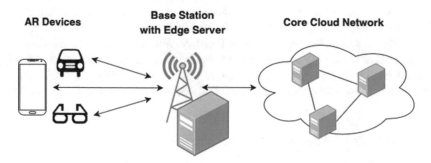

**Fig. 4** Example architecture of edge-enabled AR system

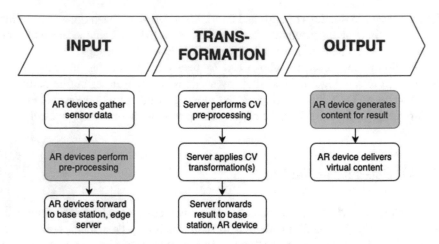

**Fig. 5** Modified AR pipeline when offloading to edge server. Highlighted steps are optional, depending on division of labor between edge server and local AR device

## 4.1  System Requirements for AR at the Edge

Once an AR system has been moved to the edge, its system requirements change slightly. Having access to additional storage and compute resources on the edge frees up local resources on the AR device, but also introduces latency while waiting for offloaded operations to return their results. It is up to the system designers to decide which operations to offload, how often, and with what data.

As discussed above, the transformation phase is the portion of the pipeline that is traditionally offloaded to the edge. This is because the computer vision or machine learning operation of choice can more or less be treated as a black box, providing results for a given set of inputs independent of anything else happening on the AR device. However, the exact data exchanged before and after offloading differs based on the type of operation being executed. Some systems may offload video clips, individual frames, or simply sets of features extracted from the local data stream. The results that the server returns could be labels, coordinates, or even encoded object files representing the content to be displayed. The degree to which data is pre-processed before offloading, or the virtual content is generated after receiving a result (designated as highlighted steps in Fig. 5) are variable depending on the system being implemented and the availability of resources on the AR device itself.

## 4.2  Quality of Augmentation at the Edge

When moving an AR system to the edge, it is important to keep in mind how the new system architecture will affect user experience, as the ultimate success of an AR system is tightly correlated with user experience and perception. It is not enough for the system to perform with low resource consumption or for the machine learning

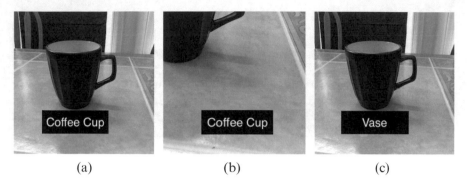

**Fig. 6** Examples of poor user experience in AR systems. (**a**) Original. (**b**) Placement error (lag). (**c**) Recognition error

components to achieve high precision and recall. The system must combine these features into a sum greater than its parts; it must not only correctly identify context, but also produce the correct output in the correct manner in a timely fashion, a characteristic that can be broadly summed up by the term **quality of augmentation**, or QoA.

Low QoA has direct impacts, not only to the general appeal of a system, but also to the physiological responses of users. Motion sickness for AR and VR systems, also called "simulator sickness", occurs when a user experiences different stimuli from the system than she expects from the real world, such as recognition errors or delays in label placement, as shown in Fig. 6. A user with simulator sickness may experience disorientation, nausea, or vomiting. AR system designers must be very careful, therefore, to ensure that users of their systems experience a high QoA.

### 4.2.1 Metrics for QoA

As important as QoA is to AR systems, it is notoriously difficult to define. While QoA is supported by traditional system metrics such as battery drain or RAM and CPU consumption, these data points cannot guarantee how users will ultimately respond. Therefore, many system designers also choose to supplement traditional system testing with more qualitative testing in the form of user studies. From gauging simulator sickness [26] to focal length [8] to facilitating device pairing in multi-user systems [54], gathering feedback directly from users in the form of questionnaires and interviews has been the traditional method of gauging the quality of a system's user experience. However, due to their open-ended nature, user studies are limited in their ability to provide quantitative system metrics, particularly when evaluating systems in the wild rather than the lab [33].

Despite the nebulous nature of QoA, there are certain measurements that an AR system designer can make to approximate it, such as measuring **how quickly content is being generated**. The speed at which the system collects data, processes it, and displays the resulting output can be measured in a number of ways, such as

**Table 1** Symbol definitions for QoA metrics

| Symbol | Description |
|---|---|
| $p$ | Precision; percentage of true positive predictions out of all positive predictions made |
| $r$ | Recall; percentage of true positive predictions made out of all ground truth positives |
| $s_i$ | Side length (in pixels) of the $i$th video frame with area $s^2$ |
| $\psi$ | Modeled complexity function (convex w.r.t. $s_i^2$) |
| $\xi$ | Modeled accuracy function (concave w.r.t $s_i^2$) |
| $c_i$ | Computational complexity of calculating the $i$th frame |
| $a_i$ | Analytical accuracy from calculating the $i$th frame |

*end-to-end latency* [9, 35–37, 44, 45, 49], and the number of *frames displayed per second* (FPS) [29, 48]. Managing these values ensures that the system can respond in real-time to changes in the user's behavior and environment, and are discussed in greater detail in Sect. 4.3.

An AR system designer can also approximate QoA by measuring the **accuracy of generated content**, typically in terms of intersection-over-union (IOU) [35, 45] and mean average precision (mAP) [36, 37]. IOU is utilized often in computer vision operations that identify specific areas of an image, such as drawing of bounding boxes. IOU is calculated as the percentage of area shared between a ground truth box and the predicted box, divided by the total area of those two boxes. A higher IOU means that the predicted box overlaps and shares more area with the ground truth box. Likewise, mAP is a common metric for computer vision operations that focus on classification, that is, identifying the subject of an image. mAP relies on the concepts of *precision*, or the rate of true positive predictions out of all predictions made (Eq. 1), and *recall*, or the rate of true positive predictions out of all ground truth positive matches (Eq. 2) (Table 1). mAP calculates the average area-under-curve when plotting precision and recall for a set of classifications. A higher mAP means that the system was able, not only to make predictions correctly, but to identify a high percentage of available items.

$$p = \frac{TP}{TP + FP} \tag{1}$$

$$r = \frac{TP}{TP + FN} \tag{2}$$

### 4.2.2 Trade-Offs for QoA

To support better quality of augmentation, AR system designers can made certain trade-offs in the **complexity** and subsequent **accuracy** of the internal computer vision operations, traditionally the most resource-intensive portions of an AR system. Figure 6 demonstrates how imbalanced complexity and accuracy can impact user experience. Figure 6b shows a system which, while accurate, takes so long to compute a label that the label placement is no longer accurate to the user's context. Figure 6c shows a system which quickly computes and places a label, but is highly inaccurate.

To understand the relationship of accuracy and complexity, the authors in [37] modeled the amount of time required to perform classification on input frames of varying sizes. They found that, for a given frame with resolution $s^2$, the time to complete the operation increases more quickly as the input resolution grows, while increases in accuracy slow and then plateau. They therefore proposed Eqs. 3 and 4, which model computational complexity and accuracy as convex and concave functions respectively with respect to the resolution of the $i$th video frame (Table 1).

This means that it is the developers' responsibility to explore the impacts of varyingly complex and accurate models to the performance of their AR systems. For example, a simpler object detection model might be preferred for a system with strict latency requirements, while a more complex bounding box model might be appropriate for a system that requires highly accurate results but can tolerate some additional latency. Ultimately, it is up to the AR system designers to make these trade-off decisions relative to the needs of their specific systems.

$$c_i = \psi(s_i^2) \tag{3}$$

$$a_i = \xi(s_i^2) \tag{4}$$

## 4.3 Impacts to Service Latency

The principal problem for AR systems trying to maintain QoA at the edge is service latency. Augmented reality systems have strict latency restrictions in order to maintain an immersive user experience; indeed, experts have calculated a maximum allowable latency of only 100ms for AR operations [5, 6]. This means that the AR system has less than a tenth of a second from the point of data collection to process that data, calculate a result, and display virtual content according to that result (Table 2).

### 4.3.1 Calculating Service Latency

For an edge-enabled AR system, **service latency** can be defined as the combination of transmission latency, or the time it takes to transmit data from the AR device to the server, and computational latency, or the time required to complete the offloaded operation. This is reflected in Eq. 5.

**Transmission latency** only applies to systems with offloaded operations, and is directly influenced by the amount of data being transported. (Because we are

**Table 2** Symbol definitions for latency metrics

| Symbol | Description |
|--------|-------------|
| $L_s$ | Service latency |
| $L_i^t$ | Transmission latency for the $i$th frame |
| $L_i^c$ | Computational latency for the $i$th frame on a shared, multi-user server |
| $\beta$ | Number of bits per pixel |
| $R_i$ | Average wireless data rate for the $i$th user |
| $N$ | Collection of possible edge servers |
| $M$ | Collection of users |
| $\sigma_{i,n}$ | Binary flag representing whether the $i$th user has been assigned to the $n$th server |
| $c_i$ | Computational complexity of processing the $i$th frame |
| $f_n$ | Set of available computational resources on the $n$th server |

assuming the target server to be on the edge, we do not consider the impact of current traffic loads within the core network to transmission time.) Equation 6 demonstrates how transmission latency can be calculated from a system level (rather than wireless link performance level) where $\beta$ is the number of bits per pixel, $s_i^2$ is the number of pixels in the $i$th frame with a side length of $s$, and $R_i$ is the $i$th user's average wireless data rate [37].

**Computational latency** is heavily influenced, not only by the amount of data, but also the complexity of the operation being performed and the amount of resources available on the server when sharing that server with other users. Equation 7 demonstrates how this type of latency can be calculated where $c_i$ is the computational complexity of the $i$th task, and where, for the $n$th server, $\sigma_{i,n}$ is a binary variable representing whether the $i$th user is assigned to this server, $f_n$ is the total set of available computational resources, and $\dfrac{f_n}{\sum_{m\in M}\sigma_{m,n}}$ is the amount of resources allocated to each user [37].

$$L_s = L_t + L_c \tag{5}$$

$$L_i^t = \frac{\beta s_i^2}{R_i} \tag{6}$$

$$L_i^c = \sum_{n\in N} \sigma_{i,n} \frac{c_i}{f_n} \sum_{m\in M} \sigma_{m,n} \tag{7}$$

### 4.3.2 Trade-Offs to Manage Latency

The degree to which service latency impacts a given AR system depends on the nature of the system and the portions of the AR pipeline being offloaded (shown in Fig. 5). For resource constrained devices such as smartphones and HMDs, the operations to be performed might be too resource-intensive to complete locally. For other devices such as "smart" vehicles with augmented windscreens, the operations to be performed might require more data than an individual device can collect on its own. Therefore, AR system designers must selectively offload certain operations either to preserve local resources or to aggregate and process larger data sets. Regardless of the purpose of the offload, the system must still respect the appropriate latency restrictions to maintain QoA for users.

While network traffic conditions and server workloads are not always under one's control, the amount of data being transmitted and the operation being performed *are*. To explore the implication of this, the authors in [37] modeled the impacts to latency and accuracy when offloading object classification operations for models of various sizes. They observed that *decreasing* either the resolution of the image being transmitted or the complexity of the model being executed dramatically decreased latency, while incurring only a minor cost to accuracy [36]. When *increasing* the resolution of the image being transmitted, they observed that computational latency increased at a gently exponential rate while accuracy tapered off after an initial increase. This led the authors to conclude that latency and accuracy of AR operations can be dynamically managed in response to changing network conditions, assuming the client application has a variety of input image sizes and computational models to choose from.

However, decreasing the frame size or complexity of the computational model has a downside, since both of these factors directly impact the system's quality of augmentation. Indiscriminately decreasing frame size or computational model complexity may decrease latency, but it will also negatively impact the accuracy of computation results. However, by selectively modifying these parameters to perform controlled approximate augmentation, clients can strike a balance between managing latency and preserving QoA.

Assuming that service latency is appropriately managed, moving an AR system to the edge gives that system access to increased computational resources, larger data sets aggregated from multiple devices, and the ability to make collaborative decisions between neighboring devices. These features can provide a number of security benefits to the various stages of the AR processing pipeline. The following sections discuss these implications in greater detail.

## 4.4   Securing AR Systems Using the Edge

### 4.4.1   Input Phase Defenses

**User Authentication**   Recall from Sect. 3.1.1 that one of the major vulnerabilities for the input phase of AR systems is authenticating the system user. Typical user authentication solutions, such as those discussed previously, assume that a user authenticates herself with a partner using wearable sensors and coordinating protocols. An example of this is the headset-pairing protocol proposed in [54], which suggests the use of gestures to create a secure communications channel directly between devices. The devices wirelessly exchange public keys, and using those keys, display abstract shapes in the air to be traced by the corresponding partner. If each user observes their partner to trace the same shape, then the pairing is successful.

However, approaches like these are limited when attempting to authenticate individual users within a larger, co-located group. Multiple users within the same physical space can create authentication interference, such as microphones collecting overlapping voice commands, or cameras collecting conflicting gestures. There is also logical ambiguity regarding connections to make between users; the system must be able to support a wide variety of communication permissions within the group, such as exclusive partner pairs, groups open to accepting new members, and closed groups with no option for outsiders to join in. The edge can help with these authentication problems by becoming the intermediary for user authentication in group settings, as demonstrated by the sample system in Fig. 7. The edge server would take responsibility for managing repositories of public keys and group membership information, mediating connections with new group members, and supplying short-term keys for use in direct communication among the group.

**Data Collection** Moving an AR system to the edge can also help control the collection of environmental data in order to preserve user and bystander privacy, as discussed in Sect. 3.1.2. In particular, edge servers are a highly advantageous place to aggregate privacy preference data from multiple users in the same geographic area, and to generate privacy policies dynamically based on environmental context in that area. One particular solution suggests the use of specific privacy markers to prevent recording of restricted 2D surfaces and 3D objects [48]. The original implementation for this solution required the authors to update their device's

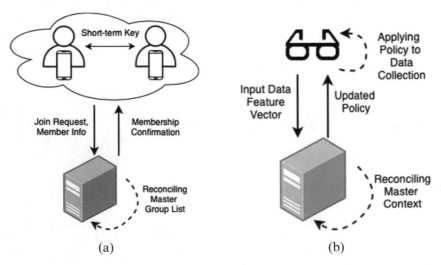

(a)                                                                              (b)

**Fig. 7** Potential uses for edge servers in (**a**) output policy generation and (**b**) display reconciliation between multiple users

hardware access layer (HAL) in order to intercept frames from the camera. This approach allowed them to sanitize each frame before applications running at less privileged layers could access them, but also incurred a significant performance hit. It is possible that, instead of modifying the device's HAL, an intermediate layer could have been created to offload the frame sanitation operation to a trusted edge server. Doing so would also allow for the aggregation of privacy markers submitted by multiple users, or for a system administrator to update the types of markers that the system can recognize without having to push an update to individual devices.

Instead of markers, systems can also leverage user- or administrator-provided policies to manage data collection. Systems such as I-Pic [1] require devices to broadcast privacy policies over BLE to prevent image capture in undesirable contexts. I-Pic assumes that each user-device pair is proactive in learning and following the privacy preferences of those nearby. Therefore, each device is responsible for storing its own local cache of privacy policies for users in the immediate area, and enforcing those policies on its captured images. Other solutions such as PrivacyManager [34] utilize data gathered from the environment (such as location, time of day, or ambient light and sound levels) to determine whether it is appropriate for a camera to record. Unlike I-Pic, PrivacyManager assumes that each user-device pair is *untrusted*, and so relies on a central authority such as a system administrator to generate and distribute policies. For both of these systems, moving data storage and policy generation to the edge is highly beneficial. Devices using I-Pic could store their privacy preferences on the edge server, without having to keep local copies. Devices using PrivacyManager could aggregate contextual information at the server, and use it to dynamically update and distribute policies.

### 4.4.2 Transformation Phase Defenses

Defenses against attacks conducted during the transformation phase of the AR pipeline borrow heavily from advancements in adversarial machine learning. Traditional defenses in this area fall into three broad categories, as described in [4]. The first option is using **modified input for training and testing**. This is considered to be a "brute force" strategy, employing approaches such as compressing the input images, adding random padding, and even training with actual adversarial examples. While these kinds of defenses can help with adversarial inputs, they can also hinder correct processing of valid inputs, since any changes to the input images would be applied regardless of the validity of the image in question. The second possibility is **adding onto the computational model**, usually in the form of a secondary network trained especially either to generate new adversarial inputs or reverse perturbations applied to existing ones. Unfortunately, the impact to service latency incurred by adding secondary networks to a computational model makes this option less attractive for AR systems.

The third option for defense is **modifying the computational model** itself, such as updating weights or loss functions within the network to reflect adversarial inputs, inserting a masking layer to the network before the classification layer, or adding a sub-network trained specifically on classification inputs. One notable example of this kind of defense is called *defensive distillation* [42]. This approach builds on the idea of "distilling" a neural network [21], or using the probability vectors generated by a pre-trained network to train a second, smaller network. This second network can then be executed on a smaller, more resource-constrained device, which is an ideal feature for AR systems. Defensive distillation uses these probability vectors, not to compress the network into a smaller architecture, but to train a new instance of the same network architecture. Training this new network instance, not on discrete labeled inputs, but on probability vectors for a group of labels, gives the network a richer understanding of the trained data sets, making it more robust and resilient to adversarial inputs.

The approaches discussed here are intended to be applied to the computer vision model before the parent application is deployed for use in the real world. However, utilizing the edge for AR systems makes these defenses, particularly modifications to an active computational model, much more feasible. Not only do edge servers provide additional processing and storage resources to AR devices, but they also provide a single point of deployment for an AR system's computational model. When changes to a model must be made, or a model must be replaced completely, it is simpler and faster to deploy the updated model to an edge server than to push that model to every subscribing AR device.

### 4.4.3 Output Phase Defenses

Recall from Sect. 3.3 that output attacks against AR systems are described in terms of who generates the content, what content is being generated, and when and where that content will be displayed. In these situations, the malicious entity may be the current application, a background application, or another user. Regardless of the source, the appropriateness of any given system output is relative to the user's current context, as output generated in an empty room will pose no threat while the same output on a crowded street may obscure crucial real-world content. Solutions like Arya [29, 30] utilize output control policies imposed at the OS-level to restrict the conditions under which output can be displayed. However, this system makes several simplifying assumptions, including manual creation of output policies and trustworthiness of fellow system users. There is no consideration for dynamically changing policies or for untrusted fellow users.

Moving output policy creation to the edge allows a system to aggregate and leverage data collected by many devices in order to determine the appropriate conditions to write into the policy. Reference [2] builds on Arya to do this, leveraging reinforcement learning on edge nodes to generate output policies dynamically. However, this solution relies on simulations of application use in order to generate policy rules, and provides no discussion on how to verify or improve the performance of that policy in the real-world. An alternative to this solution might be similar to Fig. 7a, which requires applications to register their output preferences with a designated edge server, which consolidates and assigns display parameters to each one. Alternatively, the same application being executed on multiple devices could maintain a cache of context data at the edge, such as weather conditions or network loads, that would subsequently impact how to display data for each subscribing user.

Regardless of how the output policy is generated, additional consideration needs to be paid when dealing with multi-user systems. In [31], the authors conducted a series of user studies investigating how subjects in a multi-user AR system interact with the environment and each other. They observed that users were quick to generate or manipulate content to negatively impact their partners, including modifying each other's appearance, setting up virtual barriers to block real world content, and hiding behind virtual content to obscure their own actions. Similarly, in [43], the authors observed a strong sense of ownership between users and the virtual content within their immediate space, along with adverse responses to other users modifying that content or space. In [51], the authors present several case studies of both co-located and geographically dispersed users, sharing content using both opt-out and opt-in defaults. Because these interactions are happening in real-time, this presents a unique opportunity for edge-computing to act as an intermediary between devices, as demonstrated in Fig. 7b. Instead of devices pairing up and sending content directly to each other, they can send their intended actions to an edge server, which reconciles the master view of the shared content, and delivers a sanitized view to each subscribing user. The physical proximity of an edge server to the users provides greatly improved latency over a central cloud server, allowing the system to maintain real-time reactions.

## 5   Analysis and Discussion

Today's commercial libraries for augmented reality do not currently provide explicit support for moving an AR system to the network edge. Some libraries are beginning to offer remote support in specific areas such as cloud-synced geo-located "anchors" for multi-user AR systems,[11] or leveraging cloud-based resources to execute more intensive versions of local operations.[12] However, the developer has no control over *where* these data are being stored or operations are being executed, which means that any system utilizing these functions is subject to all of the latency costs imposed by the cloud.

To explore what kind of latency costs a developer might expect when adopting offloaded operations with these commercial platforms, we conducted a series of tests comparing service latency when performing a given operation both locally and remotely. Because there are no commercially available AR libraries that explicitly support the edge, these offloaded operations are instead being executed on the

---

[11]https://developers.google.com/ar/develop/java/cloud-anchors/overview-android.

[12]https://firebase.google.com/docs/ml-kit/android/label-images.

cloud. However, should these library publishers offer edge support in the future, the performance of such offloaded operations can only improve. Leveraging the results of our experiments and the current state of commercial library offers, we conclude this section with a list of open problems for AR research and development at the network edge.

## 5.1 Experimental Results Comparing Local and Remote Processing

To conduct our experiments, we focused on two common computer vision tasks in augmented reality: text recognition and image labeling. For each computer vision task, we developed two "flavors" of a smartphone-based AR application - one which performed the task locally, and one which performed the task remotely. We then piped in frames from a pre-recorded video, so that each application would be operating on the same inputs. The CV module would operate on a single frame at a time; once the current frame completed, a new frame from the video stream would be requested, meaning slower modules received frames more infrequently. We conducted our experiments using a Samsung Galaxy GS9 smartphone running Android 10. The logic for both the local and remote computer vision operations were implemented using Firebase's Machine Learning Toolkit (ML Kit).[13]

We explored the impact of offloading a given operation based on the complexity and quality of the input video frame. For each application, we designed a collection of 20 second-long video clips with increasing complexity (volume of text, number of subjects shown on screen respectively) and quality of the video. Details of the video files used for the text recognition and image labeling apps can be found in Tables 3 and 4 respectively.

As each video frame was fed into the app's CV module, we measured the amount of time the system took to process the frame in milliseconds. Recall that some experts have calculated a maximum allowable latency of only 100ms for truly dynamic and immersive AR operations [5, 6]. With this in mind, we examined the time required to perform each selected operation (text recognition and image

**Table 3** System statistics for input videos used in **text recognition**. All videos had same resolution (1920 x 1080 px) and same running time (20 s)

|            | Volume of text | | | | | Video quality | | | |
|------------|-------|---------|---------|---------|---------|--------|------|------|----------|
|            | Lvl 1 | Lvl 2   | Lvl 3   | Lvl 4   | Lvl 5   | Low    | Med  | High | Lossless |
| File size  | 8 MB  | 11.6 MB | 10.7 MB | 12.3 MB | 12.7 MB | 500 KB | 1 MB | 8 MB | 8 MB     |
| CRF        | 0     | 0       | 0       | 0       | 0       | 51     | 37   | 23   | 0        |
| # of Frames| 608   | 614     | 607     | 613     | 610     | 608    | 608  | 608  | 608      |

---

[13]https://firebase.google.com/docs/ml-kit.

**Table 4** System statistics for input videos used in **image labeling**. All videos had same resolution (1920 x 1080 px) and same running time (20 s)

| | Number of subjects | | | | | Video quality | | | |
|---|---|---|---|---|---|---|---|---|---|
| | 1 | 2 | 3 | 4 | 5 | Low | Med | High | Lossless |
| File size | 7.6 MB | 13 MB | 12.9 MB | 16.5 MB | 12.3 MB | 518 KB | 1 MB | 7.6 MB | 7.6 MB |
| CRF | 0 | 0 | 0 | 0 | 0 | 51 | 37 | 23 | 0 |
| # of Frames | 606 | 609 | 615 | 610 | 608 | 606 | 606 | 606 | 606 |

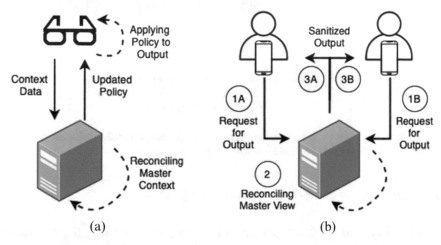

(a)                                                        (b)

**Fig. 8** Service latency when performing local and remote **text recognition** with (**a**) increasing volume of text and (**b**) increasing levels of video quality

labeling) on the appropriate video set, the results of which are shown in Figs. 8 and 9 respectively.

From these results, we can see a clear and significant impact to service latency when offloading a given operation, with minor improvements if the frame content is less complex or of low quality. With offloaded operations taking an average of 600 to 1000 ms to complete, relying on offloaded operations is a non-option for AR applications attempting to maintain a 100ms cap on service latency. However, the potential is there for AR library publishers to integrate support for the edge into their existing products and services. Performing an operation locally will always be faster than offloading, but for certain tasks, the increased compute and storage resources offered by the edge cannot be replicated or replaced on a standalone device. For those situations, integrating edge-support into these popular libraries can only be an improvement for overall system performance.

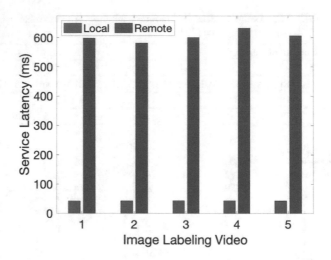

**Fig. 9** Service latency when performing local and remote **image labeling** with (**a**) increasing volume of subjects and (**b**) increasing levels of video quality

## 5.2  Impacts of Security Measures on QoA

As our AR systems inevitably grow beyond our standalone devices, the importance of security measures and their impact on system performance cannot be ignored. However, while the latency costs of offloading computer vision operations remain high, developers will be unlikely to add to that burden by incorporating security operations as well. Therefore, developers must be sensitive to the trade-off between latency and security, and manage it carefully according to their systems' needs.

One benefit of the edge that developers should be sure to leverage is the centralized storage offered by edge nodes. Edge node storage allows developers to keep large collections of images, labels, 3D models, privacy and access control policies, system state data, and computer vision modules without having to store them locally on users' devices. State data in particular can be aggregated at a high level to make better-informed decisions than a single device could make on its own. By relying on centralized storage, developers can keep the local versions of their applications small and lightweight, freeing up local resources for additional security operations.

Developers can also take advantage of the edge for redundant, multi-tiered operations with increasing levels of precision. An example of this is exhibited by Firebase's ML Kit, which recognizes 400 image labels when executing locally, but over 10,000 labels when executing in the cloud.[14] By using a more coarse-grained, lightweight model on the local device, and offloading to a more robust and

---

[14]https://firebase.google.com/docs/ml-kit/label-images.

precise model on the edge when necessary, developers can decrease the number of operations offloaded while still maintaining high levels of precision when necessary. Ultimately, it is the developer's responsibility to balance the security concerns of an AR system with QoA and user experience. As AR systems start to support multiple users and multiple applications in particular, this balance will become even more important.

## 5.3 Open Problems

In closing, we will discuss a selection of open problems that should be explored for AR systems security at the edge. The first is the development of **more holistic measurements for quality of augmentation (QoA)**. Traditional metrics for QoA such as resource consumption, frame rate, precision, and recall have been borrowed from software engineering and computer vision domains, and can help approximate QoA in standalone systems. However, these metrics fail to capture the impact of service latency and the subsequent trade-offs in computational accuracy that occur when moving AR systems to the edge. Therefore, we must develop new, more holistic metrics to reflect the changing nature of edge-enabled AR systems.

The second is the addition of **edge support to commercial AR-focused libraries and APIs**. There are many AR-focused libraries and APIs which already support the abstraction of common computer vision operations. It makes sense, therefore, for them to also include some support for offloading these operations to the edge. Some APIs, such as Firebase's ML Kit (discussed above), allow a developer to designate whether an operation should be performed locally or remotely, but do not allow the developer to stipulate what server to use. While adding such support to commercial libraries can improve overall system performance, there are risks involved, namely in controlling accidental data leakage. Any research efforts in this area would have to take this into consideration as well.

The third is for **improved user- and policy-driven I/O security for multi-user and multi-application systems**. Current approaches in AR system security focus primarily on single-user, single-application systems, with some works focusing on pair-based interactions between users. However, the capabilities of edge-enabled systems for increased processing, data aggregation, and collaboration between devices pave the way for systems supporting true multi-user, multi-application systems. It is easy to envision, for example, smart vehicles collecting data and collaborating in real-time to build wide-scale views of roadway conditions, and displaying multiple types of output on their augmented windshields, such as weather alerts, traffic updates, route suggestions, and other information. It therefore becomes crucial for these platforms to have methods of reconciling or sandboxing the input and output operations of these applications, to prevent them from compromising or interfering with each other.

# 6 Conclusions

In this chapter, we have introduced the basic concepts of AR systems and their related security concerns, as well as the implications of moving such system operations to the edge. We presented a number of potential impacts that edge computing can make on AR system security, including user authentication, data collection, transformation, and output verification. We also presented three open problems for future work: holistic metrics for quality of augmentation, edge support in commercial AR-focused libraries, and improved I/O security for multi-user-multi-app systems.

# References

1. Aditya, P., Sen, R., Druschel, P., Joon Oh, S., Benenson, R., Fritz, M., Schiele, B., Bhattacharjee, B., Wu, T.T.: I-pic: A platform for privacy-compliant image capture. In: Proceedings of the 14th Annual International Conference on Mobile Systems, Applications, and Services, pp. 235–248 (2016)
2. Ahn, S., Gorlatova, M., Naghizadeh, P., Chiang, M., Mittal, P.: Adaptive fog-based output security for augmented reality. In: Proceedings of the 2018 Morning Workshop on Virtual Reality and Augmented Reality Network, pp. 1–6 (2018)
3. Akçayır, M., Akçayır, G.: Advantages and challenges associated with augmented reality for education: A systematic review of the literature. Educ. Res. Rev. **20**, 1–11 (2017)
4. Akhtar, N., Mian, A.: Threat of adversarial attacks on deep learning in computer vision: A survey. IEEE Access **6**, 14410–14430 (2018)
5. Card, S.K.: The Psychology of Human-Computer Interaction. Crc Press (2018)
6. Card, S.K., Robertson, G.G., Mackinlay, J.D.: The information visualizer, an information workspace. In: Proceedings of the SIGCHI Conference on Human factors in computing systems, pp. 181–186 (1991)
7. Carlson, K.J., Gagnon, D.J.: Augmented reality integrated simulation education in health care. Clin. Simul. Nurs. **12**(4), 123–127 (2016)
8. Chakravarthula, P., Dunn, D., Akşit, K., Fuchs, H.: Focusar: Auto-focus augmented reality eyeglasses for both real world and virtual imagery. IEEE Trans. Vis. Comput. Graph. **24**(11), 2906–2916 (2018)
9. Chen, K., Li, T., Kim, H.S., Culler, D.E., Katz, R.H.: Marvel: Enabling mobile augmented reality with low energy and low latency. In: Proceedings of the 16th ACM Conference on Embedded Networked Sensor Systems, pp. 292–304. ACM (2018)
10. Chen, P., Liu, X., Cheng, W., Huang, R.: A review of using augmented reality in education from 2011 to 2016. In: Innovations in Smart Learning, pp. 13–18. Springer (2017)
11. Chung, N., Han, H., Joun, Y.: Tourists' intention to visit a destination: The role of augmented reality (ar) application for a heritage site. Comput. Hum. Behav. **50**, 588–599 (2015)
12. Denning, T., Dehlawi, Z., Kohno, T.: In situ with bystanders of augmented reality glasses: Perspectives on recording and privacy-mediating technologies. In: Proceedings of the 32nd Annual ACM Conference on Human Factors in Computing Systems, pp. 2377–2386. ACM (2014)
13. tom Dieck, M.C., Jung, T.: A theoretical model of mobile augmented reality acceptance in urban heritage tourism. Curr. Issues Tour. **21**(2), 154–174 (2018)

14. Egelman, S., Kannavara, R., Chow, R.: Is this thing on?: Crowdsourcing privacy indicators for ubiquitous sensing platforms. In: Proceedings of the 33rd Annual ACM Conference on Human Factors in Computing Systems, pp. 1669–1678. ACM (2015)
15. Elsayed, G., Shankar, S., Cheung, B., Papernot, N., Kurakin, A., Goodfellow, I., Sohl-Dickstein, J.: Adversarial examples that fool both computer vision and time-limited humans. Adv. Neural Inf. Process. Syst., 3910–3920 (2018)
16. Eykholt, K., Evtimov, I., Fernandes, E., Li, B., Rahmati, A., Xiao, C., Prakash, A., Kohno, T., Song, D.: Robust physical-world attacks on deep learning visual classification. In: Proceedings of the IEEE Conference on Computer Vision and Pattern Recognition, pp. 1625–1634 (2018)
17. Freina, L., Ott, M.: A literature review on immersive virtual reality in education: state of the art and perspectives. In: The International Scientific Conference eLearning and Software for Education, vol. 1, p. 133. "Carol I" National Defence University (2015)
18. Gaebel, E., Zhang, N., Lou, W., Hou, Y.T.: Looks good to me: Authentication for augmented reality. In: Proceedings of the 6th International Workshop on Trustworthy Embedded Devices, pp. 57–67 (2016)
19. Gonzalez-Franco, M., Pizarro, R., Cermeron, J., Li, K., Thorn, J., Hutabarat, W., Tiwari, A., Bermell-Garcia, P.: Immersive mixed reality for manufacturing training. Frontiers Robotics AI **4**, 3 (2017)
20. Herron, J.: Augmented reality in medical education and training. J. Electron. Resour. Med. Libr. **13**(2), 51–55 (2016)
21. Hinton, G., Vinyals, O., Dean, J.: Distilling the knowledge in a neural network. Preprint (2015). arXiv:1503.02531
22. Jana, S., Molnar, D., Moshchuk, A., Dunn, A., Livshits, B., Wang, H.J., Ofek, E.: Enabling fine-grained permissions for augmented reality applications with recognizers. In: Presented as Part of the 22nd {USENIX} Security Symposium ({USENIX} Security 13), pp. 415–430 (2013)
23. Jana, S., Narayanan, A., Shmatikov, V.: A scanner darkly: Protecting user privacy from perceptual applications. In: 2013 IEEE Symposium on Security and Privacy, pp. 349–363. IEEE (2013)
24. Joda, T., Gallucci, G., Wismeijer, D., Zitzmann, N.: Augmented and virtual reality in dental medicine: A systematic review. Comput. Biol. Med. **108**, 93–100 (2019)
25. Jung, T., tom Dieck, M.C., Lee, H., Chung, N.: Effects of virtual reality and augmented reality on visitor experiences in museum. In: Information and Communication Technologies in Tourism 2016, pp. 621–635. Springer (2016)
26. Kennedy, R.S., Lane, N.E., Berbaum, K.S., Lilienthal, M.G.: Simulator sickness questionnaire: An enhanced method for quantifying simulator sickness. Int. J. Aviat. Psychol. **3**(3), 203–220 (1993)
27. Kim, Y., Kim, H., Kim, Y.O.: Virtual reality and augmented reality in plastic surgery: a review. Arch. Plast. Surg. **44**(3), 179 (2017)
28. Lebeck, K., Kohno, T., Roesner, F.: How to safely augment reality: Challenges and directions. In: Proceedings of the 17th International Workshop on Mobile Computing Systems and Applications, pp. 45–50. ACM (2016)
29. Lebeck, K., Ruth, K., Kohno, T., Roesner, F.: Securing augmented reality output. In: 2017 IEEE Symposium on Security and Privacy (SP), pp. 320–337. IEEE (2017)
30. Lebeck, K., Ruth, K., Kohno, T., Roesner, F.: Arya: Operating system support for securely augmenting reality. IEEE Secur. Priv. **16**(1), 44–53 (2018)
31. Lebeck, K., Ruth, K., Kohno, T., Roesner, F.: Towards security and privacy for multi-user augmented reality: Foundations with end users. In: 2018 IEEE Symposium on Security and Privacy (SP), pp. 392–408. IEEE (2018)
32. Ledig, C., Theis, L., Huszár, F., Caballero, J., Cunningham, A., Acosta, A., Aitken, A., Tejani, A., Totz, J., Wang, Z., et al.: Photo-realistic single image super-resolution using a generative adversarial network. In: Proceedings of the IEEE Conference on Computer Vision and Pattern Recognition, pp. 4681–4690 (2017)

33. Lehman, S.M., Ling, H., Tan, C.C.: Archie: A user-focused framework for testingaugmented reality applications in the wild. In: 2020 IEEE Conference on Virtual Reality and 3D User Interfaces (VR). IEEE (2020)
34. Lehman, S.M., Tan, C.C.: Privacymanager: An access control framework for mobile augmented reality applications. In: 2017 IEEE Conference on Communications and Network Security (CNS), pp. 1–9. IEEE (2017)
35. Liu, L., Li, H., Gruteser, M.: Edge assisted real-time object detection for mobile augmented reality. In: The 25th Annual International Conference on Mobile Computing and Networking, pp. 1–16 (2019)
36. Liu, Q., Han, T.: Dare: Dynamic adaptive mobile augmented reality with edge computing. In: 2018 IEEE 26th International Conference on Network Protocols (ICNP), pp. 1–11. IEEE (2018)
37. Liu, Q., Huang, S., Opadere, J., Han, T.: An edge network orchestrator for mobile augmented reality. In: IEEE INFOCOM 2018-IEEE Conference on Computer Communications, pp. 756–764. IEEE (2018)
38. Makris, S., Karagiannis, P., Koukas, S., Matthaiakis, A.S.: Augmented reality system for operator support in human–robot collaborative assembly. CIRP Annals 65(1), 61–64 (2016)
39. Meyer-Lee, G., Shang, J., Wu, J.: Location-leaking through network traffic in mobile augmented reality applications. In: 2018 IEEE 37th International Performance Computing and Communications Conference (IPCCC), pp. 1–8. IEEE (2018)
40. Moosavi-Dezfooli, S.M., Fawzi, A., Fawzi, O., Frossard, P.: Universal adversarial perturbations. In: Proceedings of the IEEE Conference on Computer Vision and Pattern Recognition, pp. 1765–1773 (2017)
41. Moro, C., Štromberga, Z., Raikos, A., Stirling, A.: The effectiveness of virtual and augmented reality in health sciences and medical anatomy. Anat. Sci. Educ. 10(6), 549–559 (2017)
42. Papernot, N., McDaniel, P., Wu, X., Jha, S., Swami, A.: Distillation as a defense to adversarial perturbations against deep neural networks. In: 2016 IEEE Symposium on Security and Privacy (SP), pp. 582–597. IEEE (2016)
43. Poretski, L., Lanir, J., Arazy, O.: Normative tensions in shared augmented reality. Proc. ACM Hum. Comput. Interact. 2(CSCW), 1–22 (2018)
44. Qiu, H., Ahmad, F., Bai, F., Gruteser, M., Govindan, R.: Avr: Augmented vehicular reality. In: Proceedings of the 16th Annual International Conference on Mobile Systems, Applications, and Services, pp. 81–95 (2018)
45. Ran, X., Chen, H., Zhu, X., Liu, Z., Chen, J.: Deepdecision: A mobile deep learning framework for edge video analytics. In: IEEE INFOCOM 2018-IEEE Conference on Computer Communications, pp. 1421–1429. IEEE (2018)
46. Rauschnabel, P.A., Rossmann, A., tom Dieck, M.C.: An adoption framework for mobile augmented reality games: The case of pokémon go. Comput. Hum. Behav. 76, 276–286 (2017)
47. Raval, N., Srivastava, A., Lebeck, K., Cox, L., Machanavajjhala, A.: Markit: Privacy markers for protecting visual secrets. In: Proceedings of the 2014 ACM International Joint Conference on Pervasive and Ubiquitous Computing: Adjunct Publication, pp. 1289–1295 (2014)
48. Raval, N., Srivastava, A., Razeen, A., Lebeck, K., Machanavajjhala, A., Cox, L.P.: What you mark is what apps see. In: Proceedings of the 14th Annual International Conference on Mobile Systems, Applications, and Services, pp. 249–261. ACM (2016)
49. Ren, J., He, Y., Huang, G., Yu, G., Cai, Y., Zhang, Z.: An edge-computing based architecture for mobile augmented reality. IEEE Network 33(4), 162–169 (2019)
50. Roesner, F., Molnar, D., Moshchuk, A., Kohno, T., Wang, H.J.: World-driven access control for continuous sensing. In: Proceedings of the 2014 ACM SIGSAC Conference on Computer and Communications Security, pp. 1169–1181. ACM (2014)
51. Ruth, K., Kohno, T., Roesner, F.: Secure multi-user content sharing for augmented reality applications. In: 28th {USENIX} Security Symposium ({USENIX} Security 19), pp. 141–158 (2019)

52. Shang, J., Wu, J.: Enabling secure voice input on augmented reality headsets using internal body voice. In: 2019 16th Annual IEEE International Conference on Sensing, Communication, and Networking (SECON), pp. 1–9. IEEE (2019)
53. Shen, Y., Wen, H., Luo, C., Xu, W., Zhang, T., Hu, W., Rus, D.: Gaitlock: Protect virtual and augmented reality headsets using gait. IEEE Trans. Dependable Secure Comput. **16**(3), 484–497 (2018)
54. Sluganovic, I., Serbec, M., Derek, A., Martinovic, I.: Holopair: Securing shared augmented reality using microsoft hololens. In: Proceedings of the 33rd Annual Computer Security Applications Conference, pp. 250–261. ACM (2017)
55. Thys, S., Van Ranst, W., Goedemé, T.: Fooling automated surveillance cameras: adversarial patches to attack person detection. In: Proceedings of the IEEE Conference on Computer Vision and Pattern Recognition Workshops, pp. 0–0 (2019)
56. Tussyadiah, I.P., Jung, T.H., tom Dieck, M.C.: Embodiment of wearable augmented reality technology in tourism experiences. J. Travel Res. **57**(5), 597–611 (2018)
57. Yew, A., Ong, S., Nee, A.: Towards a griddable distributed manufacturing system with augmented reality interfaces. Robot. Comput. Integr. Manuf. **39**, 43–55 (2016)
58. Zsila, Á., Orosz, G., Bőthe, B., Tóth-Király, I., Király, O., Griffiths, M., Demetrovics, Z.: An empirical study on the motivations underlying augmented reality games: The case of pokémon go during and after pokémon fever. Personal. Individ. Differ. **133**, 56–66 (2018)

# Towards a Security-Aware Deployment of Data Streaming Applications in Fog Computing

Gabriele Russo Russo, Valeria Cardellini, Francesco Lo Presti, and Matteo Nardelli

## 1 Introduction

In the recent few years we have witnessed a worldwide explosion of the volume of daily produced data, fostered by the spread of sensors, wearable devices, and smartphones capable of collecting data about their surrounding environment and our everyday life. Today, all this information plays a key role in our society, and has become a strategical asset for institutions, companies, and scientists. The analysis of collected data allows to extract information useful for supporting decision-making, e.g., gathering new insights by identifying patterns or making predictions based on past observations.

In this context, the ability of efficiently collecting, storing, and analyzing data has become a strategic advantage. Nevertheless, the aforementioned growth has made data processing challenging from a computational point of view, and led to the development of efficient algorithms, tools, and frameworks for dealing with data (e.g., the *Map-Reduce* paradigm, and associated frameworks like *Apache Hadoop*).

Special interest has been devoted to *real-time* data analytics, which requires systems able to process data as soon as they are collected. This is critical in many application domains, e.g., in network attack detection, where monitoring information about the incoming traffic should be processed with very low latency. In this context, a primary role is played by distributed *Data Stream Processing* (DSP) systems, which allow to process unbounded sequences of data (i.e., streams),

G. Russo Russo · V. Cardellini (✉) · F. Lo Presti · M. Nardelli
Department of Civil Engineering and Computer Science Engineering, University of Rome Tor Vergata, Rome, Italy
e-mail: russo.russo@ing.uniroma2.it; cardellini@ing.uniroma2.it; lopresti@info.uniroma2.it; nardelli@ing.uniroma2.it

© Springer Nature Switzerland AG 2021
W. Chang, J. Wu (eds.), *Fog/Edge Computing For Security, Privacy, and Applications*, Advances in Information Security 83,
https://doi.org/10.1007/978-3-030-57328-7_14

flowing at very high rates, exploiting a multitude of computing nodes to spread the computation.

In the effort to further reduce processing latency with respect to data producers and consumers, which are often located at the edge of the network, recently DSP systems have been shifted from traditional cloud data centers to fog computing environments. By deploying applications in such geographically distributed infrastructures, latency reduction is achieved at the cost of handling increased heterogeneity, constrained computational and network resources, and a larger number of security concerns. These challenges especially impact the application *placement* problem, that is the problem of determining the set of computing nodes where application components are deployed and executed. This choice is indeed critical for achieving the expected Quality-of-Service (QoS), while minimizing application operating costs.

The placement problem for DSP applications has been widely investigated in literature, in the context of both traditional cloud scenarios, and geographically distributed environments (e.g., fog computing), exploiting a variety of methodologies. Existing solutions take into account performance-oriented characterizations of both the application and the computing infrastructure, in order to determine a placement scheme that optimizes one or more QoS metrics (e.g., system response time, or deployment monetary cost). Unfortunately, most the existing approaches neglect the security- and privacy-related concerns that inevitably arise when DSP applications are deployed in fog-like environments, where they may rely on a mixture of wired and wireless network links, and computing resources characterized by different software/hardware configurations, possibly acquired from multiple providers.

In this chapter, to overcome the limitations of existing placement optimization solutions, we present a simple yet quite general approach to account security related aspects. To this end, we introduce a formalism to specify application requirements and describe infrastructure features and capabilities. We also define associated metrics that capture how well different placement solutions match the specified application requirements, and allow us then to seamlessly integrate security related requirements in the overall optimization scheme. Our contributions are as follows.

- We present a formalism for specifying security-related application requirements. The idea is to represent application requirements as a forest of AND-OR requirements trees, each capturing a specific security requirement, e.g., privacy, isolation. At the same time we show how this formalism can be used to derive several requirement satisfaction metrics.
- We introduce the notion of operator and data stream configurations which define the set of security related configurations which satisfy the application requirements. This is paralleled by the notion of configurations that the infrastructure computing nodes and data links can support. The concept of configurations is the basis around which stakeholders can reason about application requirements and infrastructure characteristics and lay out the foundation of our deployment problem formulation.

- Finally, we integrate the aforementioned requirements, configurations and metrics in the placement optimization problem, which is formulated as Integer Linear Programming (ILP), and also accounts for other application non-functional requirements, e.g., response time and cost, and present the *Security-Aware DSP Placement* (SDP) problem. Focusing on a realistic case study, we show how SDP allows us to compute trade-offs between performance-based metrics, deployment cost, and security-related requirements satisfaction.

## *1.1   Organization of the Chapter*

In the next section, we will provide an overview of the basic concepts of DSP, and the challenges faced when deploying this kind of applications in fog environments, especially as regards security concerns. In Sect. 3, we present a formalism for specifying the application security-related requirements for deployment. In Sect. 4, we explain how we model the application placement problem, including the application and infrastructure model, and the associated QoS and cost metrics. The resulting problem formulation is presented in Sect. 5, along with an illustrative example of how it can be applied. We discuss the benefits and limitations of the presented approach in Sect. 6, and conclude in Sect. 7.

## 2   Background

In this section, we provide an overview of the main concepts, challenges, and research directions related to DSP, and the deployment of DSP applications in the fog environment. First, we describe in Sect. 2.1 the basic concepts and main challenges related to the deployment of DSP applications. Then, in Sect. 2.2, we focus on research works that address the placement of DSP applications. Finally, in Sect. 2.3, we describe works that deal with security and privacy issues in the DSP domain.

## *2.1   Data Stream Processing: Basic Concepts and Challenges*

A DSP application consists of a network of processing elements, called *operators*, connected by data streams. A DSP application can be represented as a directed acyclic graph (DAG), with data sources, operators, and final consumers as vertices, and streams as edges. A stream is an unbounded sequence of data items (e.g., event, tuple). Each operator is a self-contained processing element, that continuously receives incoming streams, applies a transformation on them, ranging from a simple operation (e.g., filtering, aggregation) to something more complex (e.g., applying a machine learning algorithm to detect some patterns), and generates new outgoing streams. Each data source (e.g., an IoT sensor or a message queue) generates one or more streams that feed the DSP application; differently from operators, data sources have no incoming streams. A final consumer (or sink) is a final receiver of the application streams; it can push data on a message queue, forward information to a persistent storage, or trigger the execution of some external services. Differently from operators, sinks have no outgoing streams.

An example of a DSP application related to the smart health domain, e.g., [2], is shown in Fig. 1. Data are collected by sensors on users' devices (e.g., smartphones, wrist-worn wearable devices), and sent for analysis to a DSP application. The application may carry out several kinds of processing on the data at the same time. In the example we consider, the application is used both for (i) detecting anomalies in the vital parameters monitored (e.g., skin temperature, heartbeat, and oxygen saturation in the blood) and send notifications to medical staff, and (ii) creating aggregated statistics. Users' devices push data into a message queue system, which in turn sends data as streams to the DSP system for processing. Before entering the DSP system, data can also be integrated (e.g., to merge the incoming streams into a single flow) and pre-processed (e.g., to detect duplicates). In the application DAG we can identify one source of the DSP application, four operators, and two sinks, the latter corresponding to the data consumers (medical staff and storage system). The DSP operators perform different tasks that range from aggregating data using summary statistics (the upper path in the DAG) to detecting any anomaly in the data streams (the lower path in the DAG). Although simple, this DSP application is an example of edge-native applications [56], which can take advantage of one or more of the benefits that arise from the fog/edge deployment: bandwidth scalability, low latency, enhanced privacy, and improved resiliency to WAN network failures.

DSP applications are typically deployed on either locally distributed clusters or centralized cloud data centers, which are often distant from data sources. However, pushing fast-rate data streams from sources to distant computing resources can exacerbate the load on the Internet infrastructure and introduce excessive delays experienced by DSP application users. Moreover, considering that both data sources and consumers are usually located at the network edges, a solution that allows to improve scalability and reduce network delays lies in deploying DSP applications

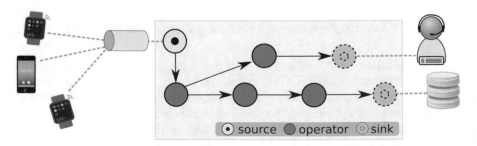

**Fig. 1** Example of a *smart health* application

not only on cloud data centers but also on edge/fog computing resources. Furthermore, the deployment on edge/fog resources can also enable users to selectively control the disclosure of sensitive information (e.g., vital parameters monitored by wrist-worn devices as in the example shown in Fig. 1).

In such a distributed scenario, a relevant problem consists in determining the computing nodes that should host and execute each operator of a DSP application, aiming to optimize some QoS attributes. This problem is known in literature as the *operator placement problem* (or scheduling problem).

Figure 2 illustrates a possible placement of the DSP application shown in Fig. 1 on the computing infrastructure. Multiple operators can also be co-located on the same computing node (e.g., $op_2$ and $op_4$).

Besides the initial placement of the DSP operators, the deployment of DSP applications can also be changed at run-time, that is during the application execution, so to self-adapt it with respect to workload changes and dynamism of the edge/fog computing environment (e.g., resource constraints, network constraints in term of latency and bandwidth, resources that join or leave the system). To this end, different approaches can be applied, ranging from the exploitation of performance-enhancing techniques (e.g., operator replication by means of elastic scale-out and scale-in operations, other types of dynamic transformation of the DAG) to the run-time adaptation of the application placement. The latter can be achieved at different grains, by placing either all the DSP operators from scratch (in this case, the placement problem is solved at regular intervals, so to update the operator location) or only a subset of operators by relying on operator migration between computing resources. Determining the operator replication degree is often addressed in literature as an independent and orthogonal decision with respect to the operator placement, but in [12] we present a problem formulation that jointly optimizes the replication and placement of DSP applications. In this chapter, we assume that the operator replication degree has been set at application design

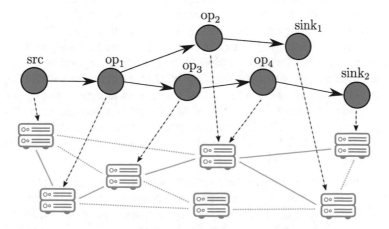

**Fig. 2** Illustration of the placement problem for a DSP application

time; so, we address the initial operator placement that is, how to place to DSP operators on the computing resources at the application start. To this end, in Sect. 2.2 we focus our literature analysis on those works that address the placement of DSP applications in the fog/edge scenario, or more generally in a geo-distributed computing environment. As regards research works dealing with the run-time self-adaptive control of DSP applications, we refer the interested reader to some surveys [20, 51] that classify and review them.

## 2.2 Placement of DSP Applications in the Fog

The DSP placement problem has been widely investigated in literature under different modeling assumptions and optimization goals (e.g., [21, 62, 63]). We review the related works organizing them along three main dimensions, that capture one or more related facets of the problem: (1) placement goals; (2) methodologies used to define the application placement; and (3) characteristics of the distributed computing infrastructure managed by the placement solution. For a deeper analysis of the state of the art, we refer the interested reader to extensive surveys, that analyze in details the research works addressing the placement problem not only in the context of DSP applications but also for other kinds of applications deployed in the fog/edge environment [7, 8, 11, 62].

**Placement Goals** Existing works consider two main classes of problems: constraint satisfaction and objective function optimization. In a *constraint satisfaction* problem, we are interested in identifying a deployment solution among all the feasible ones that satisfy some given requirements (e.g., application performance). For example, Thoma et al. [63] propose an approach to restrict the set of feasible deployment by improving the expressiveness of constraints. In most cases, not all feasible deployment result in desirable application performance; therefore, most of the existing solutions optimize (i.e., minimize or maximize) a *single-objective* function or a *multiple objective* function. A single-objective optimization considers a specific and well-defined QoS metric (e.g., response time, throughput, network usage, cost). A multi-objective optimization (or Pareto optimization) aims to combine different, possibly conflicting QoS attributes and to find the set of optimal solutions (i.e., those lying on the so-called Pareto frontier). The existing solutions aim at optimizing a diversity of objectives, such as to minimize the application response time (e.g., [6, 19, 34, 52]), the inter-node traffic (e.g., [4, 22, 27, 66, 67]), the network usage (e.g., [48, 50]), or a generic cost function that can comprise different QoS metrics (e.g., [5, 13, 21, 38, 53, 65]).

**Methodologies** The most popular methodologies used to address the operator placement problem include mathematical programming (e.g. [5, 13, 21]), graph-theoretic approaches (e.g., [23, 35]), greedy approaches (e.g., [4, 26, 33, 36, 52, 66]), meta-heuristics (e.g., genetic algorithms [60], local search [17, 61], tabu search and simulated annealing [61], steepest descent method and tabu search [31]), as well as custom heuristics (e.g., [19, 22, 40, 46, 48, 53]).

The operator placement problem, formulated as optimization problem that takes into account the heterogeneity of application requirements and infrastructural resources, turns out to be an NP-hard problem [13]. Therefore, many research efforts focus on applying different methodologies that can solve efficiently the DSP placement problem within a feasible amount of time, even for large problem instances.

**Computing Infrastructure**  Most of the existing solutions have been designed for a clustered environment, where network latencies are almost zero (e.g. [26, 34, 52]). Although interesting, these approaches might not be suitable for geo-distributed environments. Several works indirectly consider the network contribution by minimizing the amount of data exchanged between computing nodes (e.g., [4, 21–23, 32, 66]). For example, Eidenbenz et al. [21] propose a heuristic that minimizes processing and transfer cost, but it works only on resources with uniform capacity. Relying on a greedy best-fit heuristic, Aniello et al. [4] and Xu et al. [66] propose algorithms that minimize the inter-node traffic. Other works explicitly take into account network latencies, thus representing more suitable solutions to operate in a geo-distributed DSP system (e.g., [6, 13, 19, 29, 40, 48, 50]). Pietzuch et al. [48] and Rizou et al. [50] minimize the network usage, that is the amount of data that traverses the network at a given instant.

So far, only a limited number of works are specifically designed for placing DSP applications in fog/edge computing environments. SpanEdge [53] allows to specify which operators should be placed as close as possible to the data sources, while Arkian et al. [5] propose an integer non-linear formulation; to reduce resolution time, they linearize the problem; nevertheless, also linear formulations may suffer from scalability issues [40]. The work in [38] presents a Pareto-efficient algorithm to tackle the operator placement problem considering both the latency and energy consumption. Khare et al. [33] present an approach that first transforms any arbitrary DAG into an approximate set of linear chains, then uses a data-driven latency prediction model for co-located linear chains to drive a greedy heuristic, which determines the operator placement with the goal to minimize the maximum latency of all paths in the DAG. Peng et al. [47] jointly target the problems of DSP operator placement and replication in an edge scenario by proposing a two-stage approach that first employs a genetic algorithm for finding a solution and then uses a bottleneck-analysis based on the system queuing model to refine it.

The combination of cloud and edge resources have been also explored. For example, Ghosh et al. [28] propose a genetic algorithm meta-heuristic and show that their approach allows to achieve lower latency and more frequent feasible solutions than placing only on Cloud resources. Da Silva Vieth et al. [59] propose strategies that first decompose the application DAG, which is a series-parallel one, and then place its operators in a latency-aware manner. However, all these proposals focus on reducing the application latency, without taking into account any concern related to privacy and security.

A few edge-based stream processing systems support processing on fog/edge resources with the goal of reducing the need for costly data transfers. These systems include Cloud services such as AWS IoT Greengrass, Google Cloud IoT Edge, Microsoft Azure IoT, and research prototypes (e.g., Frontier [44]); however, they do not appear to place the DSP application over fog/edge resources by taking into account their peculiarities. On the other hand, this goal is pursed by some research efforts that extend existing open-source data stream processing systems such as Apache Storm (e.g., [4, 40, 53, 66]), mainly to show the feasibility of their approaches.

## 2.3 Security and Privacy in DSP

In many DSP applications, data streams carry privacy-sensitive information about users, whose confidentiality must be obviously protected throughout processing. Other applications, while not dealing with privacy-sensitive data, may carry out safety-critical tasks based on sensor-provided information (e.g., anomaly detection in a manufacturing system), where the integrity of the involved data streams must be guaranteed to avoid unintended (and possibly dangerous) application behaviors.

Guaranteeing confidentiality and integrity of data streams has become a serious challenge, especially since the availability of computing resources at the edge of the network fostered the interest for deploying DSP applications in geographically distributed infrastructures, in the aim of reducing latency. The enforcement of security and privacy policies is difficult in these environments, where the intrinsic heterogeneity, and thus the involvement of different standards and communication stacks, does not allow the application of traditional security countermeasures. This problem is particularly evident nowadays as data analytics often meet IoT scenarios, where streams originate from a multitude of potentially untrusted, distributed devices, and the need for security policy enforcement becomes critical [58].

Security and privacy issues have received limited attention in the field of distributed DSP systems so far, with research efforts being mainly devoted to application-level issues, performance, and fault-tolerance. Nonetheless, some effort has been spent investigating how to integrate *privacy-preservation* and *access control* [55] techniques in DSP systems, in order to guarantee that only authorized access to privacy-sensitive data is allowed.

Linder and Meier [37] extend the Borealis [1] streaming engine with OxR-BAC (Owner-extended Role Based Access Control), which aims at protecting the system against improper release of information, improper modification of information, and denial of service attacks. Ng et al. [43] propose a framework for privacy-preservation in data stream processing, built around the two principles of *limited disclosure* and *limited collection* of information. They design a hierarchy-based policy model and a framework to enforce privacy protection policies, and hence limit access and operation on data streams. Carminati et al. [10, 15, 16] apply Role Based Access Control [54] to DSP, relying on *secure operators* in order to replace application operators with security-aware versions. After presenting their ACStream framework in [10], in [15] they propose a query-rewriting middleware that does not target a specific underlying DSP framework.

Nehme et al. [41, 42] focus on the *continuous access control enforcement* for data streams, observing that, given the long-running nature of DSP systems, the content of the streamed data and its privacy-sensitivity may change, hence access control policies may need to be adapted dynamically as a consequence. In particular, in [42] they introduce the concept of a *security punctuation* for enforcing access control, that is a special tuple inserted directly into the data stream, allowing the data provider to attach security "metadata" to the stream. In [41], they describe FENCE, a framework for enhancing DSP systems with continuous access control enforcement through security punctuations, with limited runtime overhead.

Anh and Datta [3] focus on the problem of preserving privacy of data while stream processing is outsourced to the cloud. They present StreamForce, a framework for enforcing access control policies in presence of an untrusted cloud provider. Thoma et al. [64] propose PolyStream, a framework that allows users to cryptographically enforce access controls over streaming data on top of an unmodified DSP system. PolyStream relies on a novel use of security punctuations that enables flexible, online policy management and key distribution, with significant overhead reduction. Schilling et al. [57] focus on large-scale distributed Complex Event Processing systems, proposing access control consolidation mechanisms in order to ensure the privacy of information even over multiple processing steps in a multi-domain, large-scale application.

A different point of view on privacy-preservation is offered by Le Quoc et al. [49]. They aim at preserving users privacy, while still supporting both information high-utility and low-latency processing. Specifically, they achieve this goal by blending together two different approaches, namely, sampling (used for approximate computation) and randomized response (used for privacy-preserving analytics).

Recently, Burkhalter et al. [9] focused on the special class of applications dealing with time series data. They propose TimeCrypt, which provides scalable, real-time analytics over large volumes of encrypted time series data, by allowing users to define expressive data access and privacy policies, and enforcing them cryptographically.

The number of works that deal with system- or network-level security aspects in the context of DSP is significantly smaller. Fisher and Hancke [24] consider the network-level challenges of transmitting privacy-sensitive data streams from sensors to the processing servers. In particular, they investigate the use of the Datagram Transport Layer Security protocol, compared with the more popular Transport Layer Security protocol. Havet et al. [30] propose *SecureStreams*, a reactive framework that combines combines a high-level dataflow programming model with low-level Intel *software guard extensions* (SGX) in order to guarantee privacy and integrity of the processed data. Park et al. [45] focus on the scenario of running stream analytics on untrusted, resource-constrained devices at the edge of the network. They present StreamBox-TZ, a stream analytics engine that offers strong data security and verifiable results, by isolating computation in a Trusted Execution Environment (TEE). In particular, StreamBox-TZ relies on a data plane designed and optimized for a TEE based on ARM TrustZone.

A different approach is proposed by Chaturvedi and Simmhan [18], who apply *Moving Target Defense* (MTD) [68] techniques to protect a DSP platform; the key idea is to introduce system configuration variability at run-time so that any prior information available to an attacker becomes hardily usable. In particular, they implement several MTD mechanisms (e.g., migrating operators periodically over available computing nodes, altering the used port numbers, modifying the application graph by means of "dummy" operators), and show the feasibility of the approach by integrating them in Apache Storm.

At a higher level of abstraction, independently of the specific DSP framework in use and the possibly associated privacy-preservation mechanisms, security and privacy concerns also impact the choices made for initially deploying DSP applications over distributed infrastructures, i.e., the placement problem. As explained above, this problem has been extensively studied, but so far only performance and cost aspects have been considered in the context of DSP. Security-aware deployment and scheduling strategies have been proposed instead targeting other kinds of fog applications (e.g., [25] and [58]). In this chapter, we aim to fill the existing gap regarding the DSP placement problem, and the consideration of privacy and security for stream analytics applications in the fog. Although we specifically focus on the *initial* placement problem, the approach we will present can be applied for updating the application deployment at run-time as well, e.g., following a MTD strategy as suggested in [18].

## 3   Modeling Security-Related Requirements

Traditional strategies for deploying DSP application over distributed infrastructures aim at optimizing one or more performance metrics, e.g., application response time, or throughput. Some of them also account for the monetary cost of the computing resources chosen for running the application, assuming, e.g., a typical pay-as-you-go cost model. The recent trend of shifting the data processing applications towards the edge of the network, closer to the data producers, often forces DSP applications to be deployed in a less "trusted" environment, compared, e.g., to cloud data centers. In this new scenario, application deployment strategies should therefore account for security-related aspects in addition to the other well-known functional and non-functional metrics and stakeholders should be able to specify a set of requirements and/or objectives that involve these additional non-functional aspects.

Unfortunately, although several solutions have been proposed in literature, how stakeholders should express these requirements in a *standardized* way remains an open question. In the remainder of this section, we describe a simple yet powerful technique to formalize and organize the requirements of a DSP application with respect to the underlying computing and network infrastructure by which it is hosted.

## 3.1 Requirement Categories and Objectives

For a DSP application we might need to specify its security, privacy, or reliability requirements, in addition to commonly adopted performance and cost objectives. The application requirements could include a broad range of different aspects, including software or network configuration, hardware capabilities, location of the computing resources. Formally, we assume that application requirements are organized into different *Requirement Categories* (RCs), denoting with $\Omega_{RC}$ the set of all the considered categories. An application may hence exhibit requirements related to one or more RCs. Clearly, the relative relevance of each RC depends on the specific application. Whilst some RCs may contain critical requirements for application operation, other RCs may simply represent application "preferences" with respect to its running environment. In the following, we will show how the different importance of each requirement will impact the optimization problem we present.

With each Requirement Category $\omega$, we associate a set of *Requirement Objectives* (ROs), denoted as $\Omega_{RO}^{\omega}$. ROs represent specific properties of the computing infrastructure or the network (e.g., "type of operating system", "available encryption libraries", "wired/wireless network connectivity"). While RCs are mere abstractions, representing a collection of similar or related requirements, ROs represent concrete properties which can be evaluated in order to assess whether, e.g., a certain computing node satisfies the application needs. Specifically, given a RO $\rho$, we denote the set of values $\rho$ can be associated with as $V_{\rho}$. For example, the "type of operating system" RO might be associated with the values "Linux", "Android", "Windows", "Other". Throughout this chapter, without loss of generality, we will assume that $V_{\rho}$ is a finite set. We also find useful to define the set of the ROs comprised by all the RCs, that is $\Omega_{RO} = \bigcup_{\omega} \Omega_{RO}^{\omega}, \forall \omega \in \Omega_{RC}$.

### 3.1.1 Example

Figure 3 depicts a simple hierarchy of Requirement Categories and Objectives as a tree. We organize requirements into three categories: *Runtime Environment*, *Physical Security*, and *Network*. Each RC is associated with one or more ROs. The leaf nodes of the tree reported in the figure contain all the possible values for each RO.

## 3.2 Requirements Forest

In our approach, application requirements are expressed by means of AND-OR trees, which are widely adopted to represent security policies or requirements (e.g., in [25, 39]). AND-OR trees allow to reduce the overall requirements to

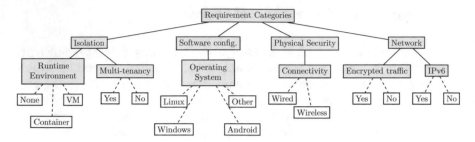

**Fig. 3** Example of a hierarchy of requirement categories and requirement objectives

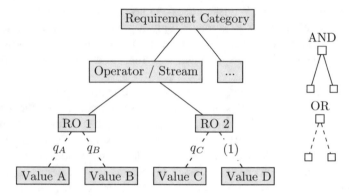

**Fig. 4** Example of the AND-OR tree used to represent application requirements with respect to a requirement category. Requirements associated with different operators (or streams) and requirements involving different requirement objectives for the same application component are in an AND relationship. Whenever the application specifies multiple accepted options for the same RO, possibly providing a preference value for each option, the corresponding nodes are in an OR relationship

the conjunctions and disjunctions of "sub-requirements" (e.g., requirements coming from specific application components).

In particular, we expect an AND-OR tree $\mathcal{T}_\omega$ to be specified by the application for each category of interest $\omega \in \Omega_{RC}$. Henceforth, the overall application requirements can be formally expressed as a *forest* $\mathcal{F} = \{\mathcal{T}_\omega : \forall \omega \in \Omega_{RC}\}$. The structure of a generic tree $\mathcal{T}_\omega$ is illustrated in Fig. 4. As shown in the figure, the root node of each tree corresponds to the RC $\omega$ itself. The next level of the tree contains nodes associated with application components (i.e., operators or streams). Each of these nodes is the root of a subtree that represents the requirements of that specific operator (or stream). Clearly, in order to satisfy the application requirements, the requirements of every component must be satisfied. Therefore, each node at this level of the tree is in conjunction with the others.

Looking at the next level of the tree, the $i$-th operator (or stream) may be associated with one or more child nodes representing ROs for which a requirement is specified. We denote the set of ROs that characterize the deployment requirements of operator $i$ as $\Omega_{RO}^{\omega,i}$. Recalling that each RO $\rho$ takes values in a finite set $V_\rho$, specifying a requirement for $\rho$ means identifying a subset of values $\tilde{V} \subseteq V_\rho$. Indeed, in our tree-based representation, RO nodes have one or more child nodes, each corresponding to a value $v \in V_\rho$ that allows to satisfy the requirement. These nodes are leaf nodes of the tree.

We further enrich our formalism by allowing leaf nodes to be associated with an optional *preference* value $q_v^i \in (0, 1]$, which is a measure of *how much* $v \in V_\rho$ satisfies the requirement of $i$ (i.e., $q_v^i = 1$ means that $v$ completely satisfies the requirement, whilst $0 < q_v^i < 1$ means that $v$ is a feasible choice for the application, but with a smaller degree of satisfaction). The preference specification is optional. Wherever the preference value is not provided, complete satisfaction of the requirement is assumed (i.e., $q_v^i = 1$). Analogously, for the values $v \in V_\rho$ not appearing in the tree, which thus do not satisfy the application requirements, we will assume the preference value to be zero.

### 3.2.1 Example

With respect to the reference DSP application depicted in Fig. 1, we show in Fig. 5 an example of the requirements that might be specified for the application. As explained, within the considered application DAG, some components are critical, because they (i) deal with privacy-sensitive information about users, and (ii) are responsible for detecting and reporting potential users' health diseases. Therefore, we expect application requirements to focus on these critical operators and streams.

For example, $op_1$ and $op_2$, which analyze users' data looking for anomalies, may require software isolation, by running either in a software container or a virtual machine. Moreover, the application may require to have those operators deployed in a dedicated node. Specifically, according to the example of Fig. 5a, $op_1$ requires a dedicated node, but also allows for deployment in a multi-tenant node, with a preference value smaller than 1; $op_2$ instead strictly requires to run in a dedicated node.

Requirements can also be specified for data streams (i.e., edges of the application DAG), imposing restrictions on the network links across which the streams can flow. Looking at Fig. 5b, we see that in the example the stream from the source to the first operator requires a network link that supports IPv6. Furthermore, all the streams in the path from the source to the first sink require data to be encrypted when traversing the network.

## 4 DSP Application Placement Modeling

Determining the *placement* of a DSP application means identifying, within the available computing infrastructure, the nodes where operators must be deployed and

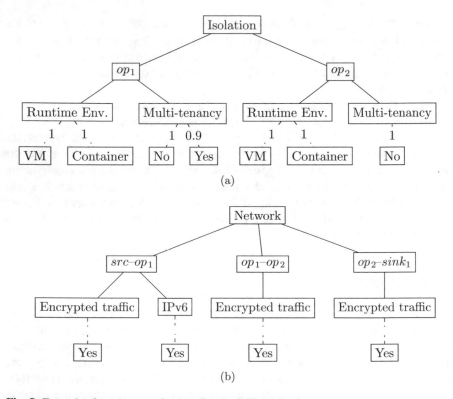

Fig. 5 Example of requirements for the reference DSP applications

executed. This choice has clearly a major impact on application performance, cost, reliability, and security, as the available nodes may (i) be equipped with different amounts of computing resources, (ii) provide different hardware or software capabilities, and (iii) be connected to each other through different network links.

Our aim is to present a linear programming formulation for determining the optimal placement of a DSP application, which takes into account performance, cost, and security metrics. To this end, in this section we will present our model of both the application and the computing infrastructure, and we will introduce the QoS metrics we want to optimize.

## 4.1 System Model

In this section, we describe the system model we consider. In particular, in Sect. 4.1.1, we present how DSP applications are modeled within our optimization framework. In Sect. 4.1.2, a model of the computing infrastructure, which hosts the applications, is introduced.

### 4.1.1 Application Model

We represent a DSP application as a labeled directed acyclic graph (DAG) $G_{dsp} = (V_{dsp}, E_{dsp})$. The nodes in $V_{dsp}$ represent the application operators as well as the data sources and sinks (i.e., nodes with no incoming and no outgoing link, respectively). The edges in $E_{dsp}$ represent the data streams that flow between nodes.

We associate each node and edge in the application graph with various non-functional attributes. Specifically, for each operator $i$, we specify: $\mu_i$, the average number of data units the operator can process per unit of time, on the reference processor;[1] $\sigma_i$, the *selectivity* of the operator (i.e., the average number of data units emitted by the operator per input data unit); $Res_i$, the amount of resources required for its execution. For the sake of simplicity, we assume $Res_i$ to be a scalar value, representing, e.g., the number of CPU cores used by the operator. Our formulation can be easily generalized to cope with a vector of required resources, including, e.g., the amount of memory needed for execution. Similarly, we characterize the stream exchanged from operator $i$ to $j$, $(i, j) \in E_{dsp}$, with its average data rate $\lambda_{(i,j)}$, and the average size (in bytes) of the data units belonging to the stream, $b_{(i,j)}$.

Moreover, we associate the application with its requirements forest $\mathcal{F}$, defined in the previous section. Each operator $\mathcal{F}$ induces a set of *feasible configurations*, i.e., the set of configurations which satisfy the operator security requirements. Formally, we define a configuration $\phi$ as a vector $\phi = (v_{\rho_1}, v_{\rho_2}, \ldots, v_{\rho_N})$, where $v_{\rho_i} \in V_{\rho_i}$ is a value associated with the RO $\rho_i$, and $\{\rho_1, \rho_2, \ldots, \rho_N\} \subseteq \Omega_{RO}$.

For each operator $i$, $i \in V_{dsp}$, the set of feasible configuration is defined as:

$$\Phi_i = \left\{ (v_{\rho_1}, v_{\rho_2}, \ldots) : v_{\rho_k} \in V_{\rho_k}, \forall \rho_k \in \bigcup_\omega \Omega_{RO}^{\omega,i} : q_{v_{\rho_k}}^i > 0 \right\} \tag{1}$$

where $\Omega_{RO}^{\omega,i}$ is the set of ROs in the RC $\omega$ that characterizes the deployment of $i$, and $q_{v_{\rho_k}}^i > 0$ identifies feasible values for $\rho$ with respect to the operator requirements. An analogous definition can be given for $\Phi_{(i,j)}$, the set of feasible configurations for every data stream $(i, j) \in E_{dsp}$.

### 4.1.2 Computing Infrastructure Model

The computing infrastructure hosting DSP applications comprises a set of computing nodes (being them powerful servers in data centers, or resource-constrained devices at the edge), and the network resources that interconnect them. Computing and network resources can be represented as a labeled, fully connected, directed

---

[1]Operator processing speed depends on the actual software/hardware architecture where it is executed. To this end, we define the operator speed with respect to a reference implementation on a reference architecture.

graph $G_{res} = (V_{res}, E_{res})$. Vertices in $V_{res}$ represent the distributed computing nodes, whereas the edges in $E_{res}$ represent the *logical links* between nodes, which result by the underlying physical network paths and routing strategies, not modeled at our level of abstraction. We also model edges of the type $(u, u)$, which capture local connectivity between operators placed in the same node $u$.

We characterize each node $u \in V_{res}$ by means of numerical attributes: $Res_u$, the amount of computing resources available at $u$; $c_u$, the monetary cost per unit of time associated with deploying an operator in the node; $S_u^{\phi}$, the processing speedup with respect to a reference processor, when using configuration $\phi$.

Each node $u \in V_{res}$ is also associated with a set of node *configurations*, $\Phi_u$. The concept of node configuration is analogous to the concept of operator configuration, introduced above. We recall that a configuration $\phi$ is a vector $\phi = (v_{\rho_1}, v_{\rho_2}, \ldots, v_{\rho_N})$, where $v_{\rho_i} \in V_{\rho_i}$ is a value associated with the RO $\rho_i$, and $\{\rho_1, \rho_2, \ldots, \rho_N\} \subseteq \Omega_{RO}$. The set of configurations available on $u$ depends on the hardware/software capabilities of $u$; it can be defined as:

$$\Phi_u \subseteq \left\{ (v_{\rho_1}, v_{\rho_2}, \ldots) : \forall v_{\rho_k} \in V_{\rho_k}^u, \forall \rho_k \in \Omega_{RO}^u \right\} \tag{2}$$

where $V_{\rho}^u \subseteq V_{\rho}$ is the set of values for the RO $\rho$ supported by $u$, and $\Omega_{RO}^u \subseteq \Omega_{RO}$ is the subset of ROs characterizing the node $u$. Note that in the definition above the equality does not necessarily hold, as some combinations of values for different ROs may be unfeasible in practice.

Similar definitions apply to the communication links $(u, v) \in E_{res}$, each being associated with a set of configurations $\Phi_{(u,v)}$, defined as:

$$\Phi_{(u,v)} \subseteq \left\{ (v_{\rho_1}, v_{\rho_2}, \ldots) : \forall v_{\rho_k} \in V_{\rho_k}^{(u,v)}, \forall \rho_k \in \Omega_{RO}^{(u,v)} \right\} \tag{3}$$

where $V_{\rho}^{(u,v)} \subseteq V_{\rho}$ is the set of values for the RO $\rho$ supported by $(u, v)$, and $\Omega_{RO}^{(u,v)} \subseteq \Omega_{RO}$ is the subset of ROs characterizing $(u, v)$. Each link $(u, v)$ and its configurations are also described by: $d_{(u,v)}^{\phi}$, the network delay between node $u$ and $v$, when using network configuration $\phi$; $B_{(u,v)}$, the network capacity available between $u$ and $v$; $c_{(u,v)}$, the monetary cost per unit of data exchanged on the link; $\alpha_{(u,v)}^{\phi}$, a coefficient that captures any data transmission overhead incurred when using link configuration $\phi$ (e.g., data encryption can lead to sending extra information).

## 4.2 DSP Placement Model

The DSP placement problem consists in determining a suitable mapping between the DSP graph $G_{dsp}$ and the resource graph $G_{res}$, as illustrated in Fig. 2. It is often the case that some operators, especially sources and sinks, are *pinned*, that is their placement is fixed. Without loss of generality, we assume that the other operators can be placed on any node of the computing infrastructure, provided that the application requirements for that operator can be satisfied when running in that node.

We can conveniently model the DSP placement problem with binary variables $x_{i,u}^{\phi}$, $i \in V_{dsp}$, $u \in V_{res}$, $\phi \in \Phi_u$. The variable $x_{i,u}^{\phi}$ has value 1 if operator $i$ is deployed on node $u$ using configuration $\phi$, and zero otherwise. A correct placement must deploy an operator on one and only one computing node; this condition can be guaranteed requiring that $\sum_u \sum_{\phi} x_{i,u}^{\phi} = 1$, with $u \in V_{res}$, $i \in V_{dsp}$, $\phi \in \Phi_u$. For each pinned operator $i_p$, being $\bar{u}$ the node it must be deployed on, clearly we have that $x_{i_p,u}^{\phi} = 0$ for every node $u \neq \bar{u}$. Furthermore, under some configurations $\Phi_u^{\epsilon} \subseteq \Phi_u$ a node $u$ becomes available for *exclusive use* of a single operator. We model the choice of exclusively acquiring nodes with additional binary variables $w_u$, $u \in V_{res}$, which have value 1 if the node $u$ is exclusively used, and zero otherwise.

We also consider binary variables associated with links, namely $y_{(i,j),(u,v)}^{\phi}$, $(i,j) \in E_{dsp}$, $(u,v) \in E_{res}$, $\phi \in \Phi_{(u,v)}$, which denotes whether the data stream flowing from operator $i$ to operator $j$ traverses the network path from node $u$ to node $v$, using network configuration $\phi$.

For short, in the following we denote by $x$ and $y$ the placement vectors for nodes and edges, respectively, where $x = \langle x_{i,u}^{\phi} \rangle$, $\forall i \in V_{dsp}$, $\forall u \in V_{res}$, $\forall \phi \in \Phi_u$ and $y = \langle y_{(i,j),(u,v)}^{\phi} \rangle$, $\forall (i,j) \in E_{dsp}$, $\forall (u,v) \in E_{res}$, $\forall \phi \in \Phi_{(u,v)}$. Similarly, we define the exclusive use vector $w = \langle w_u \rangle$, $\forall u \in V_{res}$.

Note that our model does not take into account operator replication, that is allocating multiple parallel replicas of operators to handle larger volumes of input data. The model we present can be easily extended to consider this scenario, either by tweaking the decision variables as in [14], or by introducing an intermediate DSP graph representation, which contains the replicated operators, and then computing the placement for this new graph. However, both these approaches would make the formulation less readable, and thus we do not investigate the replication problem in this chapter.

## 4.3   QoS and Cost Metrics

Placement decisions have a significant impact on both the achieved QoS and the associated monetary deployment cost. In this section, we introduce the metrics we are interested in optimizing, and will be used in our optimization problem formulation. Specifically, we are interested in optimizing the trade-offs between performance, security- and privacy-related requirements satisfaction, and deployment cost. To this end, in Sect. 4.3.1 we will consider application response time as the reference performance metric; in Sect. 4.3.2 we will show how requirement satisfaction metrics can be formulated; and in Sect. 4.3.3 we will define the deployment cost metrics we will use. Note that other metrics can be easily included

in our optimization framework, e.g., availability or network-related metrics [14]. We
do not show their formulation here because of space limitations.

### 4.3.1  Response Time

DSP applications are often used in latency-sensitive domains, where we are
interested in processing newly incoming data as soon as possible. For this reason, a
relevant performance metric for DSP application is *response time*. Given a source-
to-sink path in the application DAG, we define response time as the time it takes for
a data unit emitted by the source to reach the sink, and thus possibly producing a
result/response.

In order to formalize the above definition, we consider a source-to-sink path $\pi = (\pi_1, \pi_2, \ldots, \pi_{N_\pi})$, $\pi_i \in V_{dsp}$. We define the application response time associated
with the path $\pi$ as:

$$R^\pi(x, y) = \sum_{i=\pi_1}^{\pi_{N_\pi}} R_i(x) + \sum_{p=1}^{N_\pi - 1} D_{(\pi_p, \pi_{p+1})}(y) \tag{4}$$

where the first term accounts for the processing time spent at each operator in $\pi$,
and the second term accounts for the network delay accumulated along the path. In
presence of multiple source-to-sink paths, we may also be interested in the overall
application response time $R(x, y)$, which we define as the maximum response time
among all the paths, i.e., $R(x, y) = \max_\pi R^\pi(x, y)$.

The single operator response time in turn can be formulated as:

$$R_i(x) = \sum_{u \in V_{res}} \sum_{\phi \in \Phi_u} R_i(\lambda_i, S_u^\phi) x_{i,u}^\phi \tag{5}$$

where $R_i(\lambda, S)$ is the operator response time, evaluated with respect to the current
operator input rate $\lambda_i$, and processing speedup $S$. The total network delay along the
path $\pi$ is equal to:

$$D_{(i,j)}(y) = \sum_{(u,v) \in E_{res}} \sum_{\phi \in \Phi_{(u,v)}} d_{(u,v)}^\phi y_{(i,j),(u,v)}^\phi \tag{6}$$

### 4.3.2  Security Requirements

Performance-related metrics are not sufficient to characterize the application QoS
in highly distributed fog-like environments. Actually, it is necessary to take
into account additional, non-functional application requirements, especially those
related to security needs. To this end, in addition to the traditionally used response
time metric, we consider a set of metrics that allow quantitatively reasoning

about the satisfaction of application requirements. In particular, we will define metrics of the type $S_\alpha^\beta(x, y)$, which measure the satisfaction level of (a subset of) the application requirements, with respect to a certain deployment configuration. All these metrics take value in $[0, 1]$, with 1 indicating perfect matching of the application requirements, and 0 no satisfaction at all.

Given a placement $(x, y)$, we define an application-level satisfaction metric $S(x, y)$, which considers the requirements from every RC:

$$S(x, y) = \prod_{\omega \in \Omega_{RC}} S_\omega(x, y) \tag{7}$$

Then, we define the satisfaction of application requirements related to RC $\omega \in \Omega_{RC}$ as:

$$S_\omega(x, y) = \prod_{i \in V_{dsp}} S_\omega^i(x) \cdot \prod_{(i,j) \in E_{dsp}} S_\omega^{(i,j)}(y) \tag{8}$$

where $S_\omega^i(x)$ measures the requirements satisfaction for operator $i$, and $S_\omega^{(i,j)}(y)$ for data stream $(i, j)$, with respect to $\omega$. That is, the satisfaction of the requirements in $\omega$ implies meeting the requirements of all the operators *and* all the streams. In turn, evaluating the requirements satisfaction for an operator (or, equivalently, a stream) with respect to the current placement $x$, means evaluating how much its current configuration matches its requirements. Formally, we let $\phi(x, i)$ denote the node configuration in use by operator $i$ under placement $x$, i.e., $\phi(x, i)$ identifies the unique $\phi$ such that, for any node $u \in V_{res}$, $x_{i,u}^\phi = 1$. We have:

$$S_\omega^i(x) = S_\omega^i(\phi(x, i)) \tag{9}$$

Henceforth, for each operator $i$, we need to evaluate $S_\omega^i(\phi)$ in order to assess how much a configuration $\phi$ satisfies its requirements in $\omega$. Recalling that $\Omega_{RO}^{\omega,i}$ denotes the set of ROs in $\omega$ that characterize the deployment of $i$, which are in conjunction with each other, we get:

$$S_\omega^i(\phi) = \prod_{\rho \in \Omega_{RO}^{\omega,i}} q_{\phi_\rho}^{i,\rho} \tag{10}$$

where $q_{\phi_\rho}^{i,\rho} \in [0, 1]$ is the preference value assigned by operator $i$ to $\phi_\rho$, and $\phi_\rho$ is the value that characterizes $\rho$ in configuration $\phi$.[2]

---

[2] We note that the vector $\phi$ possibly specifies a value $\phi_\rho$ for a subset of ROs $\tilde{\Omega}_{RO} \subseteq \Omega_{RO}$. Thus, with a slight abuse of notation, we assume $q_{\phi_\rho}^{i,\rho} = 0, \forall \rho \notin \tilde{\Omega}_{RO}$, i.e., a configuration $\phi$ cannot satisfy requirements for any not specified RO.

Equivalent metrics can be formulated for data streams in a similar way. We have:

$$S_\omega^{(i,j)}(y) = S_\omega^{(i,j)}(\phi(y,(i,j)))  \tag{11}$$

where $\phi(y,(i,j))$ is the unique $\phi$ such that, for any link $(u,v) \in E_{res}$, $y_{(i,j),(u,v)}^\phi = 1$. We can thus evaluate the requirement satisfaction for $(i,j)$ under configuration $\phi$:

$$S_\omega^{(i,j)}(\phi) = \prod_{\rho \in \Omega_{RO}^{\omega,(i,j)}} q_{\phi\rho}^{(i,j),\rho}  \tag{12}$$

### 4.3.3 Deployment Cost

We model the monetary deployment cost associated with the usage of computing and network resources. We assume a typical pay-as-you-go cost model, where the monetary cost for each resource is proportional to its usage. In this scenario, we define the total deployment cost for the application, per unit of time, as $C(x,y,w)$. It accounts for both the cost of the computing resources and the network usage, as follows:

$$C(x,y,w) = \sum_{u \in V_{res}} C_u(x,w) + \sum_{(u,v) \in E_{res}} C_{(u,v)}(y)  \tag{13}$$

The computing nodes cost $C_u(x,w)$ in turn can be formulated as:

$$C_u(x,w) = \sum_{\phi \in \Phi_u \setminus \Phi_u^\epsilon} \sum_{i \in V_{dsp}} Res_i c_u x_{i,u}^\phi + Res_u c_u w_u  \tag{14}$$

where the first term accounts for the cost paid when the node is not exclusively acquired by an operator, which is proportional to $Res_i$, the amount of resources allocated to each operator $i$; the second term instead accounts for the cost paid when the node is exclusively used by an operator, which is equivalent to paying for all the resources provided by $u$, $Res_u$.

The network usage cost is defined as:

$$C_{(u,v)}(y) = c_{(u,v)} N_{(u,v)}(y)  \tag{15}$$

with $N_{(u,v)}(y)$ representing the amount of data exchanged through the link $(u,v)$, which can be computed as follows:

$$N_{(u,v)}(y) = \sum_{\phi \in \Phi_{(u,v)}} \sum_{(i,j) \in E_{dsp}} b_{(i,j)} \lambda_{(i,j)} \alpha_{(u,v)}^\phi y_{(i,j),(u,v)}^\phi  \tag{16}$$

where $b_{(i,j)}\lambda_{(i,j)}$ represents the data rate of the stream $(i, j)$, and $\alpha^{\phi}_{(u,v)}$ is a coefficient that captures any data transmission overhead incurred when using link configuration $\phi$.

# 5 Security-Aware DSP Placement Problem Formulation

When determining the placement of a DSP application, we aim at optimizing a QoS-based function $F(x, y)$ that involves one or more of the metrics presented in the previous section. Moreover, we are often expected to additionally satisfy one or more constraints, revolving around the same metrics of interest. For example, a fixed monetary budget may be allocated for running the applications; performance-related Service Level Objectives (SLOs) may have to be met; security requirements of critical operators may have to be necessarily matched in order for the application to operate.

In this work, relying on the QoS and cost metrics described above, we consider three types of constraints, respectively related to the deployment cost, application performance, and requirements satisfaction.

**Deployment Cost** Given a defined available budget $C_{max}$, the total application deployment cost must not exceed $C_{max}$.

**Performance** We assume performance-related SLO to be defined for the application, expressed in terms of application response time. In particular, an upper bound $R^{\pi}_{max}$ is defined for each source-to-sink path $\pi$. Distinct paths indeed possibly carry out processing tasks that are more or less latency-critical, and thus can be subject to different SLOs.

**Security-Related Requirements** Given a specification of the application Requirements Forest, and the requirement satisfaction metrics defined in the previous section, several constraints can be introduced in the deployment optimization problem, which allow to model both "hard" and "soft" security requirements. The generic requirement satisfaction constraint has the form:

$$S^{\beta}_{\alpha}(\cdot) \geq S^{\beta}_{\alpha,min} \tag{17}$$

By replacing $S^{\beta}_{\alpha}(\cdot)$ with the appropriate concrete metric, the constraint can be applied to a specific subset of the requirements forest. For example, by using $S_{\omega}(x, y)$, we can formulate a constraint on the satisfaction of the requirements in the RC $\omega \in \Omega_{RC}$; using $S^{i}_{\rho}(\phi(x, i))$, we can formulate a constraint associated with a specific operator $i \in V_{dsp}$, and a single RO $\rho \in \Omega_{RO}$; instead, using $S(x, y)$, the constraint applies to the whole application requirement forest.

Furthermore, by properly setting the lower bound $S^{\beta}_{\alpha,min}$, we can model different kinds of requirements. If $S^{\beta}_{\alpha,min} = 0$ and strict inequality is used, the constraint only prohibits using deployment solutions where configurations not accepted by the application are used (e.g., an operator requiring a Linux-based OS cannot be deployed in a Windows-based node); if $S^{\beta}_{\alpha,min} = 1$ we get a hard constraint, where every single involved requirement must be completely satisfied; if $S^{\beta}_{\alpha,min} \in (0, 1)$ we get a soft constraint, where only a minimum level of satisfaction is needed, based on the preference values assigned to the various configurations.

## 5.1  Problem Formulation

We formulate the Security-aware DSP Placement (SDP) problem as an Integer Linear Programming (ILP) model as follows:

$$\min_{x,y} F(x, y)$$

**subjectto** :

$$R^{\pi}_{max} \geq R^{\pi}(x, y) \qquad\qquad \forall \pi \in \Pi_{dsp} \qquad (18)$$

$$C_{max} \geq C(x, y, w) \qquad\qquad\qquad\qquad (19)$$

$$\tilde{S}^{\beta}_{\alpha,min} \leq \tilde{S}^{\beta}_{\alpha}(x, y) \qquad\qquad\qquad\qquad (20)$$

$$B_{(u,v)} \geq N_{(u,v)}(y) \qquad\qquad \forall (u, v) \in E_{res} \qquad (21)$$

$$Res_u \geq \sum_{i \in V_{dsp}} \sum_{\phi \in \Phi_u} Res_i x^{\phi}_{i,u} \qquad \forall u \in V_{res} \qquad (22)$$

$$1 = \sum_{u \in V_{res}} \sum_{\phi \in \Phi_u} x^{\phi}_{i,u} \qquad \forall i \in V_{dsp} \qquad (23)$$

$$\sum_{\phi \in \Phi_u} x^{\phi}_{i,u} = \sum_{v \in V_{res}} \sum_{\phi \in \Phi_{(u,v)}} y^{\phi}_{(i,j),(u,v)} \qquad \begin{array}{c}\forall (i,j) \in E_{dsp},\\ u \in V_{res}\end{array} \qquad (24)$$

$$\sum_{\phi \in \Phi_v} x^{\phi}_{j,v} = \sum_{u \in V_{res}} \sum_{\phi \in \Phi_{(u,v)}} y^{\phi}_{(i,j),(u,v)} \qquad \begin{array}{c}\forall (i,j) \in E_{dsp},\\ v \in V_{res}\end{array} \qquad (25)$$

$$w_u \geq \sum_{i \in V_{dsp}} \sum_{\phi \in \Phi^{\epsilon}_u} x^{\phi}_{i,u} \qquad \forall u \in V_{res} \qquad (26)$$

$$(1 - w_u)M \geq \sum_{i \in V_{dsp}} \sum_{\phi \in \Phi_u \setminus \Phi^{\epsilon}_u} x^{\phi}_{i,u} \qquad \forall u \in V_{res} \qquad (27)$$

$$w_u \in \{0, 1\} \qquad \forall u \in V_{res}$$

$$x^{\phi}_{i,u} \in \{0, 1\} \qquad \begin{array}{c}\forall i \in V_{dsp},\\ u \in V_{res},\\ \phi \in \Phi_u\end{array}$$

$$y^{\phi}_{(i,j),(u,v)} \in \{0, 1\} \qquad \begin{array}{l} \forall (i,j) \in E_{dsp}, \\ (u,v) \in E_{res}, \\ \phi \in \Phi_{(u,v)} \end{array}$$

where $M \gg 1$ is a large constant, and $\tilde{S}(\cdot) = \log S(\cdot)$ is used in order to obtain linear expressions for the requirements satisfaction metrics. Constraints (18)–(20) model the QoS- and cost-related bounds described above, where—as explained—(20) may actually be replaced by several constraints involving different security require-ments. Constraints (21) and (22) are capacity constraints, modeling, respectively, the limited capacity of network links, and the limited amount of computational resources available at nodes. Equation (23) reflects the fact that a single node and a single configuration must be chosen for each operator. Equations (24) and (25) are flow conservation constraints, which model the logical AND relationship between placement variables. Constraints (26) and (27) model the relationship between exclusive use variables and placement configurations.

It is easy to realise that SDP is a NP-hard problem. To this end, it suffices to observe that SDP is a generalization of the optimal DSP placement problem presented in [13], which has been shown to be NP-hard.

## 5.2 Example

We formulate the SDP problem for the smart health application presented in the previous sections. The requirements for this application, described in Sect. 3, were reported in Fig. 5. We further assume that a fixed budget $C_{max}$ is allocated for deploying the application, and the application is expected to meet SLOs formulated in terms of response time. Specifically, denoting as $\pi_1$ the operators path from the source to $sink_1$, we assume $\pi_1$ to have strict latency requirements, and set the maximum response time along the path, $R^{\pi_1}_{max}$, to 10 ms. For the other path, $\pi_2$, we set $R^{\pi_2}_{max} = 100$ ms. Moreover, as we want none of the application requirements to be ignored, we require $S(x, y)$ to be strictly greater than zero.

We consider different scenarios for formulating the SDP problem, as follows:

- *Scenario A*: we solve SDP maximizing the satisfaction of application require-ments, i.e., $F(x, y) = \tilde{S}(x, y)$;
- *Scenario B*: we solve SDP minimizing the worst-case application response time, i.e., $F(x, y) = \max_{\pi \in \Pi_{dsp}} R_{\pi}(x, y)$. We also consider three cases, considering an additional constraint on overall requirements satisfaction: (i) $S(x, y) > 0$; (ii) $S(x, y) \geq 0.9$; (iii) $S(x, y) \geq 0.99$.

The computing infrastructure we consider for deployment is depicted in Fig. 6. The infrastructure is composed of 15 geographically distributed computing nodes: 3 edge nodes, 4 fog nodes distributed across two micro-data centers, and 8 cloud nodes distributed across two data centers. We assume that the application data source is pinned on the first edge node. The operators and the sinks can be freely

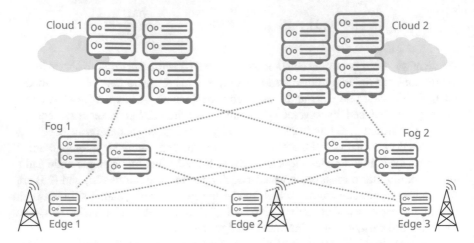

**Fig. 6** Illustration of the computing infrastructure used in the example, comprising edge nodes, fog micro-data centers, and cloud data centers

placed in any node of the infrastructure. Compared to a reference cloud node, processing at the fog nodes is 10% slower, and 20% slower at the edge. Conversely, cloud nodes are cheaper than those in the fog, which in turn are cheaper than edge nodes. We assume network delay to be negligible within the same geographical region (i.e., the same (micro-)data center). We further assume that fog and cloud nodes support operator execution either as bare processes, or within containers or within VMs, whereas operators cannot be deployed in edge nodes using VMs. Moreover, we assume edge nodes cannot be acquired exclusively by a single operator in this scenario.

**Results in Scenario A** In this scenario, SDP aims at maximizing the security requirements satisfaction. To illustrate the results, we solve SDP considering different choices for the monetary budget $C_{max}$, namely 5, 6, 7, and 10 \$/h. The optimization results in this scenario are reported in Table 1. In Fig. 7, we show the deployment computed by SDP for $C_{max} \in \{5, 10\}$ \$/h. In all the considered experiments, traffic encryption is enforced for all the data streams in the application. The minimum budget $C_{max} = 5$ \$/h leads to a requirements satisfaction degree equal to 0.9. As illustrated in Fig. 7a, in this case fog and cloud nodes are used for deploying application operators (except for the pinned data source), with $op_2$ running in an exclusively acquired node. Conversely, $op_1$ is deployed in a multi-tenant node, leading to partial requirements satisfaction. When the cost budget is raised to 10 \$/h, we can note that (i) $op_1$ is deployed in a dedicated node, completely satisfying its isolation requirement, (ii) the application path to $sink_2$ is deployed in the fog, without relying on any cloud node, and (iii) the other application path is deployed across 3 geographical regions instead of 4 as in the previous case. The total application response time along the two paths is thus reduced, respectively,

**Table 1** Optimization results in *Scenario A* of the example

| $C_{max}$ ($/h) | $C$ ($/h) | $R^{\pi_1}$ (ms) | $R^{\pi_2}$ (ms) | $S$ |
|---|---|---|---|---|
| 5 | 4.6 | 9.1 | 6.0 | 0.9 |
| 6 | 5.9 | 9.1 | 12.0 | 1.0 |
| 7 | 6.5 | 9.1 | 9.0 | 1.0 |
| 10 | 9.7 | 6.1 | 4.0 | 1.0 |

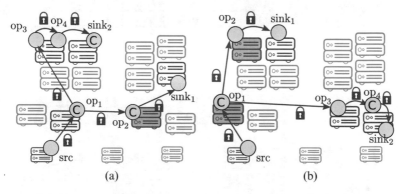

**Fig. 7** Deployment computed by SDP in example Scenario A. (a) $C_{max}$ = 5 $/h. (b) $C_{max}$ = 10 $/h

from 9 ms to 6 ms and from 6 ms to 4 ms. However, the deployment cost is doubled, from 4.6 $/h to 9.7 $/h.

In the other configurations, instead, with slightly larger allocations for $C_{max}$, SDP is able to perfectly match application requirements. In all the cases, the response time along both the paths is within the SLO bound.

**Results in Scenario B** In this scenario we aim at minimizing the application response time, while requiring a minimum level of requirements satisfaction, $S_{min}$. We again consider different values for the maximum cost $C_{max} \in \{5, 6, 7, 10\}$, and solve SDP varying the requirements satisfaction constraint. In Table 2 we report the optimization results, while in Fig. 8 we show the deployment computed by SDP for two illustrative cases.

Requiring $S(x, y) > 0$ means that only configurations with non-zero requirements satisfaction can be adopted. Whatever the monetary budget, in this case SDP computes placement solutions where requirements satisfaction is rather low. Figure 8a shows the solution for $C_{max} = 5$ $/h. In order to minimize latency, differently from Scenario A, for this scenario SDP does not use cloud nodes. Traffic encryption is only used on the "critical" path towards $sink_1$, and only $op_2$ is deployed in a dedicated node, as it does not admit other configurations.

When we require $S(x, y) \geq 0.9$, SDP computes solutions characterized by higher requirements satisfaction (0.9 in all the considered settings), with negligible

**Table 2** Optimization results in *Scenario B* of the example. The case where $S_{min}$ is set to 0.99 is not feasible when a budget of 5 \$/h is allocated

| $C_{max}$ (\$/h) | $S_{min}$ | $C$ (\$/h) | $R^{\pi_1}$ (ms) | $R^{\pi_2}$ (ms) | $S$ |
|---|---|---|---|---|---|
| 5 | > 0 | 4.6 | 6.1 | 6.0 | 0.11 |
| 6 | > 0 | 5.2 | 1.1 | 6.0 | 0.23 |
| 7 | > 0 | 5.2 | 1.1 | 6.0 | 0.45 |
| 10 | > 0 | 9.4 | 1.1 | 1.0 | 0.23 |
| 5 | 0.90 | 4.6 | 6.1 | 6.0 | 0.90 |
| 6 | 0.90 | 5.2 | 1.1 | 6.0 | 0.90 |
| 7 | 0.90 | 5.2 | 1.1 | 6.0 | 0.90 |
| 10 | 0.90 | 9.4 | 1.1 | 1.0 | 0.90 |
| 5 | 0.99 | - | - | - | - |
| 6 | 0.99 | 5.9 | 6.1 | 6.0 | 1.00 |
| 7 | 0.99 | 6.5 | 4.1 | 6.0 | 1.00 |
| 10 | 0.99 | 9.1 | 2.1 | 4.0 | 1.00 |

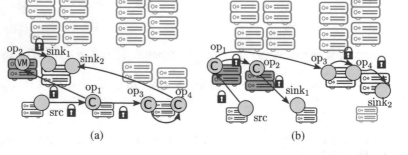

(a)                                                        (b)

🔒 Traffic encryption  🖥 Exclusively used node ◯ Bare process ©Container ⓋⓂVirtualMachine

**Fig. 8** Deployment computed by SDP in example Scenario B

impact on application response time and deployment cost. Interestingly, when we require $S(x, y) \geq 0.99$, SDP is not able to find a feasible solution if the monetary budget is 5 \$/h. Indeed, it would need to exclusively acquire more than one computing node, and that is expensive. With a higher budget instead SDP is able to perfectly match the application requirements. We note that this happens at the cost of slightly higher response time, especially on the critical path $\pi_1$. Figure 8b shows the deployment determined when $C_{max} = 10$ \$/h. Again, "far" cloud nodes are not used in this scenario. In this setting, traffic encryption is enabled for all the data streams, and an additional fog node is exclusively acquired for deploying $op_1$ to satisfy the security requirements.

# 6 Discussion

State-of-the-art solutions for the DSP placement problem allow to effectively optimize several QoS metrics, taking into account both functional and non-functional characterizations of the application operators and the underlying computing infrastructure. However, when we move applications out of the traditionally used data centers, we are forced to look at security, privacy, and data integrity concerns as major issues. Unfortunately, existing techniques often completely neglect these aspects.

SDP aims at overcoming this limitation by reserving a primary role to application deployment requirements. The forest-based approach we presented for specifying application requirements allows to easily integrate satisfaction metrics into the placement optimization problem; moreover, it provides large flexibility for formalizing application needs. In particular, compared to similar works in the literature, our approach allows—when needed—to apply different policies to different types of requirements (e.g., to multiple RCs), and to different operators/data streams, instead of necessarily collapsing "security" to a single numerical indicator. It is worth observing that the requirements formalism and the associated metrics presented in this chapter might not perfectly suit every application domain. Nevertheless, SDP can be easily adapted to work with a different toolbox for quantitative modeling of security aspects (e.g., a probabilistic model like that presented in [25]).

The concept of operator (data stream) and node (link) configurations we introduced in SDP provides additional degrees of freedom in the placement optimization. Compared to previous work on the topic, SDP goes beyond simply determining a mapping from operators to computing nodes, by also identifying the specific software configuration to be adopted for each operator. The integration of this configuration optimization approach with existing software orchestration platforms will be subject of further investigations, in order to have a system capable of automatically choosing and *applying* necessary configuration for the deployed application.

As noted above, some possible extensions of the presented problem formulation were intentionally not tackled in this work, and deferred to future research. First of all, the formulation can be easily generalized to take into account the operator replication problem, e.g., adopting the approach used in [14]. Moreover, we have not specifically covered here the issues related to run-time deployment adaptation, focusing instead on the initial application placement. Nonetheless, adaptation overhead metrics can be readily introduced in SDP, e.g., following our previous work [12].

# 7 Conclusion

In this chapter, we looked at how security aspects impact the placement problem for DSP applications. The recent trend of shifting data analytics services from traditionally used cloud data centers to fog computing environments, in order to reduce network latency between applications and data sources, forces us to deal with a broader range of security and privacy issues.

We presented an approach, based on a forest of AND-OR trees, for specifying additional non-functional application requirements, which are hardily captured by existing techniques for placement optimization. Relying on this formalism, we also defined a set of metrics that allow to quantitatively reason about requirements satisfaction, especially as regards security aspects. We included these metrics in the ILP-based SDP problem, which determines the optimal DSP application placement according to several QoS constraints. By means of an illustrative case study of a smart health application in the fog, we provided insights about the trade-offs between performance, cost, and security computed by SDP. Our approach provides great flexibility for specifying requirements, as well as optimization objectives and constraints. We pointed out some open research directions, especially as regards the integration of SDP with existing DSP frameworks. As future work, we plan to investigate this direction, by complementing our modeling effort with experimental validations, where concrete cybersecurity issues must be fitted within SDP.

# References

1. Abadi, D.J., Ahmad, Y., Balazinska, M., Çetintemel, U., et al.: The design of the Borealis stream processing engine. In: Proc. CIDR '05, pp. 277–289 (2005)
2. Agbo, C.C., Mahmoud, Q.H., Mikael Eklund, J.: A scalable patient monitoring system using Apache Storm. In: Proc. 2018 IEEE Canadian Conf. on Electrical Computer Engineering, pp. 1–6, CCECE '18 (2018)
3. Anh, D.T.T., Datta, A.: Streamforce: Outsourcing access control enforcement for stream data to the clouds. In: Proc. ACM CODASPY '14, pp. 13–24 (2014)
4. Aniello, L., Baldoni, R., Querzoni, L.: Adaptive online scheduling in Storm. In: Proc. ACM DEBS '13, pp. 207–218 (2013)
5. Arkian, H.R., Diyanat, A., Pourkhalili, A.: MIST: Fog-based data analytics scheme with cost-efficient resource provisioning for IoT crowdsensing applications. J. Parallel Distrib. Comput. **82**, 152–165 (2017)
6. Backman, N., Fonseca, R., Çetintemel, U.: Managing parallelism for stream processing in the cloud. In: Proc. HotCDP '12, pp. 1:1–1:5. ACM (2012)
7. Bellendorf, J., Mann, Z.A.: Classification of optimization problems in fog computing. Future Gener. Comput. Syst. **107**, 158–176 (2020)
8. Brogi, A., Forti, S., Guerrero, C., Lera, I.: How to place your apps in the fog: State of the art and open challenges. Softw. Pract. Exp. (2019)
9. Burkhalter, L., Hithnawi, A., Viand, A., Shafagh, H., Ratnasamy, S.: TimeCrypt: Encrypted data stream processing at scale with cryptographic access control. In: Proc. USENIX NSDI '20, pp. 835–850 (Feb 2020)

10. Cao, J., Carminati, B., Ferrari, E., Tan, K.L.: ACStream: Enforcing access control over data streams. In: Proc. IEEE ICDE '09, pp. 1495–1498 (2009)
11. Cardellini, V., Lo Presti, F., Nardelli, M., Rossi, F.: Self-adaptive container deployment in the fog: A survey. In: Proc. ALGOCLOUD '19. LNCS. Springer (2020)
12. Cardellini, V., Lo Presti, F., Nardelli, M., Russo Russo, G.: Optimal operator deployment and replication for elastic distributed data stream processing. Concurr. Comput. Pract. Exp. **30**(9) (2018)
13. Cardellini, V., Grassi, V., Lo Presti, F., Nardelli, M.: Optimal operator placement for distributed stream processing applications. In: Proc. ACM DEBS '16, pp. 69–80 (2016)
14. Cardellini, V., Grassi, V., Lo Presti, F., Nardelli, M.: Optimal operator replication and placement for distributed stream processing systems. ACM SIGMETRICS Perfom. Eval. Rev. **44**(4), 11–22 (May 2017)
15. Carminati, B., Ferrari, E., Cao, J., Tan, K.L.: A framework to enforce access control over data streams. ACM Trans. Inf. Syst. Secur. **13**(3) (Jul 2010)
16. Carminati, B., Ferrari, E., Tan, K.: Enforcing access control over data streams. In: Proc. ACM SACMAT '07, pp. 21–30 (2007)
17. Chandramouli, B., Goldstein, J., Barga, R., Riedewald, M., Santos, I.: Accurate latency estimation in a distributed event processing system. In: Proc. IEEE ICDE '11, pp. 255–266 (2011)
18. Chaturvedi, S., Simmhan, Y.: Toward resilient stream processing on clouds using moving target defense. In: Proc. IEEE ISORC '19, pp. 134–142 (2019)
19. Chatzistergiou, A., Viglas, S.D.: Fast heuristics for near-optimal task allocation in data stream processing over clusters. In: Proc. ACM CIKM '14, pp. 1579–1588 (2014)
20. de Assunção, M.D., da Silva Veith, A., Buyya, R.: Distributed data stream processing and edge computing: A survey on resource elasticity and future directions. J. Netw. Comput. Appl. **103**, 1–17 (2018)
21. Eidenbenz, R., Locher, T.: Task allocation for distributed stream processing. In: Proc. IEEE INFOCOM '16 (2016)
22. Eskandari, L., Mair, J., Huang, Z., Eyers, D.: T3-Scheduler: A topology and traffic aware two-level scheduler for stream processing systems in a heterogeneous cluster. Future Gener. Comput. Syst. **89**, 617–632 (2018)
23. Fischer, L., Scharrenbach, T., Bernstein, A.: Scalable linked data stream processing via network-aware workload scheduling. In: Proc. 9th Int'l Workshop Scalable Semantic Web Knowledge Base Systems (2013)
24. Fisher, R., Hancke, G.: DTLS for lightweight secure data streaming in the Internet of Things. In: Proc. 2014 9th Int'l Conf. on P2P, Parallel, Grid, Cloud and Internet Computing, pp. 585–590 (Nov 2014)
25. Forti, S., Ferrari, G.L., Brogi, A.: Secure cloud-edge deployments, with trust. Future Gener. Comput. Syst. **102**, 775–788 (2020)
26. Gedik, B., Özsema, H., Öztürk, O.: Pipelined fission for stream programs with dynamic selectivity and partitioned state. J. Parallel Distrib. Comput. **96**, 106–120 (2016)
27. Ghaderi, J., Shakkottai, S., Srikant, R.: Scheduling storms and streams in the cloud. ACM Trans. Model. Perform. Eval. Comput. Syst. **1**(4), 14:1–14:28 (2016)
28. Ghosh, R., Simmhan, Y.: Distributed scheduling of event analytics across edge and cloud. ACM Trans. Cyber Phys. Syst. **2**(4) (Jul 2018)
29. Gu, L., Zeng, D., Guo, S., Xiang, Y., Hu, J.: A general communication cost optimization framework for big data stream processing in geo-distributed data centers. IEEE Trans. Comput. **65**(1), 19–29 (2016)
30. Havet, A., Pires, R., Felber, P., Pasin, M., Rouvoy, R., Schiavoni, V.: SecureStreams: A reactive middleware framework for secure data stream processing. In: Proc. ACM DEBS '17, pp. 124–133 (2017)
31. Janßen, G., Verbitskiy, I., Renner, T., Thamsen, L.: Scheduling stream processing tasks on geo-distributed heterogeneous resources. In: Proc. IEEE Big Data '18, pp. 5159–5164 (2018)

32. Jiang, J., Zhang, Z., Cui, B., Tong, Y., Xu, N.: StroMAX: Partitioning-based scheduler for real-time stream processing system. In: Proc. DASFAA '17, pp. 269–288. Springer (2017)
33. Khare, S., Sun, H., Gascon-Samson, J., Zhang, K., Gokhale, A., Barve, Y., Bhattacharjee, A., Koutsoukos, X.: Linearize, predict and place: Minimizing the makespan for edge-based stream processing of directed acyclic graphs. In: Proc. ACM/IEEE SEC '19, pp. 1–14 (2019)
34. Lakshmanan, G.T., Li, Y., Strom, R.: Placement of replicated tasks for distributed stream processing systems. In: Proc. ACM DEBS '10, pp. 128–139 (2010)
35. Li, J., Deshpande, A., Khuller, S.: Minimizing communication cost in distributed multi-query processing. In: Proc. IEEE ICDE '09, pp. 772–783 (2009)
36. Li, T., Tang, J., Xu, J.: A predictive scheduling framework for fast and distributed stream data processing. In: Proc. 2015 IEEE Int'l Conf. on Big Data, pp. 333–338 (2015)
37. Lindner, W., Meier, J.: Securing the Borealis data stream engine. In: Proc. 10th Int'l Database Engineering and Applications Symp., pp. 137–147. IDEAS '06 (2006)
38. Loukopoulos, T., Tziritas, N., Koziri, M., Stamoulis, G., Khan, S.U.: A Pareto-efficient algorithm for data stream processing at network edges. In: Proc. IEEE CloudCom '18, pp. 159–162 (2018)
39. Luna Garcia, J., Langenberg, R., Suri, N.: Benchmarking cloud security level agreements using quantitative policy trees. In: Proc. 2012 ACM Workshop on Cloud Computing Security, pp. 103–112. CCSW '12 (2012)
40. Nardelli, M., Cardellini, V., Grassi, V., Lo Presti, F.: Efficient operator placement for distributed data stream processing applications. IEEE Trans. Parallel Distrib. Syst. 30(8), 1753–1767 (2019)
41. Nehme, R.V., Lim, H., Bertino, E.: FENCE: continuous access control enforcement in dynamic data stream environments. In: Proc. IEEE ICDE '10, pp. 940–943 (2010)
42. Nehme, R.V., Rundensteiner, E.A., Bertino, E.: A security punctuation framework for enforcing access control on streaming data. In: Proc. IEEE ICDE '08, pp. 406–415 (2008)
43. Ng, W.S., Wu, H., Wu, W., Xiang, S., Tan, K.: Privacy preservation in streaming data collection. In: Proc. IEEE ICPADS '12, pp. 810–815 (Dec 2012)
44. O'Keeffe, D., Salonidis, T., Pietzuch, P.: Frontier: Resilient edge processing for the Internet of Things. Proc. VLDB Endow. 11(10), 1178–1191 (Jun 2018)
45. Park, H., Zhai, S., Lu, L., Lin, F.X.: Streambox-TZ: Secure stream analytics at the edge with trustzone. In: Proc. USENIX ATC '19, pp. 537–554 (2019)
46. Peng, B., Hosseini, M., Hong, Z., Farivar, R., et al.: R-Storm: Resource-aware scheduling in Storm. In: Proc. Middleware '15, pp. 149–161. ACM (2015)
47. Peng, Q., Xia, Y., Wang, Y., Wu, C., Luo, X., Lee, J.: Joint operator scaling and placement for distributed stream processing applications in edge computing. In: Proc. ICSOC '19, pp. 461–476. LNCS. Springer (2019)
48. Pietzuch, P., Ledlie, J., Shneidman, J., Roussopoulos, M., et al.: Network-aware operator placement for stream-processing systems. In: Proc. IEEE ICDE '06 (2006)
49. Quoc, D.L., Beck, M., Bhatotia, P., Chen, R., Fetzer, C., Strufe, T.: PrivApprox: Privacy-preserving stream analytics. In: Proc. USENIX ATC '17, pp. 659–672 (Jul 2017)
50. Rizou, S., Durr, F., Rothermel, K.: Solving the multi-operator placement problem in large-scale operator networks. In: Proc. ICCCN '10, pp. 1–6 (2010)
51. Röger, H., Mayer, R.: A comprehensive survey on parallelization and elasticity in stream processing. ACM Comput. Surv. 52(2), 36:1–36:37 (2019)
52. Rychly, M., Koda, P., Pavel: Scheduling decisions in stream processing on heterogeneous clusters. In: Proc. 8th Int'l Conf. Complex, Intelligent and Software Intensive Systems (2014)
53. Sajjad, H.P., Danniswara, K., Al-Shishtawy, A., Vlassov, V.: SpanEdge: Towards unifying stream processing over central and near-the-edge data centers. In: Proc. IEEE/ACM SEC '16, pp. 168–178 (2016)
54. Sandhu, R.S., Coyne, E.J., Feinstein, H.L., Youman, C.E.: Role-based access control models. Computer 29(2), 38–47 (1996)
55. Sandhu, R.S., Samarati, P.: Access control: Principle and practice. IEEE Commun. Mag. 32(9), 40–48 (1994)

56. Satyanarayanan, M., Klas, G., Silva, M., Mangiante, S.: The seminal role of edge-native applications. In: Proc. IEEE EDGE '19, pp. 33–40 (2019)
57. Schilling, B., Koldehofe, B., Rothermel, K., Ramachandran, U.: Access policy consolidation for event processing systems. In: Proc. NetSys '13, pp. 92–101. IEEE Computer Society (2013)
58. Sicari, S., Rizzardi, A., Grieco, L., Coen-Porisini, A.: Security, privacy and trust in Internet of Things: The road ahead. Comput. Netw. **76**, 146–164 (2015)
59. da Silva Veith, A., de Assunção, M.D., Lefèvre, L.: Latency-aware placement of data stream analytics on edge computing. In: Proc. ICSOC '18, pp. 215–229. LNCS. Springer (2018)
60. Smirnov, P., Melnik, M., Nasonov, D.: Performance-aware scheduling of streaming applications using genetic algorithm. Procedia Comput. Sci. **108**, 2240–2249 (2017)
61. Stanoi, I., Mihaila, G., Palpanas, T., Lang, C.: WhiteWater: Distributed processing of fast streams. IEEE Trans. Softw. Eng. **19**(9), 1214–1226 (2007)
62. Starks, F., Goebel, V., Kristiansen, S., Plagemann, T.: Mobile distributed complex event processing – Ubi sumus? Quo vadimus? In: Mobile Big Data: A Roadmap from Models to Technologies, pp. 147–180. Springer (2018)
63. Thoma, C., Labrinidis, A., Lee, A.J.: Automated operator placement in distributed data stream management systems subject to user constraints. In: Proc. IEEE ICDEW '14, pp. 310–316 (2014)
64. Thoma, C., Lee, A.J., Labrinidis, A.: PolyStream: Cryptographically enforced access controls for outsourced data stream processing. In: Proc. ACM SACMAT '16, pp. 227–238 (2016)
65. Tian, L., Chandy, K.M.: Resource allocation in streaming environments. In: Proc. 7th IEEE/ACM Int'l Conf. Grid Computing, pp. 270–277 (2006)
66. Xu, J., Chen, Z., Tang, J., Su, S.: T-Storm: Traffic-aware online scheduling in Storm. In: Proc. IEEE ICDCS '14, pp. 535–544 (2014)
67. Zhou, Y., Ooi, B.C., Tan, K.L., Wu, J.: Efficient dynamic operator placement in a locally distributed continuous query system. In: On the Move to Meaningful Internet Systems 2006, LNCS, vol. 4275, pp. 54–71. Springer (2006)
68. Zhuang, R., DeLoach, S.A., Ou, X.: Towards a theory of moving target defense. In: Proc. 1st ACM Workshop on Moving Target Defense, pp. 31–40. MTD '14 (2014)

# Blockchain of Finite-Lifetime Blocks for Edge-IoT Applications

Shravan Garlapati 🆔

## 1 Introduction

The Internet of Things (IoT) industry continues to grow rapidly, and it is forecasted that, by 2030, at least 50 billion devices will have internet connectivity [1]. IoT devices and sensors generate, process and exchange huge amounts of safety-critical and privacy-sensitive data and hence the security of these devices and data are a major concern. Considering the explosive number of devices, it is not a simple task to address issues in security, privacy and data integrity. To be specific, many of the state-of-the art security solutions are centralized and they may not fit well for IoT due to the massive scale of the generated data, single point of failure and many-to-one nature of the traffic. To overcome these issues, contrary to the traditional centralized approaches, numerous recent research works proposed the use of decentralized approaches such as Blockchain [2–5].

Blockchain is a peer-to-peer (P2P) distributed system that offers improved security and privacy of data [6]. In the last decade, blockchain technology has gained a lot of attention as it is widely used in cryptocurrencies such as bitcoin, Ethereum, ripple etc. In recent years, efforts are underway to adopt blockchain technology in different sectors such as financial services, supply chain, healthcare, IoT etc. [7–9]. This work considers a scalable blockchain architecture targeted at IoT applications in edge computing environment. As the number of connected IoT devices continue to explode exponentially, data storage and processing pose serious scalability problems to centralized cloud architectures. The *Edge computing*—a distributed computing paradigm alleviates these problems by bringing computation, data processing and storage closer to IoT devices to improve response times and

S. Garlapati (✉)
IEEE, San Diego, CA, USA
e-mail: gshra09@vt.edu

© Springer Nature Switzerland AG 2021
W. Chang, J. Wu (eds.), *Fog/Edge Computing For Security, Privacy, and Applications*, Advances in Information Security 83,
https://doi.org/10.1007/978-3-030-57328-7_15

save bandwidth [10–12]. With Edge computing, before sending it to the central cloud, IoT data can be pre-processed at the edge servers to compress and summarize the collected data. Thus, edge application services reduce the amount of data that must be moved, and the distance data must travel, resulting in lower latency and reduced transmission costs. Edge computing offers perfect middle-layer services between cloud and IoT device layer i.e. it not only handles data processing tasks for cloud but also takes care of computational offloading of resource constrained IoT devices for real-time applications such as facial, audio and video recognition, autonomous cars, smart cities, industry 4.0 and home automation systems etc. [13, 14]. Thus, Edge computing has the potential to reduce the computational load and power consumption of IoT devices [15–17].

In this context, few recent research works proposed *Edge-IoT* systems where edge-servers form a distributed network and employ blockchain to not only manage the allocation of resources at the edge to resource constrained IoT devices but also to support storing and sharing of IoT data. In [18], Edge-chain presented a credit-based resource management system to allocate edge server resource pool to IoT devices based on pre-defined set of rules, historical usage and application types. Edge-chain uses blockchain to manage all the IoT activities and transactions for secure data logging and auditing. Also, [19] proposed a cognitive Edge-IoT framework that hosts, and processes offloaded geo-tagged multimedia payload and transactions from IoT nodes and stores results in a blockchain and decentralized cloud repositories to support secure and privacy-oriented sharing economy services in a smart city. A decentralized storage, access control, data management and sharing system employing blockchain is presented in [20] to manage time-series IoT data at the edge.

As discussed above, many of the previous research works focused on using blockchain to manage the allocation of edge resources to IoT devices, distributed access control, secure storage and data management at the edge, but except [21], none of them focused on *storage scalability*, a major challenge for *Edge-IoT*. In Bitcoin Network (BCN), as of today, more than 250 GB of storage is required to store full blockchain. The throughput of the bitcoin is around 10 transactions per sec. On the contrary, considering the massive scale, transaction rate of IoT devices can be significantly higher compared to monetary transactions of bitcoin. Hence, the storage capacity needs of *Edge-IoT* can also be significantly higher compared to bitcoin and it is possible that the edge servers can ultimately run out of space to store the full chain. To address this issue, an existing work proposed LiTiChain—a scalable and lightweight blockchain architecture [21]. In cryptocurrency systems, monetary transactions and blocks are stored permanently on the blockchain. On the other hand, IoT data has finite-lifetime and hence expired transactions and blocks can be deleted from the blockchain. LiTiChain exploited this idea and proposed a novel blockchain architecture to minimize the storage requirements. The disadvantage of LiTiChain is, instead of deleting the blocks immediately upon expiration, it is possible that some blocks are retained longer to validate remaining blocks, which results in *additional storage cost*. To address this issue, $\mu$-*LiTiChain*—a novel architecture, which is a generalized version of

*LiTiChain* is presented in this chapter. Also, *p-LiTiChain* and *s-LiTiChain*, which are variants of *μ-LiTiChain* are presented. These novel architectures aim at reducing the *additional storage cost* incurred by LiTiChain and improving security of the chain. In this process, *μ-LiTiChain*, *p-LiTiChain* and *s-LiTiChain* provide a trade-off in the blockchain design (see Sect. 4) in terms of storage cost, security and computational cost.

The remainder of this chapter is organized as follows: Section 2 describes the preliminaries that lay the foundation to understand the subsequent sections. LiTiChain architecture is described in Sect. 3. A generalized version of LiTiChain architecture i.e. *μ-LiTiChain* and its two variants i.e. *p-LiTiChain* and *s-LiTiChain* architectures are presented in Sect. 4. Also, in Sect. 4, trade-offs in blockchain design in terms of storage cost, security and computational cost are discussed. Section 5 presents the simulation results and analysis. Section 6 concludes the chapter.

## 2 Preliminaries

This section describes the basic concepts of blockchain, and system model used in this study.

### 2.1 Blockchain

This sub-section describes the basic concepts of blockchain that are derived from the Bitcoin [6, 21]. Blockchain can be defined as a specific type of distributed ledger that records transactions between two parties in an efficient and secure manner. In simple terms, blockchain is a growing list of records called as *blocks*, which are linked using cryptography [22]. As the blocks are connected using cryptographic hash, a block in a blockchain cannot be easily altered and hence the data is resistant to modification. In order to operate as a distributed ledger, a blockchain based system is typically managed by a P2P network. For example: in *Edge-IoT* systems (see Fig. 1), edge-servers form a P2P network. In order to transfer money, a bitcoin node generates and broadcasts a new transaction to the network. Every new transaction in a BCN is validated by the bitcoin nodes in the network. In BCN, certain users with large computational resources known as *miners*, participate in mining process. Every miner node independently aggregates transactions into a new block and when the new block is full, it is appended to blockchain by mining process i.e. miners solve a cryptographic puzzle with a certain difficulty known as *Proof of Work* (*PoW*). A miner node that first solves the *PoW* appends the mined block to the blockchain and broadcasts the solution to the network. All miners in the network validate the solution, accept the updated blockchain and re-broadcasts the solution. The miner

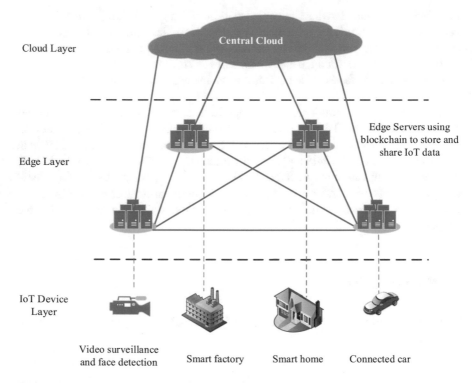

Cloud Layer

Edge Layer

IoT Device
Layer

Edge Servers using
blockchain to store and
share IoT data

Video surveillance
and face detection        Smart factory        Smart home        Connected car

**Fig. 1** Usage of blockchain in edge-IoT systems

node that first solved the *PoW* is rewarded with bitcoins. The details of *Blockheader,*
*Blockheight, and Nonce Computation* are explained below.

### 2.1.1 Structure of a Block

As per [23], a block consists of header, metadata and long list of transactions. The
*Blockheader* consists of three sets of metadata. First, *previous blockhash*, a pointer
to the previous block. The second set, *timestamp, Nonce* and *difficulty* are related
to the mining process. The third part is the *merkle tree root*. A block is identified
by two identifiers, they are: *Blockhash* and *Blockheight*. *Blockhash* is the primary
identifier and it is obtained by hashing the blockheader twice using the SHA256
algorithm. Another way to identify a block is by its position in the blockchain,
known as *blockheight*. The first block, known as *Genesis block*, is at a *blockheight*
of zero.

### 2.1.2 Nonce Computation

It is aforementioned that, in BCN, every miner node independently aggregates transactions into a new block and when the blocksize reaches a threshold (ex: 1 MB), it tries to solve a *PoW* cryptographic hash computation problem. *PoW* takes *previous blockhash, merkle root and timestamp* as input and computes hash using SHA256. The computational challenge involves finding a *Nonce* that results in an output hash with certain number of leading bits as zero. The difficulty level in *Nonce computation* i.e. the number of leading bits as zeros in output hash is given by the *Difficulty Target* parameter specified in the *blockheader*. In bitcoin, for every new block, depending on the *Difficulty Target,* miners may test *billions or trillions* of *Nonce* before the requirements are met. The block is valid only if the miner succeeds in finding a *Nonce* that meets the target. The miner node broadcasts the valid block to the neighbors, which is further propagated to the BCN.

## 2.2 System Model

The system model assumed in this study is as shown in Fig. 1. It consists of three layers. They are IoT device layer, Edge layer and cloud layer. IoT device layer consists of IoT devices that communicate with the Edge servers in the Edge layer. Generally, every IoT device is associated with an edge server in its vicinity. It is also possible that an IoT device can communicate with multiple edge servers. Edge layer contains a network of edge servers that can communicate with each other in a P2P manner *via* internet. Every edge server is responsible for collecting, processing and storing of data from the group of IoT devices associated with it. Also, in the case of resource constrained IoT devices, edge servers manage allocation of resources for resource intensive tasks processed by the IoT devices. Cloud layer employs central storage entity. The edge servers scan, pre-process and compress the raw data collected from IoT devices before sending it to central storage at the cloud layer.

In Edge layer, edge servers employ blockchain to maintain the security and privacy of IoT devices and data. Any form of the activity on IoT data such as computation, access, transfer, storage etc., is considered as a transaction on the blockchain. Unlike bitcoin blockchain, edge servers employ a permissioned blockchain. This ensures that only the edge servers with permission can be a full node in the blockchain network. A full node is a node that maintains a full blockchain database with all transactions. Hence, a full node can independently and authoritatively validate and verify any transaction without relying on any other node in the network [23]. Also, permissioned blockchain has the potential to meet the high throughput requirements of IoT data transactions. As edge servers employ permissioned blockchain, they use Practical Byzantine Fault Tolerance (PBFT) algorithm for distributed consensus [24]. Block creation and deletion are validated and verified through PBFT.

# 3 LiTiChain

As mentioned earlier, LiTiChain proposed a scalable and lightweight architecture
for blockchain of finite-lifetime blocks targeted at edge-IoT applications. The details
of the LiTiChain are discussed in the following sub-sections:

## 3.1 Expiration Time Ordering Graph

In LiTiChain, *lifetime* of a block is defined as the difference between creation and
expiration times of the block. As in conventional blockchain, if the finite-lifetime
blocks are chained in the order of their arrival times, when an expired block is
deleted, it is possible that the blockchain can be disconnected. As shown in Fig. 2,
when block $b_2$ is deleted at the end of its expiration time, it results in disconnected
blockchain. Hence, in order to ensure chain connectivity even after the expired
blocks are deleted, LiTiChain proposed a graph structure based on the expiration
time of the blocks. Construction of Expiration time Ordering Graph (EOG) i.e.
inserting new blocks and deleting expired blocks from the blockchain is as shown
in Fig. 3. The number inside a block represents expiration time $e_i$. A new block
is added to the blockchain by creating a directed edge from the new block to an
existing block. The direction of the edge indicates that the hash of a block from
which the edge is emanating is a function of the hash of the block it is pointing to.
Let us assume that the block created at the $i$th time instant is denoted by $b_i$ for $i = 1$,
$2, \ldots$. Also, let $t_i$ and $e_i$ respectively denote the creation and expiration time of a
block $b_i$. According to [21], the procedure to construct EOG is as follows:

- If There Exist a Set of Blocks Whose Expiration Time is Later than the new block
  $b_i$. In this set, block $b_i$ is connected to the block with the earliest expiration time
  by a directed edge. Expiration time of the *Genesis block G* is infinity.
- If expiration time of the new block $b_i$ is later than all the existing blocks, $b_i$ is
  connected to *G* via a directed edge from $b_i$ to *G*.

The directed graph i.e. EOG constructed based on the above rules will result in
a tree topology with *Genesis block* G as the root node and the remaining blocks are
connected based on the precedence relation of expiration times. For any directed
edge, the node from which the edge is emanating from is called as a child node and
the node to which the edge is pointing to is known as a parent node. As shown in
Fig. 3, for every node, expiration time of a parent node is later than its child nodes.

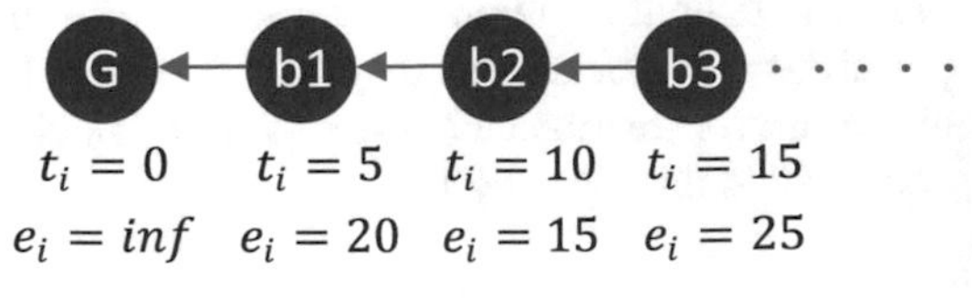

**Fig. 2** Blockchain based on
linear order of arrival times.
Genesis block (G), creation
time ($t_i$) and expiration time
($e_i$)

$t_i = 0 \quad\quad t_i = 5 \quad\ t_i = 10 \quad t_i = 15$

$e_i = inf \quad e_i = 20 \quad e_i = 15 \quad e_i = 25$

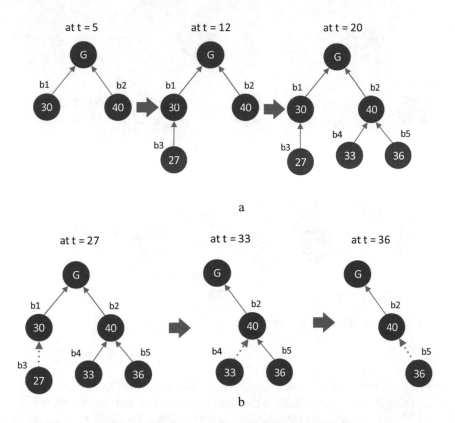

**Fig. 3** Construction of expiration time ordering graph (**a**) inserting blocks and (**b**) deleting blocks

Hence, unlike the conventional blockchain shown in Fig. 2 that can get disconnected upon the deletion of blocks, at all times, EOG remains a connected graph. But, the problem with EOG is, *blockheight* i.e. the distance measured from a block to the *Genesis block* can be short. In Fig. 3, *blockheight* of $b_5$ is 2 but in conventional blockchain the *blockheight* of $b_5$ would be 5. Additionally, as the expired blocks are deleted in EOG, unlike conventional blockchain, size of the chain may not grow, which results in shallow branches.

## 3.2 Structure of LiTiChain

As per the *longest* chain rule in the bitcoin blockchain i.e. longer the chain the harder it is for the attacker to undo the chain. In other words, longer chains are assumed to be more secure and hence are preferred [6, 21]. Therefore, to overcome the shallow EOG-based chains, another graph based on arrival time of blocks known as Arrival Ordering Graph (AOG) is coupled with the EOG to form LiTiChain, it is as shown

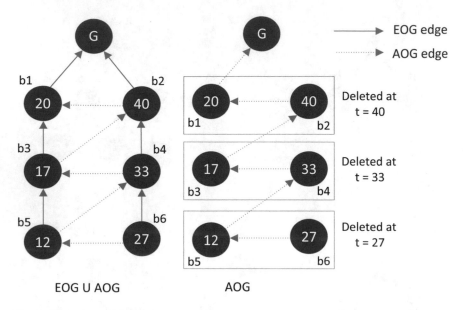

**Fig. 4** Illustration of LiTiChain structure

in Fig. 4. In conventional blockchain, a block $b_i$ is connected to its previous block $b_{i-1}$ using *previousblockhash*. On the other hand, in LiTiChain, as shown in Fig. 5, every block is connected to two blocks—a parent block $b_{i*}$ *via* EOG and previous block $b_{i-1}$ *via* AOG. Therefore, for a given block $b_i$, there will be two directed edges, one connecting to $b_{i*}$ and other connecting to $b_{i-1}$. In other words, hash $h_i$ of block $b_i$ depends on both *previous blockhash* $h_{i-1}$ and *parent blockhash* $h_{i*}$. Also, height of a block is now defined as the *maximum* distance measured from the block to the *Genesis block* along EOG and AOG. Therefore, as shown in Fig. 4, with the addition of AOG edges, *blockheight* of $b_6$ is increased from 4 to 6, which offers improved security as per the longest chain rule.

## 3.3  Retention Cost

When the lifetime of a block $b_i$ is expired, ideally the block should be deleted. But, if a block $b_{i+1}$ with the expiration time $e_{i+1} > e_i$ is connected to $b_i$, there exists an AOG edge from $b_i$ to $b_{i+1}$. Therefore, as per the rules of LiTiChain, block $b_i$ is not deleted immediately upon its expiration as it is needed to verify the validity of $b_{i+1}$. Hence, the lifetime of $b_i$ is extended to $e_{i+1}$ i.e. block $b_i$ is retained beyond its expiration time $e_i$. The *retention* or the *lifetime extension* mechanism aids in maintaining the "hash-chained" property of the blockchain. But the disadvantage of

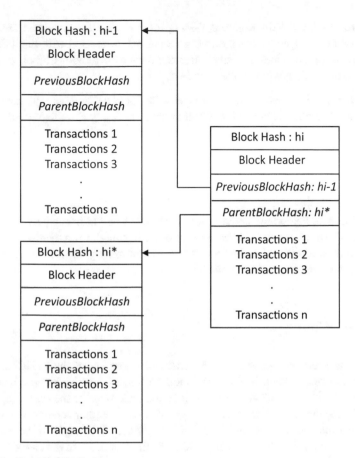

**Fig. 5** *Blockhash* in LiTiChain

retention mechanism is, it incurs additional storage cost, called as *retention cost*. For example, in Fig. 4, block $b_1$ should be ideally deleted at time instance 20. As there exists an AOG edge from $b_2$ to $b_1$ i.e. hash of $b_2$ depends on $b_1$, to validate $b_2$ i.e. verify that block $b_2$ is not corrupted by an attacker, $b_1$ needs to be stored until the expiration time of $b_2$ i.e. until time instance 40. Assuming the size of each block is 1 MB and the time count is in seconds, the retention cost of $b_1$ in Fig 4 is 20 MB-Sec. Section 4 presents two new architectures that aim at reducing the retention cost.

## 3.4   Limiting AoG Edges and Blockheight

Retention cost of the LiTiChain is directly proportional to the number of AOG edges i.e. retention cost increases with the increase in the number of AOG edges. Hence, in order to reduce the retention cost, the number of AOG edges needs to be reduced.

In this process, LiTiChain proposed *R-Height* Block Insertion scheme to reduce the number of AOG edges. Let us assume a new block $b_i$ created with two edges, one connected to parent block $b_{i*}$ and other connected to previous block $b_{i-1}$. The procedure for *R-Height* block insertion is as follows:

- If the height of the parent block $b_{i*}$ is less than or equal to $R$, as explained in Sect. 3.2, an AOG edge is created as usual from $b_i$ to $b_{i-1}$. So, this incurs retention cost.
- If the height of the parent block $b_{i*}$ is greater than $R$, AOG edge is not created from $b_i$ to $b_{i-1}$ and hence there is no *retention cost* for block $b_i$. In this case, in Fig. 5, both *PreviousBlockHash* and *ParentBlockHash* are set to the hash $h_i^*$ of $b_i^*$.

The rationale behind the above rules is, once the height of a newly added block reaches a threshold, it is considered to be sufficient enough from security perspective and hence for its child nodes, the focus shifts to minimizing the overall retention cost of the chain by removing AOG edges.

# 4  $\mu$-LiTiChain

This section presents $\mu$-*LiTiChain,* a generalized version of the LiTiChain architecture presented in Sect. 3. As mentioned earlier, as per the *longest chain rule,* from the security perspective, higher values are preferred for the *blockheight.* But, in LiTiChain, higher values of *blockheight* results in higher retention cost. Let $K$ denote the *blockheight* threshold. In LiTiChain, if the *blockheight* of the parent of a newly added block $b_i$ is less than or equal to $K$, then an AOG edge is added between block $b_i$ and $b_{i-1}$, which incurs retention cost. On the other hand, if the *blockheight* of the parent is greater than $K$, AOG edge is not added and the retention cost of block $b_{i-1}$ is zero. As shown in Fig. 6, in LiTiChain, as the value of $K$ is varied as 1, 2 and 4, overall retention cost increases as 20, 36 and 51. $\mu$-*LiTiChain* architecture provides an opportunity to reduce the retention cost in the design of blockchain of finite-lifetime blocks targeted at Edge-IoT applications

If an attacker intends to corrupt as many transactions as possible in a blockchain, he has to undo the total blockchain i.e. compute new hash for all the blocks. In this process, majority of the attacker's time and resources are spent in *Nonce computation* (see Sect. 2.1.2) i.e. finding a new *Nonce* for each block in the chain. Therefore, *if the number of AOG edges reduces while the height of the blocks increases or remains same as in LiTiChain, overall retention cost of the chain can be reduced without lowering attacker's difficulty in undoing the chain i.e. security of the chain is maintained at the same level.* The proposed $\mu$-*LiTiChain* architecture aims at this.

As discussed in Sect. 2.1, in *public blockchain* systems like bitcoin, for every mined block, miners are rewarded with bitcoins. To reduce the number of bitcoins spent as a reward, BCN aims at reducing the number of mined blocks and hence the

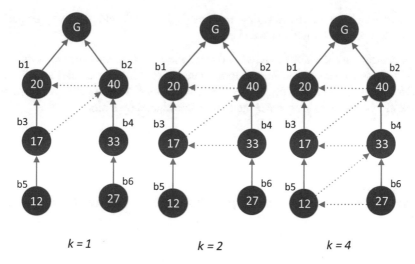

**Fig. 6** Increase in number of AOG edges in LiTiChain with $K$

*blocksize* used in bitcoin is generally higher. Unlike bitcoin, as discussed in Sect. 2.2, Edge-IoT employs *permissioned blockchain*. Hence, miners are not rewarded for solving PoW. Therefore, if edge-servers have enough *Nonce computation* resources; it is ok to reduce the *blocksize* to increase the number of mined blocks, which increases the height of blocks in a blockchain. Hence, *when LiTiChain branches are shallower, blocksize can be reduced to increase the number of mined blocks added to the chain to increase the depth of LiTiChain branches.* As depth increases i.e. as the *blockheight* of the newly added blocks is greater than the threshold $K$, *blocksize* can be increased i.e. the number of mined blocks can be reduced to reduce the number of *Nonce computations*. The proposed schemes exploit this principle to reduce the number of AOG edges and minimize the *retention cost*. For the same number of transactions to be processed, the reduction in *blocksize* increases the number of mined blocks and the number of *Nonce computations*, which leads to higher computational cost. So, there exists a trade-off between *retention cost* i.e. additional storage cost, *security* i.e. height of the blocks in the chain and *Nonce computational* cost. *The proposed $\mu$-LiTiChain architecture provides with an opportunity to explore this trade-off in the design of blockchain.*

As mentioned earlier, if the edge-servers are equipped with enough *Nonce computation* resources; when the *blocksize* is reduced, the number of mined blocks increases, which leads to the increase in the height of the blocks in a blockchain. For a given block, there are two ways in which a block can be replaced by multiple mined blocks, they are:

- When *blocksize* can be varied, a regular block in a blockchain can be split into multiple sub-blocks of same size.

- When *blocksize* cannot be varied, multiple light-weight blocks can be connected to a regular block to increase its height.

Depending on the above steps, two new architectures that are variants of $\mu$-LiTiChain are presented in this chapter. They are: *p-LiTiChain* and *s-LiTiChain*. Let us define block expansion factor ($\mu$) as the number of blocks replacing a regular block in a blockchain to increase the height introduced by a block.

$$\mu = \begin{cases} p \ for \ p - LiTiChain \\ s \ for \ s - LiTiChain \end{cases}$$

In LiTiChain, $\mu = 1$ for all the blocks. If $\mu = p \ or \ s$ for at least a block in the chain, it is respectively called as *p-LiTiChain* or *s-LiTiChain*. The details are as follows.

## 4.1   *s-LiTiChain*

As discussed above, reducing *blocksize* is equal to splitting regular block of a LiTiChain into multiple sub-blocks, together known as *s-block*. Let $s$ be the number of sub-blocks. As shown in Fig. 7b, replacing a regular block with an *s-block* increases the number of EOG edges and the height introduced by a block from 1 to $s$, which has the potential to reduce the number of AOG edges and overall retention cost of the chain for a given $K$. Fig. 7c shows a diamond shaped box representing an *s-block*. The parameters $e_i$ and $\mu_i$ inside the *s-block* respectively represent expiration

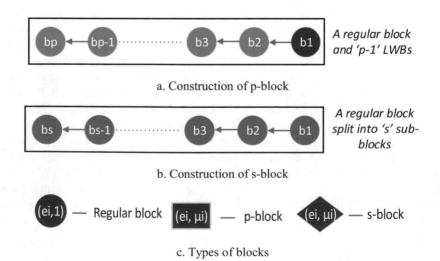

a. Construction of p-block

b. Construction of s-block

c. Types of blocks

**Fig. 7** Construction of *p-block* and *s-block*

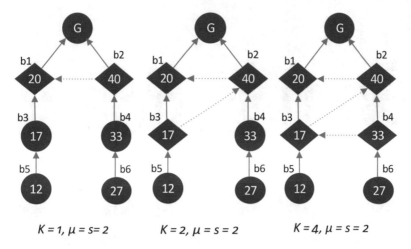

**Fig. 8** Construction of s-LiTiChain

time of the block and block expansion factor used in the construction of this block. If multiple block expansion factors are used in the design of blockchain, it is possible that each *s-block* may employ different block expansion factor. Hence, the subscript $i$ for the block expansion factor $\mu$ inside the *s-block* shown in Fig. 7c. On the other hand, if only two values of block expansion factor are used in the blockchain design i.e. $\mu \in \{\mu_0 = 1, \ \mu_1 > 1\}$, as in Fig. 8, $\mu_i$ is not shown inside the *s-block*. Also, as shown in Fig. 8, $\mu = s = 2$ indicates that $\mu$ can take two values i.e. $\mu \in \{1, \ 2\}$. In other words, $\mu = 1$ is implicit. In Fig. 8, $\mu = s = 2$ for *s-block* and $\mu = s = 1$ for regular block. In Fig. 8, for $\mu = s = 2$, when $K$ is varied as 1, 2 and 4, overall retention cost of the chain increases as 20, 20 and 36 i.e. compared to the LiTiChain discussed in Fig. 6, retention cost is reduced or unchanged for the same $K$. A $\mu$-*LiTiChain* that employs at least one *s-block* is known as *s-LiTiChain*. In this work, to simplify the process of building *s-LiTiChain*, it is assumed that $\mu = s$ takes only two values ($s = 1$ and another fixed value $s > 1$). In other words, *s-LiTiChain* can only have two types of blocks i.e. a regular block ($s = 1$) as in LiTiChain and an *s-block* ($s > 1$).

## 4.2 p-LiTiChain

Let us assume that for some reason, it is not a best option, or it is inconvenient to split a block into multiple sub-blocks. In this case, *s-LiTiChain* cannot be implemented and hence as an alternative, *p-LiTiChain* architecture is presented. In bitcoin, the number of transactions in a block are in the range of 1000–2000 and *blocksize* is around 1 MB. Let's assume a *Lightweight Block* (LWB) with few empty transactions (e.g.: 10) and size of around 10 KB i.e. approximately 100 times lighter than

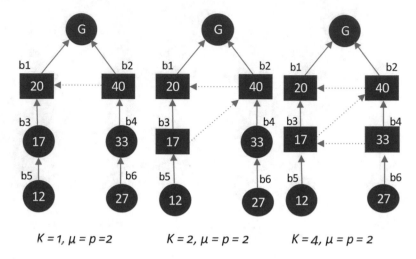

**Fig. 9** Construction of p-LiTiChain

a regular block. As shown in Fig. 7a, a *p-block* contains a regular block as in LiTiChain and *p*-1 LWBs. Hence, if a regular block in a LiTiChain is replaced with a *p-block*, the number of EOG edges and height contribution of a block increases from 1 to *p*, which reduces the number of AOG edges and the overall retention cost of the chain. Figure 7c shows a rectangular box representing a *p-block*. The parameters $e_i$ and $\mu_i$ inside the *p-block* respectively represent expiration time of the block and block expansion factor used in the construction of the *p-block*. As discussed in Sect. 4.1, like *s-LiTiChain*, if multiple block expansion factors are used in the design of blockchain, it is possible that each *p-block* may employ different block expansion factor. Hence, the subscript $i$ for the block expansion factor $\mu$ inside the *p-block* shown in Fig. 7c. On the other hand, if only two values of block expansion factor are used in the blockchain design i.e. $\mu \in \{\mu_0 = 1, \ \mu_1 > 1\}$, as in Fig. 9, $\mu_i$ is not shown inside the *s-block*. Also, as shown in Fig. 9, $\mu = 2$ indicates that $\mu$ can take two values i.e. $\mu \in \{1, \ 2\}$. In other words, $\mu = 1$ is implicit. In Fig. 9, $\mu = p = 2$ for *p-block* and $\mu = p = 1$ for regular block. In Fig. 9, for $\mu = p = 2$, when $K$ is varied as 1, 2 and 4, retention cost varies as 20, 20 and 36 i.e. compared to the LiTiChain in Fig. 6, retention cost is reduced or unchanged for a given $K$. A $\mu$-*LiTiChain* that contains at least one *p-block* is known as *p-LiTiChain*. In this work, to reduce the complexity of *p-LiTiChain*, it is assumed that $p$ takes only two values ($p = 1$ and another fixed value $p > 1$). In other words, a *p-LiTiChain* can only have two types of blocks i.e. a regular block ($p = 1$) as in LiTiChain and a *p-block* (with fixed $p > 1$).

The process of constructing *s-LiTiChain* and *p-LiTiChain* i.e. inserting and deleting blocks is as follows:

- When the *Blockheight* of the Parent of a Newly Added Block is Less than or Equal to $K$, the New Block Added to the Chain is an *s/p-block* and an AOG edge is connected between block $b_i$ and $b_{i-1}$.

**Table 1** Pros and Cons of *s-LiTiChain* and *p-LiTiChain*

| s-LiTiChain | p-LiTiChain |
| --- | --- |
| It requires the blockchain design parameter *blocksize* to be varied | It is not required to vary the blockchain design parameter *blocksize* |
| There is no storage overhead due to block splitting | There is storage overhead due to that addition of light-weight blocks. The amount of overhead increases with block expansion factor |
| When block expansion factor is greater than blockheight threshold, the retention cost of s-LiTiChain is zero. Hence, the storage overhead is zero | When block expansion factor is greater than blockheight threshold, the retention cost of p-LiTiChain is zero. But the overhead due to LWBs is non-zero and for higher values of block expansion factor, it is significant. Hence, the storage overhead of p-LiTiChain can be non-zero |

- When the *blockheight* of the parent is greater than $K$, a *regular block is inserted without an AOG edge.*

For a given number of AOG edges, both *s-LiTiChain* and *p-LiTiChain* have the potential to increase the average *blockheight* of a LiTiChain, respectively by at least $s$ and $p$. In other words, for the same number of AOG edges i.e. for same *retention cost*, both *s-LiTiChain* and *p-LiTiChain* offer better security compared to LiTiChain i.e. the effort required by an attacker to undo the chain increases compared to LiTiChain. In *p-LiTiChain*, even though the size of an LWB is assumed to be negligible i.e. 100 times smaller than a regular block, for higher values of $p$, the overhead due to LWBs is considerable i.e. for $\mu = p = 10$, LWB overhead is around 100 KB. Hence, for higher values of $\mu$, *s-LiTiChain* can be preferred over *p-LiTiChain*. On the other hand, as shown in Fig. 9, *p-LiTiChain* simplifies the blockchain design as it offers the same level of security as *s-LiTiChain* without varying the *blocksize*. Pros and cons of *s-LiTiChain* and *p-LiTiChain* are summarized in Table 1.

## 4.3  Algorithms

Given *blockheight* threshold $K$ and valid values of block expansion factor $\mu$, the steps for adding a new block to the $\mu$-*LiTiChain* is as shown in Algorithm 1. The procedure to delete a block in $\mu$-*LiTiChain* is as shown in Algorithm 2. The steps given in Algorithm 1 and Algorithm 2 applies to both *p-LiTiChain* and *s-LiTiChain*. In the case of *p-LiTiChain* and *s-LiTiChain*, respectively set $\mu = p$ and $\mu = s$. The format of a block in a $\mu$-*LiTiChain* is assumed to be similar as in conventional blockchain with few additional fields in the *blockheader*. They are *ParentBlockHash* and *ExpiryTime*. Please refer to Sect. 3 for the details of these fields. In addition to these parameters, a new parameter i.e. *DeletionTime* is stored outside the blockchain for every block.

Retention mechanism discussed in Sect. 3.3 is illustrated in line 32 of Algorithm 1 i.e. the value of the *Deletiontime* of previous block $b_{i-1}$ is set to the *Expirytime* of new block $e_i$. This ensures that the block $b_{i-1}$ is not deleted until the validity of the block $b_i$ is expired. Also, *DeletionTime* of the newly added block $b_i$ is set to $e_i$. But, at a later time instance, when another new block $b_{i+1}$ is added, it is possible that the *DeletionTime* of the block $b_i$ can be set to $e_{i+1}$. So, unlike *ExpirtyTime* which is fixed, *DeletionTime* is a varying field. Hence, *DeletionTime* is not a part of the *blockheader* as a block cannot be modified after adding it to the blockchain.

*Algorithm 1  Insertion of a block*

1: Output: Updated Blockchain with newly created block $b_i$
with endtime $e_i$ and block expansion factor $\mu$
2: Input: Existing Blockchain, height constraint $K$, allowed
range of values of block expansion factor $\mu$, time
instance of update $D$.
3: /* Determining the parent block according to EOG */
4: $l^* \leftarrow \arg\min_l \{l|e_i \le e_l\}$, $e_l$ is the endtime of the existing
blocks in the blockchain.
5: $b_{i*} \leftarrow b_{l*}$
6: Let $d$ denote the height of the parent block $b_i^*$
7: /* Determining the previous block according to AOG */
8: $m^* \leftarrow \arg\max_m \{m|e_m > D\}$, $e_m$ is the endtime of the
blocks in the blockchain
9: /* Constraining the height of the new block */
10: **if** $(d == 0)$ **then** /* Parent is Genesis Block */
11: $\mu \leftarrow \mu_{max}$
12: $b_{i-1} \leftarrow b_{i*}$
13: **else** /* Parent is a non-Genesis Block */
14: **if** $(d \le K)$ **then** /* Height of parent <= K */
15: $b_{i-1} \leftarrow b_{m*}$
16: **if** $((d + \mu_{max}) > K)$ **then**
17: $j^* \leftarrow \arg\min_j \left(j \mid (d + \mu_j) > K\right)$
18: $\mu \leftarrow \mu_{j*}$
19: **else**
20: $\mu \leftarrow \mu_{max}$
21: **endif**
22: **else** /* Height of parent > K */
23: $\mu \leftarrow 1$
24: $b_{i-1} \leftarrow b_{i*}$
25: **endif**
26: **endif**
27: Create block $b_i$. The header of $b_i$ is updated as below:

28: Assign current time to the creation timestamp $t_i$
29: Assign the value $Hash(b_i^*)$ to the field *ParentBlockHash*
30: Assign the value $Hash(b_i^-)$ to the field *PreviousBlockHash*
31: Assign the value $e_i$ to the field *ExpirtyTime*
32: Set *DeletionTime* for blocks $b_i$ and $b_i^-$ to $e_i$

*Algorithm 2  Deletion of blocks*

1: Output: Updated blockchain after the removal of expired blocks.
2: Input: Time instance of update D, Number of blocks N in the blockchain, $e_i$ - end time of block $b_i$
2: **for** $i = 1 : N$
3: **if** $(e_i = = D)$
4: delete block $b_i$
5: **endif**
6: **endfor**

# 5  Performance Evaluation

In this section, performance of *LiTiChain* and *μ-LiTiChain* are evaluated using simulations. *Total Retention cost*, *average* and *maximum blockheight, total number of Nonce computations* are used as the performance metrics.

## 5.1  Performance Metrics

As mentioned earlier, $e_i$ is the expiration time of a block. Let $d_i$ be the deletion time of a block and $N$ denote the total number of blocks. Total Retention Cost $(\delta_K)$ of all the blocks for a given value of *blockheight* threshold $K$ is given as follows:

$$\delta_K = \sum_{i=0}^{N} (d_i - e_i)$$

The *average blockheight H* and the *maximum blockheight M* are computed at the time of creating a new block and deleting an expired block. $H$ is obtained by taking average of *blockheight* over all the blocks that are alive in the chain. Similarly, $M$ is the maximum of the *blockheight* of all the blocks that are not expired in the chain. Let $\overline{H}$ and $\overline{M}$ denote the time averages of $H$ and $M$ during the lifetime of a blockchain. $\overline{H}$ *and* $\overline{M}$ are used as the performance metrics. The *number of Nonce computations* $(\epsilon)$ of p-LiTiChain and s-LiTiChain are generally higher compared to LiTiChain. Hence, *number of Nonce computations* $(\epsilon_K)$ for a given $K$ is used as a performance metric.

## 5.2 Simulation Setup

In [21], realistic IoT data published by the New York City Taxi and Limousine Commission (TLC) with trip record data for yellow taxis is analyzed. The trip duration is considered as the lifetime of the transactions. Based on the simulations, it was concluded that, if the lifetime of blocks in a blockchain has bimodal distributions with small and large lifetime values, it would result in worst retention cost. The reason is that, compared to unimodal distributions, in bimodal distributions the short lifetime blocks suffer relatively more due to the time held back by the long lifetime blocks. For the purpose of simulations, in order to generate the lifetime data with a bimodal distribution, similar to [21], lifetime data is sampled from $Z$ which is a mixture of the following two Gaussian distributions:

$$Z = \begin{cases} Z_1 & w.p. \ 0.5 \\ Z_2 & w.p. \ 0.5 \end{cases}$$

where $Z_1 \sim N(300, 110^2)$ and $Z_2 \sim N(1200, 110^2)$. In order to study the variation in performance metrics w.r.t mixing proportion, similar to [21], the following two distributions are also considered.

$$\hat{Z} = \begin{cases} Z_1 & w.p. \ 0.1 \\ Z_2 & w.p. \ 0.9 \end{cases}$$

$$\tilde{Z} = \begin{cases} Z_1 & w.p. \ 0.9 \\ Z_2 & w.p. \ 0.1 \end{cases}$$

Around 10000 lifetime data are sampled from the above distribution. The value of $\mu$ is varied between 1 and 500 and the values of $K$ considered are 10, 50, 100, 300, 400 and 500. For the ease of explanation, let's call $Z$, $\hat{Z}$ and $\tilde{Z}$ as *case1*, *case2* and *case3*.

## 5.3 Simulation Results

MATLAB is used to evaluate the performance of *LiTiChain* and *$\mu$-LiTiChain*. Simulation results are averaged over 10 iterations.

### 5.3.1 LiTiChain

As discussed above, in *LiTiChain*, as $K$ increases, the number of AOG edges increase and hence $\delta_K$ increases. Fig. 10 shows the variation of relative $\delta_K$ w.r.t $\delta_{10}$ for different values of $K$. As expected, it is obvious from the simulations that $\delta_K$ increases with $K$ in LiTiChain. The same behavior is observed for different mixing proportions i.e. $\delta_K$ increases with $K$ for $Z$, $\hat{Z}$ and $\tilde{Z}$. But, the amount of increase in $\delta_K$ varies with mixing proportions. Compared to $Z$, in the case of $\hat{Z}$, the decrease in relative retention cost is around 0.2 for all values of $K$. On the other hand, in the case of $\tilde{Z}$, compared to $Z$, the decrease in relative retention cost is in the range of 0.2–0.3 for $K \leq 100$ and for $K > 100$, it is in the range of 0.5–0.7. The variation in the average absolute values of retention cost w.r.t $K$ for $Z$, $\hat{Z}$ and $\tilde{Z}$ is as shown in Fig. 11. The following conclusions can be drawn from Fig. 11:

- In *case1*, the proportion of $Z_1$ and $Z_2$ is split exactly into half. Hence, on average the number of blocks connecting to *Genesis block* via EOG edge are higher in $Z$ when compared to $\hat{Z}$ and $\tilde{Z}$. Therefore, as per the rules given in Sect. 3, the number of AOG edges in $Z$ are higher when compared to $\hat{Z}$ and $\tilde{Z}$, resulting in higher retention costs for $Z$.
- In *case2* and *case3*, *retention cost* contribution from the majority population i.e. $Z_2$ in *case2* and $Z_1$ in *case3* almost remains same but the distinguishing factor is the *retention cost* from the minority population i.e. $Z_1$ in *case2* and $Z_2$ in *case3*.
- In *case2*, the proportion of $Z_2$ is higher than $Z_1$. The mean of $Z_1$ is 300 which is less than the mean of $Z_2$ i.e. 1200 and the difference in the mean is 900. Based on the rules given in Sect. 3 for the construction of LiTiChain, the probability of a block with a lifetime value selected from $Z_1$ (*minority population*) forming an AOG edge with a block whose lifetime value is selected from $Z_2$ (*majority population*) is higher than the probability of forming an AOG edge with a block whose lifetime value is selected from $Z_1$. This is as shown in Fig 12a. AOG edge between blocks $b_i$ and $b_{i+1}$ doesn't incur *retention cost*. Hence, the *retention cost* contribution from most of the blocks with the lifetime values selected randomly from $Z_1$ is close to zero. Therefore, compared to *case3*, *retention cost* incurred by *case2* is lower.
- On the other hand, in *case3*, the proportion of $Z_1$ is higher than $Z_2$. Based on the rules given in Sect. 3 for the construction of LiTiChain, the probability of a block with a lifetime value selected from $Z_2$ (*minority population*) forming an AOG edge with a block whose lifetime value is selected from $Z_1$ (*majority population*) is higher than the probability of forming an AOG edge with a block whose lifetime value is selected from $Z_2$. This is as shown in Fig 12b. AOG edge between blocks $b_i$ and $b_{i+1}$ incurs non-zero *retention cost*. Hence, the *retention cost* incurred by most of the blocks with the lifetime values selected randomly from $Z_2$ is non-zero and it is around 900 MB-sec (i.e. 1200–300) for each block. Therefore, compared to *case2*, *retention cost* incurred by *case3* is

**Fig. 10** Increase in relative retention cost w.r.t *blockheight* threshold $K$ in LiTiChain: (**a**) case 1 — Z i.e. Z1 w.p. 0.5, Z2 w.p. 0.5, (**b**) case 2 — $\hat{Z}$ i.e. Z1 w.p. 0.1, Z2 w.p. 0.9, and (**c**) case 3 — $\check{Z}$ i.e. Z1 w.p. 0.9, Z2 w.p. 0.1

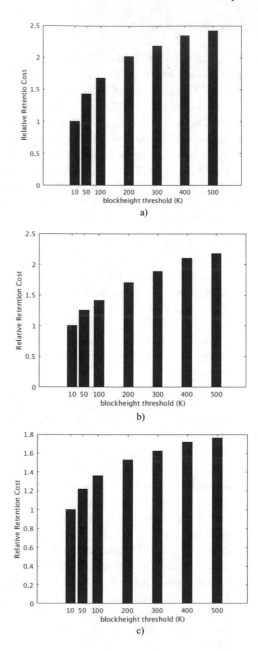

a)

b)

c)

higher. In *case1*, the proportion of $Z_1$ and $Z_2$ is split exactly into half. Hence, the frequency of occurrence of the scenario shown in Fig 12b is higher in *case1* when compared to *case3*. So, the *retention cost* in *case3* is lower when compared to *case1*.

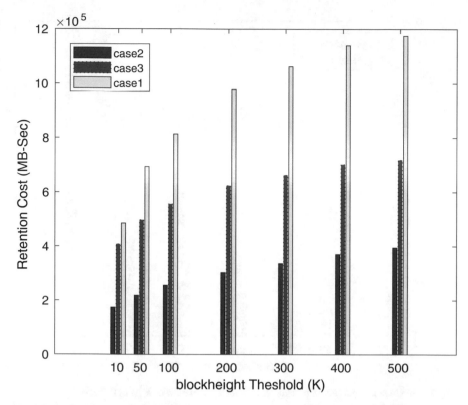

**Fig. 11** Variation in average absolute retention cost w.r.t *blockheight* threshold $K$ in LiTiChain

a)                                              b)

**Fig. 12** Scenarios resulting in retention cost difference between case 2 and case 3: (**a**) case 2, doesn't incur retention cost and (**b**) case 3, incurs retention cost

Figure 13 shows the variation in $\overline{H}$ and $\overline{M}$ for different values of $K$ in LiTiChain. As expected, in LiTiChain, both $\overline{H}$ and $\overline{M}$ increase as the value of $K$ is increased. But, as shown in Fig. 10, $\delta_K$ i.e. the additional storage cost incurred by the LiTiChain increases with increase in $K$. $\mu$-LiTiChain architecture aims at increasing $\overline{H}$ and $\overline{M}$ while reducing $\delta_K$.

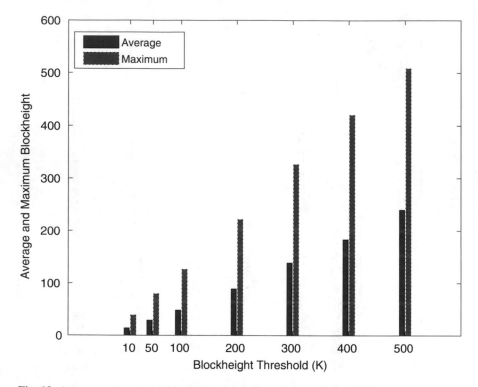

**Fig. 13** Average and maximum blockheight for different values of $K$ in LiTiChain

### 5.3.2  $\mu$-LiTiChain

*case1–Z*: To study the performance of $\mu$-*LiTiChain* i.e. *p-LiTiChain* and *s-LiTiChain*, two values of $K$, a lower value i.e. $K = 50$ and a higher value i.e. $K = 500$ are considered. Fig. 14 shows the variation in $\overline{H}$ and $\overline{M}$ w.r.t $\mu$ for $K = 50$ and *500*. As the value of $\mu$ increases, $\overline{H}$ and $\overline{M}$ increase for both the values of $K$. For $K = 50$, $\mu \geq 10$ results in an *average blockheight* of $\overline{H} > 50$. Hence, $\mu \geq 10$ offers the required security as per the *longest chain rule* in the average sense. Figure 15 shows the variation in relative $\delta_K$ w.r.t increase in $\mu$. As expected, *retention cost* $\delta_{50}$ decreases with increase in $\mu$ and it is zero when $\mu > K$ i.e. *when the value of block expansion factor is greater than the blockheight threshold, it is not required to extend the lifetime of the expired blocks and hence the additional storage costs are zero.* Hence, in Fig. 15, for $K = 50$, $\mu > = 60$ results in *zero retention cost*.

The lower *retention cost* and higher *average blockheight* $\overline{H}$ are achieved at the expense of increase in the total number of *Nonce Computations* ($\epsilon$). In other words, *the reduction in storage cost and improved security are obtained at the expense of*

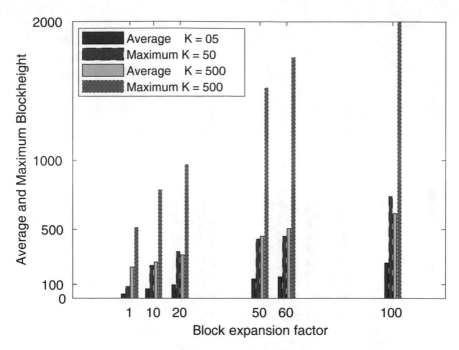

**Fig. 14** Average and maximum blockheight w.r.t block expansion factor $\mu$ for $K = 50$ and $500$

*higher computational costs.* Figure 16 presents the relative increase in $\epsilon$ w.r.t $\mu$ for $K = 50$ and $500$. As mentioned above, for $K = 50$, according to Fig. 14, $\mu \geq 10$ has the potential to offer the required security for the chain and as per Fig. 15, $\mu = 60$ results in *zero retention cost*. Also, as shown in Fig. 16, $\epsilon_{50}$ i.e. the *total number of Nonce computations* for $\mu = 60$ are around 6.14 times compared to $\mu = 1$. Hence, for $K = 50$, $\mu = 60$ offers the best performance in terms of *storage cost*, *security* and *computational cost*. On the other hand, as shown in Fig. 14, for $K = 500$, $\overline{H}$ is greater than $K$ for $\mu = 60$. But, according to Fig. 15, $\delta_{500}$ decreases with $\mu$ but it does not reach zero as the value $\mu$ is not high enough. Similar to $K = 50$, there exists a value of $\mu$ for every $K$ that offers optimal performance in terms of *storage cost*, *security* and *computational cost*.

The difference in the *retention cost* savings offered by *p-LiTiChain* and *s-LiTiChain* is as shown in Fig. 15. For a given $K$, the *retention cost* is same for both *p-LiTiChain* and *s-LiTiChain* when $\mu = 1$ for all the blocks. But, for $\mu \geq 1$ and $\mu \leq K$, as it employs LWBs, *p-LiTiChain* always results in higher *retention cost* compared to *s-LiTiChain*. For $\mu \geq K$, *retention cost* is zero for both *p-LiTiChain* and *s-LiTiChain*. Also, unlike *s-LiTiChain*, where the $\delta_K$ always decreases with increase in $\mu$, after a certain threshold value of $\mu$, the *retention cost* of *p-LiTiChain* tends to increase rather than decreasing. The reason for this is that, as explained in Sect. 4, the LWB overhead becomes significant for higher values of $\mu$. Hence, as shown in

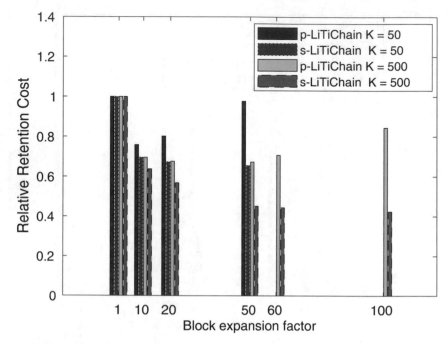

**Fig. 15** Variation in relative retention cost w.r.t block expansion factor $\mu$ for $K = 50$ and 500 for *p-LiTiChain* and *s-LiTiChain*

Fig. 15, for $K = 500$ and $\mu = 100$, the retention cost of *p-LiTiChain* approximately doubles when compared to *s- LiTiChain*. As mentioned earlier in Sect. 4, the size of LWBs is assumed to be 100 times lighter than regular blocks.

Figure 16 highlights another important point i.e. for $K = 50$, the *total number of Nonce Computations* decreases when $\mu$ increases from 50 to 60 and then again increases for $\mu = 100$. The reason is illustrated by Fig. 17. When $\mu = 50$, blockheight of $b_1$ is not greater than the threshold $K = 50$. Hence, a $\mu$-block is used for $b_2$. Therefore, the *total number of Nonce computations* $\epsilon = 101$. On the other hand, when $\mu = 60$ and 100, blockheight of $b_1$ is greater than the threshold $K = 50$. Hence, a regular block is used for $b_2$ and the *total number of Nonce computations* are 62 and 102, respectively for $\mu = 60$ and $\mu = 100$.*case2-$\hat{Z}$ and case3-$\tilde{Z}$*: The variation in average and maximum blockheight w.r.t block expansion factor for $K = 50$ and $K = 500$ is as shown in Fig. 18. The following conclusions can be drawn from Fig. 18:

- In *case2*, the proportion of $Z_2$ is higher than $Z_1$. Also, the mean lifetime of $Z_2$ i.e. 1200 is higher than $Z_1$ which is 300. Hence, in steady state, at any given time instance, the number of blocks alive in the blockchain are more in *case2* when compared to *case1*. Therefore, the *average* and *maximum blockheight* shown in Fig. 18a for *case2* are higher than the respectively values show in Fig. 13 for *case1*.

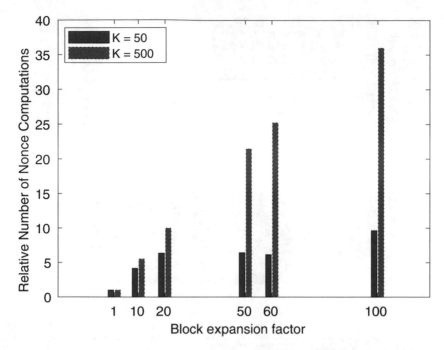

**Fig. 16** Increase in relative number of nonce computations w.r.t block expansion factor $\mu$ for $K =$ 50 and $K = 500$

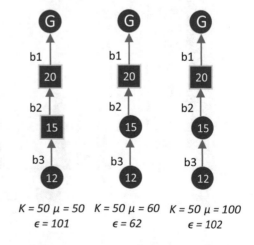

**Fig. 17** Illustration of difference in number of *nonce computations* w.r.t $\mu$

- On the other hand, in *case3*, the proportion of $Z_1$ with a mean lifetime value of 300 is higher than $Z_2$ whose mean lifetime value of 1200. Hence, in steady state, at any given time instance, the number of blocks alive in the blockchain are less in *case3* when compared to *case2* and *case1*. Therefore, the *average* and *maximum*

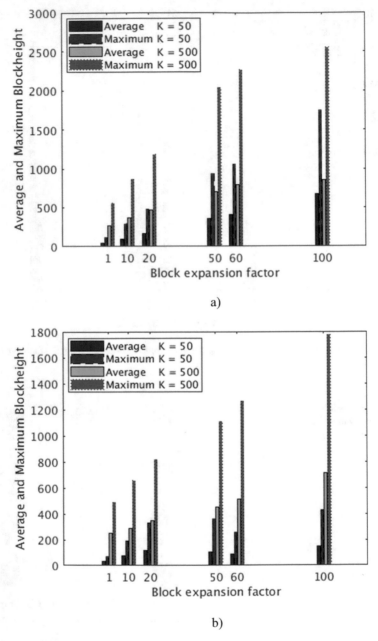

a)

b)

**Fig. 18** Average and maximum blockheight w.r.t block expansion factor $\mu$ for $K = 50$ and 500: (**a**) case 2 and (**b**) case 3

*blockheight* shown in Fig. 18b for *case3* are lower than the respectively values show in Figs. 13 and 18a for *case1* and *case2*.

The difference in the absolute *retention cost* between *case2* and *case3* for different values of *blockheight* threshold and block expansion factor is as shown in Fig. 19 for both *p-LiTiChain* and *s-LiTiChain*. The conclusions drawn from Fig. 19 are similar to Fig. 11, they are reiterated below, and these conclusions hold for all combinations of $\mu$ and $K$ shown in Fig. 19:

- In both *case2* and *case3*, the *retention cost* contribution from the majority population i.e. $Z_2$ in *case2* and $Z_1$ in *case3* is almost same. But the major differentiating factor is the *retention cost* contribution from minority population i.e. $Z_1$ in *case2* and $Z_2$ in *case3*.
- In *case2*, based on the rules given in Sect. 3 for the construction of LiTiChain, the probability of a block with a lifetime value selected from $Z_1$ (*minority population*) forming an AOG edge with a block whose lifetime value is selected from $Z_2$ (*majority population*) is higher than the probability of forming an AOG edge with a block whose lifetime value is selected $Z_1$. This is as shown in Fig 12a. Hence, the *retention cost* contribution from most of the blocks with the lifetime values selected randomly from $Z_1$ is close to zero. Therefore, compared to *case3*, retention cost incurred by *case2* is lower.

- On the other hand, in *case3*, based on the rules given in Sect. 3 for the construction of LiTiChain, the probability of a block with a lifetime value selected from $Z_2$ (*minority population*) forming an AOG edge with a block whose lifetime value is selected from $Z_1$ (*majority population*) is higher than the probability of forming an AOG edge with a block whose lifetime value is selected from $Z_2$. This is as shown in Fig. 12b. Hence, the *retention cost* incurred by most of the blocks with the lifetime values selected randomly from $Z_2$ is non-zero and it is around 900 MB-sec (i.e. 1200–300) for each block. Therefore, compared to *case2*, retention cost incurred by *case3* is higher.

The difference in the *total number of Nonce Computations* between *case2* and *case3* for different values of *blockheight* threshold and block expansion factor is as shown in Fig. 20. The following conclusions can be drawn from Fig. 20:

- In *case2*, it was observed from the simulations that, for $\mu = 50$, the number of blocks connected to *Genesis block* via EOG edge are 1190. All these blocks are $\mu$-blocks i.e. *s/p*-blocks. The number of $\mu$-blocks connected to *non-Genesis block* are 20. For the same distribution of lifetime data i.e. $\hat{Z}$, the change in $\mu$ does not vary the number of $\mu$-blocks connected to *Genesis block* but the number of $\mu$-blocks connected to *non-Genesis block* varies. So, when $\mu$ is increased from 50 to 60, the number of $\mu$-blocks connected to *Genesis block* remained at 1786

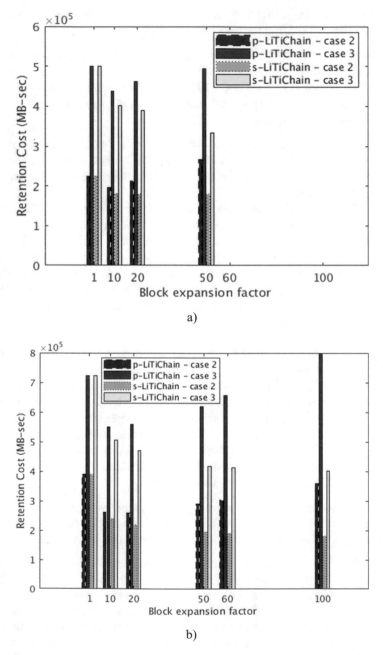

**Fig. 19** Comparison of retention cost between *case2* and *case3* for different values of blockheight threshold ($K$) and block expansion factor ($\mu$): (**a**) $K = 50$ and (**b**) $K = 500$

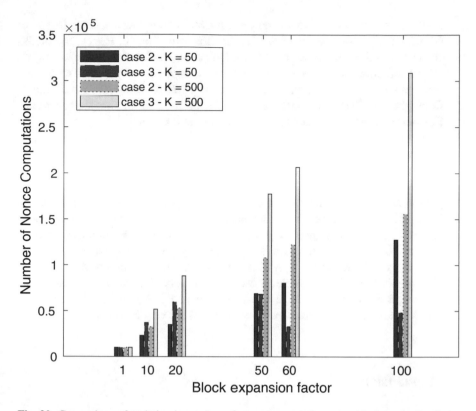

**Fig. 20** Comparison of variation in number of nonce computations w.r.t block expansion factor for different values of *blockheight* threshold $K$

but the number of $\mu$-blocks connected to *non-Genesis block* are reduced to zero. Hence, for $\mu = 50$, the *number of Nonce Computations* is $((1190 + 20)*50)$ whereas for $\mu = 60$, *number of Nonce Computations* are $(1190 * 60)$. Hence, when $\mu$ is varied from 1 to 60, the *number of Nonce Computations* increases linearly with $\mu$.

- In *case3*, it was observed from the simulations that, for $\mu = 50$, the number of blocks connected to *Genesis block* via EOG edge are 387, majority of them belongs to $Z_2$. All these blocks are $\mu$-blocks i.e. *s/p*-blocks. The number of $\mu$-blocks connected to *non-Genesis block* are 797. For the same distribution of lifetime data i.e. $\tilde{Z}$, the change in $\mu$ does not vary the number of $\mu$-blocks connected to *Genesis block* but the number of $\mu$-blocks connected to *non-Genesis block* varies. So, when $\mu$ is increased from 50 to 60, the number of $\mu$-blocks connected to the *Genesis block* remained at 387 but the number of $\mu$-blocks connected to *non-Genesis block* are reduced to zero. The reason being, all the $\mu$-blocks connected to the *Genesis block* introduced a *blockheight* of 60, which is greater than the *blockheight* threshold $K = 50$. Hence, the blocks connected to these $\mu$-blocks via EOG edge are regular blocks. Therefore, when $\mu$ is varied from 1 to 50, the *number of Nonce Computations* increases with $\mu$ but when $\mu$ is varied from 50 to 60, the *number of Nonce Computations* decreases from $((387 + 797)_* 50)$ to $(580_* 60)$. When $\mu > 60$, the *number of Nonce Computations* again increases as $(387_* \mu)$ increases with $\mu$.

- From above two points, it can be concluded that the variation in the *number of Nonce Computations* w.r.t $\mu$ depends significantly on the mixing proportion of $Z_1$ and $Z_2$. Comparing *case2* and *case3*, the number of $\mu$-blocks connected to *Genesis block* are very high in *case2*. The reason for this is: In *case2*, the proportion of $Z_2$ is 90% which is significantly higher than $Z_1$. Hence, as shown in Fig. 21, in *case2*, block $b_3$ is connected to *Genesis block* whereas in *case3*, as the proportion of $Z_2$ significantly lower than $Z_1$, block $b_3$ is connected to *non-Genesis block*. In *case2*, the probability of the average lifetime of a block (e.g. $b_3$) being close to 1200 is very high whereas in *case3*, the probability of the average lifetime of a block (e.g. $b_3$) being close to 300 is very high. So, the probability of $b_3$ being a $\mu$-block is very high in *case2* whereas it is less in *case3*. Hence, the number of $\mu$-blocks connected to *Genesis block* are very high in *case2*.

# 6  Conclusion

This chapter started by reviewing the LiTiChain architecture presented in the literature for Edge-IoT applications. LiTiChain architecture has the potential to reduce IoT data storage costs at Edge servers but it results in higher than ideally expected storage costs. In order to reduce the *retention cost* incurred by the LiTiChain, novel $\mu$-*LiTiChain*—a generalized version of the LiTiChain architecture is presented. Two variants of $\mu$-*LiTiChain* i.e. *p-LiTiChain* and *s-LiTiChain* architectures are also presented. Pros and cons of *p-LiTiChain* and *s-LiTiChain* are also discussed. With extensive simulations, it was shown that the $\mu$-*LiTiChain* i.e. *p-LiTiChain* and *s-LiTiChain* architectures have the potential to reduce storage cost and offer better security when compared to LiTiChain at the expense of computational costs. In conclusion, unlike the LiTiChain architecture, $\mu$-*LiTiChain, p-LiTiChain* and *s-LiTiChain* architectures presented in this chapter offer a tradeoff between storage cost, security and computational cost which is worth exploring when designing blockchain for finite-lifetime data applications.

**Fig. 21** Illustration of difference in number of nonce computations between case 2 and case 3: (**a**) case 2 and (**b**) case 3

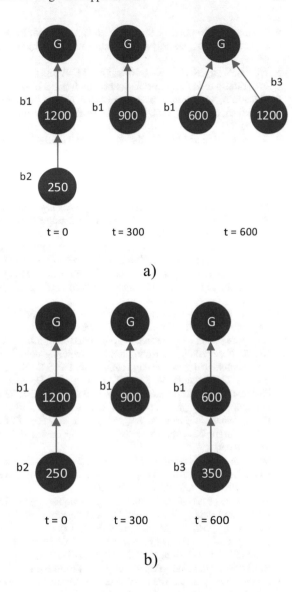

a)

b)

# References

1. Number of internet of things (IoT) connected devices worldwide in 2018, 2025 and 2030. https://www.statista.com/statistics/802690/worldwide-connected-devices-by-access-technology/ (2020)
2. Ferrag, M.A., Derdour, M., Mukherjee, M., Derhab, A., Maglaras, L., Janicke, H.: Blockchain technologies for the internet of things: Research issues and challenges. IEEE Internet Things J. **6**(2), 2188–2204 (2019)
3. Lo, S.L., Liu, Y., Chia, S.Y., Xu, X., Lu, Q., Zhu, L., Ning, H.: Analysis of blockchain solutions for IoT: a systematic literature. IEEE Access. **7**, 58822–58835 (2019)

4. Hassija, V., Chamola, V., Saxena, V., Jain, D., Goyal, P., Sikdar, B.: A survey on IoT security: application areas, security threats, and solution architectures. IEEE Access. **7**, 82721–82743 (2019)
5. Wu, M., Wang, K., Cai, X., Guo, S., Guo, M., Rong, C.: A comprehensive survey of blockchain: from theory to IoT applications and beyond. IEEE Internet Things J. **6**(5), 8114–8154 (2019)
6. Nakamoto, S.:Bitcoin: A peer-to-peer electronic cash system, 2008.
7. Treleaven, P., Brown, R.G., Yang, D.: Blockchain technology in finance. Computer. **50**(9), 14–17 (2017)
8. Saberi, S., Kouhizadeh, M., Sarkis, J., Shen, L.: Blockchain technology and its relationships to sustainable supply chain management. Int. J. Prod. Res. **57**(7), 2117–2135 (2019)
9. Mettler, M.: Blockchain technology in healthcare: the revolution starts here. In IEEE 18th International Conference on e-Health Networking, Applications and Services (Healthcom), Munich, Germany, (2016).
10. Edge computing. https://en.wikipedia.org/wiki/Edge_computing
11. Shi, W., Cao, J., Zhang, Q., Li, Y., Xu, L.: Edge computing: Vision and challenges. IEEE Internet Things J. **3**(6), 854–864 (2016)
12. Pan, J., McElhannon, J.: Future edge cloud and edge computing for internet of things applications. IEEE Internet Things J. **5**(1), 439–449 (2017)
13. Ndikumana, A., Nguyen, T.H., Kim, D.H., Kim, K.T., Hong, C.S.: Deep learning based caching for self-driving cars in multi-access edge computing. IEEE Trans. Intell. Transp. Syst. 1–16 (2020)
14. Li, H., Ota, K., Dong, M.: Learning IoT in edge: deep learning for the internet of things with edge computing. IEEE Network. **32**(1), 96–101 (2018)
15. Satyanarayanan, M.: The emergence of edge computing. Computer. **50**(1), 30–39 (2017)
16. Kumar, N., Zeadally, S., Rodrigues, J.J.: Vehicular delay-tolerant networks for smart grid data management using mobile edge computing. IEEE Commun. Mag. **54**(10), 60–66 (2016)
17. Ayoade, G., Karande, V., Khan, L. Hamlen, K.: Decentralized IoT data management using blockchain and trusted execution environment. In IEEE International Conference on Information reuse and Integration (IRI), Salt Lake City, UT, (2018).
18. Pan, J., Wang, J., Hester, A., AlQerm, I., Liu, Y., Zhao, Y.: Edgechain: An edge-IoT framework and prototype based on blockchain and smart contracts. IEEE Internet Things J. **6**(3), 4719–4732 (2019)
19. Rahman, M.A., Rashid, M.M., Hossain, M.S., Hassanain, E., Alhamid, M.F., Guizani, M.: Blockchain and IoT-based cognitive edge framework for sharing economy services in a smart city. IEEE Access. **7**, 18611–18621 (2019)
20. Shafagh, H., Burkhalter, L., Hithnawi, A., Duquennoy, S.: Towards blockchain-based auditable storage and sharing of IoT data. In the proceedings of the 2017 on cloud computing security workshop, pp. 45–50. ACM, New York, (2017)
21. Pyoung, C.K., Baek, S.J.: Blockchain of finite-lifetime blocks with applications to edge-based IoT. IEEE Internet Things J. **7**(3), 2102–2116 (2020)
22. Blockchain. https://en.wikipedia.org/wiki/Blockchain
23. Antonopoulos, A.M.: The blockchain. In: Mastering bitcoin: unlocking digital cryptocurrencies. O'Reilly Media, Newton, MA (2014)
24. Mingxiao, D., Xiaofeng, M., Zhe, Z., Xiangwei, W., Qijun, C.: A review on consensus algorithm of blockchain. In IEEE international conference on systems, man, and cybernetics (SMC), Banff, AB, Canada, (2017)

Printed in the United States
by Baker & Taylor Publisher Services